BETA DECAY AND MUON CAPTURE

BETA DECAY
AND
MUON CAPTURE

MASATO MORITA

Osaka University
Toyonaka, Osaka, Japan

1973

W. A. BENJAMIN, INC.
ADVANCED BOOK PROGRAM
Reading, Massachusetts

London · Amsterdam · Don Mills, Ontario · Sydney · Tokyo

Library of Congress Cataloging in Publication Data

Morita, Masato.
Beta decay and muon capture.

Includes bibliographies.
1. Beta decay. 2. Muons—Capture. I. Title.
QC795.M63 539.7'523 73-4520

ISBN 0-8053-7152-4
ABCDEFGHIJ-MA-7987654

Printed in the United States of America

CONTENTS

PREFACE

As is well known, understanding the weak processes in the low energy region involves two problems. One is the study of the nature of weak interactions, and the other is the study of nuclear structure from various phenomena, typically beta decay and muon capture, induced by these weak interactions. About sixteen years ago, properties of the weak interactions were almost unknown, although these interactions are one of the four fundamental (strong, electromagnetic, gravitational, and weak) interactions among elementary particles. In this book we describe how the properties of weak interactions were revealed only in the last sixteen years, and introduce many exciting findings on these properties, in particular those for parity nonconservation, by using examples in beta decay. In addition to various kinds of violation of the symmetry principles in weak processes, we emphasize the universality of the weak coupling constant and the $V-A$ interactions. Furthermore, there probably exists a certain basic similarity among the different interactions, which may be called the universality of interactions. As evidence of this we cite the conserved vector currents for the electromagnetic and weak interactions.

Once the nature of weak interactions becomes well known, we can study nuclear structure from beta decays and muon captures, as we can from gamma decays or nuclear reactions. This field of low energy nuclear physics has grown in the last decade, and important topics are introduced here. Investigation of nuclear structure from the muon capture reactions is relatively new, but is promising, since several meson factories are being completed. The muon is closely bound to the nucleus, and it has no strong interactions. Therefore, muon capture reactions, supplemented by x-ray experiments, supply information on nuclear shape, nuclear moment distributions, and various properties of nuclear levels without disturbing nuclear states appreciably.

In Chapter 1, we give an outline of the investigation of beta decay before the discovery of parity nonconservation. In Chapter 2, we describe the Fermi theory, the beta-ray spectrum, the $ft_{1/2}$ value, and electron–neutrino angular correlation for determining the $V-A$ interaction. In Chapter 3, the great discovery of parity nonconservation is introduced. Transformation properties of the Dirac wave functions and of the relativistic covariants under discrete transformations P, C, and T are discussed. The two-component neutrino theory is presented. The beta-ray angular distributions and beta–gamma angular correlations are derived from the theory of angular momentum, which is given in the Appendix. In Chapter 4, theoretical grounds are described for the parity-nonconserving $V-A$ interaction. This is also generalized to include the meson cloud effects, as in electromagnetic interactions. Universality of the $V-A$ interaction in weak processes is emphasized. In Chapter 5, the related weak processes are studied in the low as well as high energy regions. In Chapter 6, the forbidden beta decays are discussed. A number of formulas are collected for use in analyzing experimental data in the first forbidden transitions. Higher-order corrections to the allowed transitions are also given. In Chapter 7, nuclear structure problems that have recently been studied from the beta decays are presented. The muon capture in complex nuclei is a process similar to that of the orbital electron capture, but the difference is the mass of the relevant leptons. In Chapter 8, the weak interaction responsible for muon capture in hydrogen is studied, and is found to be similar to that of beta decay. A formalism is, therefore, developed for the allowed and forbidden transitions of the muon capture reactions. In Chapter 9, a possibility of studying nuclear structure from the muon capture reactions is shown. In particular, nuclear structure of C^{12} and O^{16} is studied in detail. In Chapter 10, we introduce the recent developments in the study of beta decays and related processes. The Appendix includes the theory of angular momentum and discussion of the Dirac equation.

The book is based on the lectures given at Osaka University in the first-year graduate courses on nuclear physics, quantum mechanics, and advanced nuclear physics. The contents are presented in a simple way, demand minimal prerequisites, and are self-contained. In several cases, however, it is difficult to derive the results concisely, so the results are simply presented. References are provided for readers whose interests are deeper or more extensive. (This does not mean, however, that a complete list of references for the weak interactions is included.) Square root symbols are replaced by the corresponding power symbols.

The book was written at the suggestion of Professor E. Segré, to whom I am deeply appreciative. I was at Columbia University, New York, on the occasion of the discovery of parity nonconservation and during the following years, when the most exciting and stimulating investigations in this field of physics were being conducted. For their part in making possible my stay at Columbia, I am greatly indebted to Professors R. Serber and W. W. Havens, Jr. I am also grateful to Professors C. S. Wu and T. D. Lee for their invaluable and enlightening discussions in those days. The contribution to the preparation of sections 7.4, 7.5, and 7.9 of Professors M. Yamada and J. I. Fujita is gratefully acknowledged, as is Dr. R. Morita's assistance.

The author expresses his sincere thanks to the individual authors of whose figures and tables appear in this book for their permission to use them. He also acknowledges the same permissions obtained from Annual Reviews Inc.;

the Clarendon Publishing Company; the *Journal of the Physical Society of Japan*; the North-Holland Publishing Company; *Physical Review*; *Physical Review Letters*; and *Progress of Theoretical Physics.*

Many formulas and figures in sections A.1–A.7 come from the author's article "*Theory of Nuclear Orientation*" in *Lectures in Theoretical Physics,* Vol. VI (W. E. Britten *et al.*, eds.), p. 358 (Interscience, 1962). He is greatly indebted to John Wiley & Sons, Inc., for their permission to reproduce these materials.

Osaka, Japan

MASATO MORITA

Chapter One

INTRODUCTION

1.1 BACKGROUND

One of the fundamental problems in nuclear and particle physics is to find interactions among elementary particles and to describe the physical system that obeys the physical laws by using these interactions. At the present stage of our knowledge, there are four interactions in nature: the strong, electromagnetic, and weak interactions among elementary particles; and the gravitational force between two masses. These interactions are characterized by the magnitude and dimension of their coupling constants. For example, the coupling constant f of the pion–nucleon interaction is derived from the pion–nucleon scattering data, and we have $f^2/\hbar c \sim 15$. ($\hbar$ is Planck's constant over 2π and c is the velocity of light.) The pion–nucleon interaction is classified as a strong interaction. In the case of the electromagnetic interactions, the coupling constant is the electric charge e, which is universal and is given by the fine structure constant $e^2/\hbar c = 1/137$. The weak interactions are responsible for decay processes of elementary particles or atomic nuclei. The strength of these interactions is relatively weak and the coupling constant C is given by $(CM^2c/\hbar^3)^2 \sim 10^{-10}$, M being the nucleon mass. (See Eq. (2.173).) Finally, the gravitational force constant $G_{\mathbf{grav}}$ is extremely small: $G_{\mathbf{grav}}M^2/\hbar c \sim 10^{-38}$. Among these four interactions, we confine ourselves to the weak interactions, in particular those for beta decays and muon captures.

Study of the nucleus began with the discovery of the natural radioactivity of uranium compounds by Becquerel in 1896 [1]. Since then, the radiations from nuclei have been supplying us with a tremendous amount of information concerning nuclear structure. As is shown in Fig. 1.1 [2], radioactive nuclei emit successively alpha, beta, and gamma rays, and the nuclei mutate to other species. Studying Fig. 1.1, we are tempted to assume that the interactions responsible for the alpha, beta, and gamma transitions have a certain basic relationship, since all these radiations come from the same point. It has since been revealed that there is, in fact, an important similarity between the electromagnetic interaction for gamma decays and the weak interaction for beta decays. Alpha decays are also induced by the Coulomb interaction. (And the strong interaction has to be taken into consideration if a nuclear model, such as an alpha-cluster model, is adopted.)

Figure 1.1 anticipates the historic development of the study of these interactions. The study of electromagnetic interaction was steady, culminating in the quantum electrodynamics of Tomonaga, Schwinger, and Feynman. This theory has wide application in atomic as well as nuclear physics. Its validity holds even at the high energies that can be produced by the accelerators available today. In contrast with this, the study of weak interactions has been controversial. Alpha and gamma rays have line spectra, whereas beta rays have a continuous distribution of energy. This was first noticed by Chadwick in

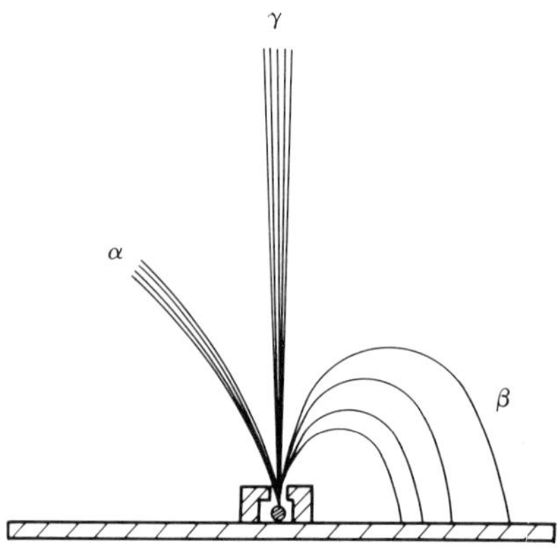

Fig. 1.1 Radioactivity. A magnetic field is applied perpendicular to the plane of the sheet.

1914[3]. The continuous spectrum was once interpreted by Bohr [4] as an indication that the laws of the conservation of energy and momentum do not hold when one deals with a nuclear process where light particles play an essential part. The difficulty was, however, removed by the introduction of the neutrino by Pauli [5]. The conservations of energy, momentum, angular momentum, and statistics were recovered again in the processes of beta decay. The next big problem concerned the linear combination of five relativistic covariants in beta interaction. It was once understood as a combination of scalar, tensor, and pseudoscalar. When precise data on the electron–neutrino angular correlation became available, the true interaction was, however, found to be a linear combination of vector and axial vector (see section 2.15). Finally, the most exciting finding came to us. It is the parity nonconservation in weak interactions in 1957 [6], which was suggested by Lee and Yang [7]. After these findings, important properties of the weak interactions were mostly established in the early 1960s.

As a consequence, we are ready to use weak processes, as well as electromagnetic processes, for the study of nuclear structure. That is, we can discuss nuclear structure problems by using information obtained from beta decays. These data are supplemented by our knowledge of the related processses, such as gamma decay, Coulomb excitations, and (n, p) reactions. An interesting field of nuclear physics is now rapidly developing.

In this book, properties of weak interactions are studied mainly from the beta decays. Universality of the weak interactions is observed in the related processes. Furthermore, the usefulness of a similarity between electromagnetic and weak interactions is emphasized. By using two typical weak processes, beta decay and muon capture, we shall study nuclear structure problems.

1.2 CONSERVATION OF ENERGY AND ANGULAR MOMENTUM

Beta decay is the spontaneous emission of an electron or a positron. The latter was first observed in 1934 by Curie and Joliot [8]; this was two years after the discovery of the positron in cosmic radiation. An inverse process of beta decay is electron capture, in which the nucleus absorbs one of its orbital electrons. This possibility was first suggested by Yukawa and Sakata [9]. The experimental proof was performed by Alvarez for the case of Ga^{67} [10]. All three electron processes are closely related theoretically and are called beta decay.

As was pointed out in section 1.1, Chadwick noticed in 1914 that the spontaneously emitted electrons in beta decay have a continuous distribution in energy. The reason for this was not ascertained in the next twenty years. The continuous nature of the beta spectrum suggests that individual nuclei do not always emit electrons of the same energy when they decay, while conservation of energy should hold for individual beta disintegrations. In 1922, Meitner studied this problem in detail [11]. Since the nuclear states are quantized, the emitted electrons are monoenergetic. There must be some reason for the reduction in the electron energy after emission. If this is the case, all electrons have the same disintegration energy, which is equal to the maximum energy of the electron spectrum. On the other hand, if electrons are emitted with various energies, then the average energy of disintegration is equal to the mean energy of the continuous spectrum. In this connection, the disintegration energy was measured by a calorimetric method. All beta particles were stopped in the calorimeter and delivered their kinetic energy to it in the form of heat. Ellis and Wooster performed this experiment in beta decay of $RaE(Bi^{210})$, and they found (350 ± 40) keV per disintegration [12]. This result agrees very well with the average energy of the beta spectrum, 390 keV, and it does not agree at all with the maximum energy 1050 keV. The result was also confirmed by Meitner and Orthmann [13]. This finding is in an apparent nonconservation of energy in the beta decay.

At that time, the constituents of the nucleus were supposed to be the protons and electrons. For example, the N^{14} nucleus was thought to consist of 14 protons and 7 electrons. If so, it would obey the Fermi–Dirac statistics, whereas spectroscopic analysis of this nucleus indicated that N^{14} obeys the Bose–Einstein statistics. The determination of nuclear spin led to the same conclusion.

Pauli noticed that neither energy nor spin and statistics were apparently conserved in beta decay. For example, the beta decay of RaE is

$$\mathrm{RaE}^{210} \rightarrow \mathrm{RaF}^{210} + \beta^-,$$

where the spins of RaE and RaF are $\hbar$ and 0, respectively. Since the spin of the electron is $\frac{1}{2}\hbar$, the total angular momentum of the system in the right-hand side is $\frac{1}{2}\hbar$. This is different from the spin in the left-hand side.

1.3 PAULI'S NEUTRINO HYPOTHESIS

To resolve these difficulties, Pauli proposed a new particle, the neutrino, which is created and emitted simultaneously with the electron in beta decay. He assumed that the neutrino has a very small rest mass, a spin of $\frac{1}{2}\hbar$, and no charge. This particle does not interact strongly with matter, and it would be very hard to detect. In the calorimetric experiments, the neutrinos are presumably carried part of the disintegration energy away through the calorimeter walls. The heat effects were due only to electrons. This idea was presented by Pauli at the American Physical Society meeting in Pasadena in 1931.

Meanwhile, neutrons were discovered by Chadwick [14], and the proton–neutron hypothesis of the nuclear building stones was proposed and fitted into the picture [15]. Pauli was greatly encouraged by this favorable development and emphasized his neutrino hypothesis at the Solvay Congress in Brussels in 1933 [5]. Fermi, who was also at the Congress, was immediately attracted by the idea of Pauli's neutrino. Thus, the famous Fermi theory of beta decay appeared soon after the Congress [16].

1.4 CLASSIFICATION OF BETA DECAYS

The beta-decay rate has an enormous range of variation from isotope to isotope. The nuclei of B^{12} have a half life of 0.0027 sec, while K^{40} has a half life of 1.5×10^9 years. Theoretically, the decay rates depend on disintegration energies, nuclear matrix elements, and details of the beta-decay interaction. In order to investigate various beta decays systematically, it is convenient to classify the beta decay by the change of the spin J and parity π of the initial and final states, as in Table 1.1, where Yes refers to $\pi_i \neq \pi_f$ and No refers to $\pi_i = \pi_f$. The allowed transition is subdivided into two parts; the Fermi transition ($\Delta J = 0$), and the Gamow-Teller transition ($\Delta J = 0, \pm 1$). As is shown in Chapters 2, 3, and 4, all important information on the interaction of beta decays can be obtained from the data in the allowed transition, while the nuclear structure problems can be obtained from both allowed and forbidden transitions. The forbiddenness n is used for the nth forbidden transition, and it is zero for the allowed transition.

Table 1.1 Classification of Beta Decay

Name of Transition	$\Delta J = J_i - J_f$	Parity Change
Allowed transition	0, ±1	No
First forbidden transition	0, ±1, ±2	Yes
Second forbidden transition	±2, ±3	No
Third forbidden transition	±3, ±4	Yes
etc.		

1.5 DISINTEGRATION ENERGIES

The disintegration energy Q is the total kinetic energy of the final system in the reaction. The electron decay is the nuclear transformation,

$$(A, Z) \to (A, Z+1) + \beta^- + \bar{\nu}. \qquad (1.1)$$

The disintegration energy is given by

$$Q = ({}_Z M^A - {}_{Z+1} M^A)c^2. \qquad (1.2)$$

Here, M is the atomic mass for (A, Z), Z and A being the atomic number and the mass number, respectively. A small recoil energy of the nucleus is neglected in Eq. (1.2). The positron decay is

$$(A, Z) \to (A, Z-1) + \beta^+ + \nu, \qquad (1.3)$$

and its Q value is

$$Q = ({}_Z M^A - {}_{Z-1} M^A) - 2mc^2. \qquad (1.4)$$

The positron decay always competes with orbital electron captures. The K-electron capture takes place most frequently. The reaction is

$$(A, Z) + e^- \to (A, Z-1) + \nu, \qquad (1.5)$$

and its Q value is

$$Q = ({}_{Z}M^{A} - {}_{Z-1}M^{A})c^{2} - \text{B.E.} \tag{1.6}$$

where B.E. is the binding energy of the orbital electron to be captured.

REFERENCES

1 H. Becquerel, *Compt. rend.* **122,** 420 (1896).
2 M. Curie, *Recherches sur les substances de radioactivité,* Université de Paris, 1903.
3 J. Chadwick, *Verhandl. deut. physik.* **16,** 383 (1914).
4 N. Bohr, Chemistry and the quantum theory of atomic constitution, *J. Chem. Soc.* 1932, p. 349.
5 W. Pauli, Noyaux atomiques, *Proc. Solvay Congr.,* Brussels (1933), p. 324.
6 For beta decay, C. S. Wu, E. Ambler, R. W. Hayward, D. D. Hoppes, and R. P. Hudson, *Phys. Rev.* **105,** 1413 (1957).
7 T. D. Lee and C. N. Yang, *Phys. Rev.* **104,** 254 (1956).
8 I. Curie and F. Joliot, *Compt. rend.* **198,** 254 (1934).
9 H. Yukawa and S. Sakata, *Proc. Phys. Math. Soc. Japan* **17,** 467 (1935); **18,** 128 (1936).
10 L. W. Alvarez, *Phys. Rev.* **54,** 486 (1938).
11 L. Meitner, *Z. Phys.* **9,** 131 and 145 (1922).
12 C. D. Ellis and W. A. Wooster, *Proc. Roy. Soc. London* **A117,** 109 (1927).
13 L. Meitner and W. Orthmann, *Z. Phys.* **60,** 143 (1930).
14 J. Chadwick, *J. Proc. Roy. Soc.* **A136,** 692 (1932).
15 W. Heisenberg, *Z. Phys.* **77,** 1 (1932).
16 E. Fermi, *Ric. Sci.* **2,** Pt. 12 (1933); *Z. Phys.* **88,** 161 (1934).

Chapter Two

FERMI THEORY OF BETA DECAY

2.1 FERMI THEORY OF BETA DECAY

In 1934, Fermi published a theory of beta decay that was constructed in an analogous way to the theory of electromagnetic radiation. He posited that the emission and absorption of the photon are induced by the time-dependent interaction between the radiation system and the electromagnetic field. The interaction is the product of the electric current **j** due to the charged particles and the vector potential **A**. For example, the emission of a gamma ray from the nucleus which has the mass number A and the atomic number Z is represented by

$$(A, Z)^* \rightarrow (A, Z) + \gamma. \tag{2.1}$$

The parentheses signify the ground state of the nucleus, and the superscript asterisk denotes its excited state. The interaction becomes

$$\mathscr{H}_\gamma = -\sum_n \frac{e}{M_\mathrm{p} c} \mathbf{p}_n(\mathbf{r}_n) \cdot \mathbf{A}(\mathbf{r}_n, t). \tag{2.2}$$

Here $\mathbf{p}_n$ and $\mathbf{r}_n$ are the momentum and the position of the nth proton, and its mass is M_p. In the quantum mechanical treatment, **A** can create or annihilate a photon, so that

$$\mathbf{A}(\mathbf{r}, t) = \mathbf{A}(\mathbf{r})\, e^{-i\omega t} + \mathbf{A}^*(\mathbf{r})\, e^{i\omega t}, \tag{2.3}$$

with

$$\mathbf{A}(\mathbf{r}) = \text{const. } \boldsymbol{\epsilon} e^{i\mathbf{k}\cdot\mathbf{r}}.$$

Here $\mathbf{A}(\mathbf{r})$ annihilates a photon that has the wave number **k** and the polarization $\boldsymbol{\epsilon}$. Similarly, $\mathbf{A}^*(\mathbf{r})$ creates a photon. The emission rate for the interaction Eq. (2.3) is given by

$$\frac{2\pi}{\hbar} |H_{\mathrm{fi}}|^2 \frac{dn}{dE}, \tag{2.4}$$

with

$$H_{\mathrm{fi}} = -\frac{e}{M_{\mathrm{p}}c}\int \psi_{\mathrm{f}}^{*}\sum_{n}\exp(-i\mathbf{k}\cdot\mathbf{r}_n)\boldsymbol{\epsilon}\cdot\mathbf{p}_n\,\psi_{\mathrm{i}}\,d\tau$$

and

$$d\tau = d\mathbf{r}_1\,d\mathbf{r}_2\cdots d\mathbf{r}_A\,.$$

Here dn/dE is the final state density in the photon energy domain E and $E + dE$. ψ_{i} and ψ_{f} are the nuclear wave functions of the initial and final states, respectively. The sum n runs over all protons.

In nuclear beta decay, the electron and neutrino are emitted simultaneously. We call the neutrino accompanied by the negative electron an antineutrino. This is required from the law of lepton number conservation, as we will see later. Therefore, in electron decay,

$$(A, Z) \to (A, Z+1) + \beta^{-} + \bar{\nu}. \tag{2.5}$$

In positron decay,

$$(A, Z) \to (A, Z-1) + \beta^{+} + \nu. \tag{2.6}$$

Now, let us consider a correspondence of the photon emission and beta decay (more specifically, electron decay). Comparing Eqs. (2.1) and (2.5), we see that the photon is replaced by the system of the electron and antineutrino.

$$\mathbf{A}^{*}(\mathbf{r}) \to \psi_{\mathrm{e}}^{*}(\mathbf{r})\,\psi_{\bar{\nu}}^{*}(\mathbf{r}). \tag{2.7}$$

Here ψ^{*} is the wave function of the created lepton, which has the form

$$\psi^{*}(\mathbf{r}) \sim \exp(-i\mathbf{p}\cdot\mathbf{r}/\hbar) \tag{2.8}$$

where $\mathbf{p}$ is the lepton momentum. (The lepton refers to either the electron, the neutrino, or their antiparticles.) Next, in Eq. (2.4), the differential operator acting on the initial wave function is replaced by an operator τ_{+}, which changes the neutron into the proton and annihilates the proton in the initial state. However, τ_{+} does not change the spatial and spin state of the nucleon.

$$-\frac{\mathbf{p}}{M_{\mathrm{p}}c} \to \tau_{+}. \tag{2.9}$$

Finally, the interaction of beta decay has a characteristic strength that corresponds to the strength of the electromagnetic interaction, i.e., the electric charge

$$e \to g. \tag{2.10}$$

The replacements in Eqs. (2.7), (2.9), and (2.10) introduce an interaction Hamiltonian for electron decay,

$$\mathscr{H}_{\beta^-} = g \sum_n \psi_e^*(\mathbf{r}_n)\psi_{\bar{\nu}}^*(\mathbf{r}_n)\tau_+^{(n)}. \tag{2.11}$$

The emission of the antineutrino can be interpreted as the absorption of the neutrino in the negative energy state, since the neutrino is a Dirac particle. For this reason, we write the electron decay in the form

$$(A, Z) + \nu \rightarrow (A, Z+1) + \beta^-, \tag{2.12}$$

and the interaction Hamiltonian for beta decay is

$$\mathscr{H}_\beta = g \sum_n [\psi_e^*(\mathbf{r}_n)\psi_\nu(\mathbf{r}_n)\tau_+^{(n)} + \psi_e(\mathbf{r}_n)\psi_\nu^*(\mathbf{r}_n)\tau_-^{(n)}]. \tag{2.13}$$

Here, the first terms induce the electron decay, and the second terms induce the positron decay. The operator τ_- changes the proton into a neutron and annihilates the initial neutrons. The sum n runs over all nucleons. Moreover, the emission of the antineutrino with energy $E_{\bar{\nu}}$ and momentum $p_{\bar{\nu}}$ corresponds to the capture of the neutrino from a state of energy $-E_{\bar{\nu}}$ and momentum $-p_{\bar{\nu}}$.

2.2 STATISTICAL SHAPE OF THE BETA-RAY SPECTRUM

The transition probability of emission of the beta particles with energy E_e per unit time is given by

$$P(E_e)\, dE_e = \frac{2\pi}{\hbar} |H_{fi}|^2 \frac{dn}{dE_o}, \tag{2.14}$$

with the matrix element of the interaction Hamiltonian

$$H_{fi} = \int \psi_f^* \mathscr{H}_\beta \psi_i \, d\tau. \tag{2.15}$$

Here ψ_i and ψ_f are the wave functions of the initial and final nuclear states. The lepton wave functions are normalized in a volume V,

$$\int \psi^* \psi \, d\mathbf{r} = 1, \tag{2.16}$$

and

$$\psi = \left(\frac{1}{V}\right)^{1/2} \exp(i\mathbf{p}\cdot\mathbf{r}/\hbar). \tag{2.17}$$

The rapidity of the variation of ψ can be measured by the de Broglie wavelength $\lambdabar = \hbar/p$, which is around 10^{-11} cm for a usual value of the lepton momentum p. On the other hand, the nuclear radius is less than 10^{-12} cm.

This means that the lepton wave functions are almost constant in the nucleus. We replace them by the value at the origin $r = 0$,

$$\psi_e(0) = \psi_\nu(0) = (1/V)^{1/2}. \tag{2.18}$$

The integration (2.15) is now proportional to

$$\int \psi_f^* \sum_n \tau_+^{(n)} \psi_i \, d\tau, \tag{2.19}$$

which is nothing but the overlapping integral of the nuclear wave functions. It is called the Fermi matrix, and is written briefly as

$$\int 1 \equiv \int \psi_f^* \sum_n \tau_+^{(n)} \psi_i \, d\tau. \tag{2.20}$$

Hereafter, we also use the symbol $\int \Omega$ if the integral on the right-hand side includes the operator $\Omega\tau_\pm^{(n)}$. From Eqs. (2.18) and (2.20) we have

$$H_{fi} = \frac{g}{V} \int 1. \tag{2.21}$$

The number of the plane wave states in the momentum domain p and $p + dp$ with the particle anywhere in V is

$$dn = \frac{4\pi p^2 V \, dp}{h^3}. \tag{2.22}$$

The total number of final states in the resultant lepton energy domain E_o and $E_o + dE_o$ is

$$\frac{dn}{dE_o} = \frac{dn_e \, dn_{\bar\nu}}{dE_o} = \frac{16\pi^2 V^2}{h^6} \frac{p_e^2 \, dp_e \, p_{\bar\nu}^2 \, dp_{\bar\nu}}{dE_o}. \tag{2.23}$$

If the neutrino mass is zero, we have

$$E_{\bar\nu} = cp_{\bar\nu} \qquad \text{and} \qquad dE_o = dE_{\bar\nu} \tag{2.24}$$

for a constant energy of the electron. Here

$$E_o = E_{\bar\nu} + E_e \tag{2.25}$$

corresponds to the maximum energy of the electron in the beta decay. Finally we have

$$\begin{aligned} \frac{dn}{dE_o} &= \frac{16\pi^2 V^2}{(2\pi\hbar)^6 c} \, p_e^2 p_{\bar\nu}^2 \, dp_e \\ &= \frac{V^2}{4\pi^4 \hbar^6 c^5} \, p_e E_e (E_o - E_e)^2 \, dE_e. \end{aligned} \tag{2.26}$$

From Eqs. (2.14), (2.21), and (2.26), we have the probability of the emission of the electrons with energy E_e as

$$P(E_e)\, dE_e = \frac{g^2}{2\pi^3 c^5 \hbar^7} \left| \int 1 \right|^2 p_e E_e (E_o - E_e)^2\, dE_e. \tag{2.27}$$

In many references, the atomic units

$$\hbar = c = m = 1 \tag{2.28}$$

are adopted, where m is the electron mass. This is the case in our book. Then, the energy and momentum are measured in units of mc^2 and mc (and the length is given in units of $\hbar/mc$). We put

$$E = \frac{E_e}{mc^2}, \qquad p = \frac{p_e}{mc}, \qquad E_o = \frac{E_o}{mc^2}. \tag{2.29}$$

Here we use the same symbol E_o in the atomic units. We also introduce a nondimensional weak coupling constant G which is given by $g = G(\hbar^3/m^2c)$. Consequently, the beta-ray spectrum becomes

$$P(E)\, dE = \frac{mc^2}{\hbar} \frac{G^2}{2\pi^3} \left| \int 1 \right|^2 pE(E_o - E)^2\, dE. \tag{2.30}$$

The factor $mc^2/\hbar$ is for transformation from atomic units to centimeter-gram-second (cgs) units. In fact,

$$\frac{\hbar}{mc^2} = 1.287 \times 10^{-21} \text{ sec}. \tag{2.31}$$

As shown in Fig. 2.1, the formula (2.30) can well reproduce the characteristic nature of the continuous beta-ray spectrum. The factor $pE(E_o - E)^2$

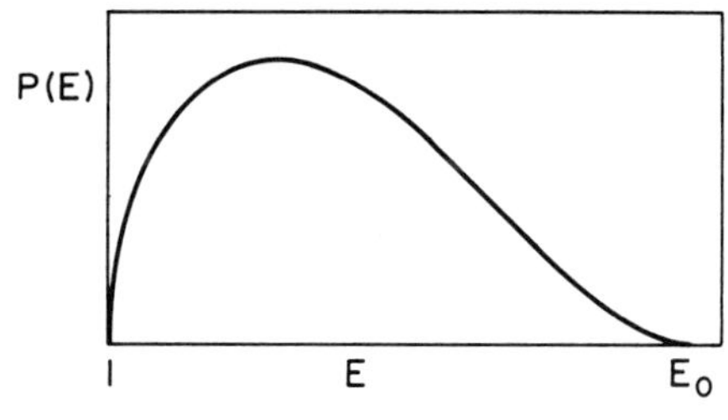

Fig. 2.1 A sketch of the statistical spectrum, Eq. (2.30).

in Eq. (2.30) called the statistical shape is, however, subject to at least two major alterations. The first is the selection rule. Since the Fermi matrix (2.20) is the overlapping integral of the initial and final wave functions, it has nonzero value only if the initial and final nuclear states have the same spin and parity, $J_i = J_f$ and $\pi_i = \pi_f$. This class of beta decays is called the Fermi transition. The beta decay

$$O^{14} \rightarrow N^{14} + \beta^+ + \nu, \qquad 0^+ \rightarrow 0^+ \tag{2.32}$$

is a typical example. On the other hand, there is a spin-change beta decay, such as

$$He^6 \rightarrow Li^6 + \beta^- + \bar{\nu}. \qquad 0^+ \rightarrow 1^+ \tag{2.33}$$

Transitions of this type cannot be induced by the interaction (2.13) originally introduced by Fermi. A generalization of the interaction will be done with inclusion of the lepton spins, so that the spin-change transitions are properly taken into account. This will be seen in the next section.

The second point to be considered is the effect of the Coulomb fields of the nucleus as well as the orbital electron clouds on the emitted beta particles. In order to derive Eq. (2.30), we simply assumed that the wave function of the electron is the plane wave, and this assumption will be modified, in section 2.7, by taking into account those Coulomb interactions.

2.3 SPINS OF LEPTONS

The interaction of beta decay introduced in section 2.1 is simply the product of the four fermions. A very slight generalization of this form can be performed if we take into account the spinor character of the lepton wave function. Since the lepton has an intrinsic spin $\frac{1}{2}\hbar$, its wave function is given by

$$\psi = \left(\frac{1}{V}\right)^{1/2} \begin{pmatrix} \psi^1 \\ \psi^2 \end{pmatrix} \exp\left[i(\mathbf{p}\cdot\mathbf{r} - Et)/\hbar\right] \tag{2.34}$$

in nonrelativistic theory. ψ^1 and ψ^2 are the probability amplitudes representing the spin along and opposite to the axis of quantization, respectively, and

$$|\psi^1|^2 + |\psi^2|^2 = 1. \tag{2.35}$$

We can drop the exponential factor in the nucleus, as before. We construct the bilinear form of lepton wave functions, and similarly for nucleons. The interaction Hamiltonian density is now the product of these two bilinear forms in such a way that the product is rotationally invariant. This is because the beta decay does not depend on a special choice of the coordinate system. And the law of nature is the same in all directions in space.

We have two components for each lepton, thus, four different bilinear forms,

$$\psi_e^\dagger \psi_\nu \qquad \text{and} \qquad \psi_e^\dagger \sigma_i \psi_\nu, \qquad i = x, y, z. \tag{2.36}$$

Here, $\psi^\dagger$ is the hermitian conjugate of ψ:

$$\psi = \begin{pmatrix} \psi^1 \\ \psi^2 \end{pmatrix} \qquad \text{and} \qquad \psi^\dagger = (\psi^{1*}, \psi^{2*}). \tag{2.37}$$

The Pauli spin matrices σ_i are given by

$$\sigma_x = \begin{pmatrix} 0 & 1 \\ 1 & 0 \end{pmatrix}, \qquad \sigma_y = \begin{pmatrix} 0 & -i \\ i & 0 \end{pmatrix}, \qquad \sigma_z = \begin{pmatrix} 1 & 0 \\ 0 & -1 \end{pmatrix}. \tag{2.38}$$

Since the electron and antineutrino have an intrinsic spin $\frac{1}{2}\hbar$, the emitted electron-antineutrino system can carry either the resultant intrinsic spin of zero or one unit of $\hbar$. (See Appendix A.2 on the addition of angular momenta.) The wave function $\psi_e^\dagger \psi_\nu$ corresponds to the system with spin zero and remains invariant under the rotation. That is scalar. The other components $\psi_e^\dagger \sigma_i \psi_\nu$ transform like the ith component of a vector quantity under the rotation. These components do not change sign under the space inversion $\mathbf{r} \to -\mathbf{r}$ because ψ's do not depend on $\mathbf{r}$. Therefore, $\psi_e^\dagger \sigma_i \psi_\nu$ are components of the axial vector. The bilinear forms for nucleons also make scalar and axial vector parts. The interaction Hamiltonian density invariant under the rotation can be made with the product of a lepton scalar and a nucleon scalar or similarly for axial vectors.

$$(\psi_p^\dagger \tau_+ \psi_n)(\psi_e^\dagger \psi_\nu), \qquad (\psi_p^\dagger \tau_+ \sigma \psi_n)(\psi_e^\dagger \sigma \psi_\nu). \tag{2.39}$$

2.4 FERMI AND GAMOW-TELLER TRANSITIONS

We classified beta decays by the spin and parity changes between the initial and final nuclear states in section 1.4. The allowed transition is also subdivided into two parts:

$$\begin{array}{lll} \text{Fermi transition: } \Delta J = 0, & & \text{parity change No.} \\ \text{Gamow–Teller transition: } \Delta J = 0, \pm 1, & & \text{parity change No.} \\ \qquad\qquad (\text{But no } 0 \to 0.) & & \end{array} \tag{2.40}$$

These transitions are now induced by the interaction (2.39).

Fermi Transition

The Fermi transitions are induced by

$$H_\beta = \psi_f^\dagger \mathscr{H}_\beta \psi_i = g_F \left(\psi_f^\dagger \sum_n \tau_+^{(n)} \psi_i\right) \left(\psi_e^\dagger \psi_\nu\right) + \text{h.c.} \tag{2.41}$$

The conservation of angular momentum requires

$$\mathbf{J}_i = \mathbf{J}_f + \mathbf{J}_{e\nu} = \mathbf{J}_f + 0, \tag{2.42}$$

where $\mathbf{J}_{e\nu}$ is the resultant spin carried by the emitted lepton system. Equation (2.42) is satisfied only if $J_i = J_f$. The spatial integral $\int \psi_f^\dagger\ \Sigma_n\ \tau_+^{(n)}\ \psi_i$ will also vanish, unless the ψ_f and ψ_i have the same parity, $\pi_i = \pi_f$.

Gamow–Teller Transition

The Gamow-Teller transitions are induced by

$$H_\beta = g_{GT} \left(\psi_f^\dagger \sum_n \tau_+^{(n)} \sigma^{(n)} \psi_i\right) \cdot (\psi_e^\dagger \sigma \psi_\nu) + \text{h.c.} \tag{2.43}$$

In this case, the lepton system has the resultant spin one, and angular momentum conservation requires

$$\mathbf{J}_i = \mathbf{J}_f + \mathbf{J}_{e\nu} = \mathbf{J}_f + 1. \tag{2.44}$$

From Appendix A.2 we see that Eq. (2.44) is satisfied by

$$J_f = J_i + 1, J_i, J_i - 1. \tag{2.45}$$

That is, $\Delta J = 0, \pm 1$. As a special case, Eq. (2.45) does not hold for $J_i = J_f = 0$. This means that the $0 \to 0$ transition is forbidden for the Gamow–Teller interaction, Eq. (2.43). The integral $\int \psi_f^\dagger\ \Sigma_n \tau_+^{(n)} \sigma^{(n)} \psi_i\ d\tau$, called the Gamow–Teller matrix, vanishes unless ψ_i and ψ_f have the same parity. This is because the operator $\sigma^{(n)}$ acting on the nth nucleon does not change its sign under the space inversion.

The beta-ray spectrum for the allowed transitions now becomes

$$P(E)\,dE = \frac{mc^2}{\hbar} \frac{1}{2\pi^3} PE(E_o - E)^2 \left(g_F^2 |\textstyle\int 1|^2 + g_{GT}^2 |\textstyle\int \sigma|^2\right) dE \tag{2.46}$$

in the units $\hbar = m = c = 1$. The Fermi and Gamow–Teller matrices are defined by

$$\int 1 = \int \psi_f^\dagger \sum_n \tau_+^{(n)} \psi_i\, d\tau,$$

$$\int \sigma = \int \psi_f^\dagger \sum_n \tau_+^{(n)} \sigma^{(n)} \psi_i\, d\tau, \tag{2.47}$$

and

$$|\textstyle\int \sigma|^2 = |\int \sigma_x|^2 + |\int \sigma_y|^2 + |\int \sigma_z|^2.$$

Until now, the calculation was done for nonrelativistic leptons. This treatment is not accurate enough, since the neutrino and beta particle are moving at or near the velocity of light. In order to formulate the relativistic theory of beta decay, we need information about the relativistic wave equation, which is given in Appendices A.9–A.12.

2.5. RELATIVISTIC FORMULATION FOR BETA DECAY

In the following, we adopt four-component spinors for the fermion wave functions ψ, which are solutions of the Dirac equation

$$\left(\gamma_\mu \frac{\partial}{\partial x_\mu} + \frac{mc}{\hbar}\right)\psi = 0. \tag{2.48}$$

Choosing the explicit forms for the Dirac matrices

$$\gamma_k = -i\beta\alpha_k, \qquad k = 1, 2, 3, \qquad \text{and} \qquad \gamma_4 = -\beta, \tag{2.49}$$

we have

$$(E + c\boldsymbol{\alpha}\cdot\mathbf{p} + \beta mc^2)\psi = 0 \tag{2.50}$$

from Eq. (2.48). We use the expression (2.50) as the Dirac equation throughout this book. The other choice,

$$\gamma_\mu = -i\beta\alpha_k, \qquad \gamma_4 = \beta, \tag{2.51}$$

is equally often used in the literature.

In the relativistic case, a generalization of the interaction Eq. (2.11) can be obtained again from the analogy of beta decay with electromagnetic radiations. A part of the electromagnetic interactions is

$$-\frac{e}{c}j_\mu A_\mu = -ei(\bar{\psi}_{\rm p}\gamma_\mu\psi_{\rm p})A_\mu. \tag{2.52}$$

As we studied, the quantity $i(\bar{\psi}_{\rm p}\gamma_\mu\psi_{\rm p})$ transforms under the Lorentz transformation as a four-vector, which is now the proton current density. A_μ is the four-vector potential. It is evident that this interaction (2.52) is Lorentz invariant. Now, we replace

(a) electric charge e by the vector coupling constant $C_{\rm V}$;
(b) electric current $i(\bar{\psi}_{\rm p}\gamma_\mu\psi_{\rm p})$ by weak current $i(\bar{\psi}_{\rm p}\gamma_\mu\tau_+\psi_{\rm n})$;
(c) electromagnetic potential A_μ by a bilinear form of leptons $i(\bar{\psi}_{\rm e}\gamma_\mu\psi_\nu)$.

As a result, we have an interaction Hamiltonian density

$$C_{\rm V}(\bar{\psi}_{\rm p}\gamma_\mu\tau_+\psi_{\rm n})(\bar{\psi}_{\rm e}\gamma_\mu\psi_\nu), \tag{2.53}$$

which is relativistically invariant. This is the one that Fermi originally introduced. The operator τ_+ will be dropped in the following few expressions. This is because the quantity, e.g., $(\psi_p^\dagger \tau_+ \psi_n)$, in the isospin formalism is given by

$$\psi_p^\dagger \tau_+ \psi_n = (\psi_p^\dagger, 0)\begin{pmatrix} 0 & 1 \\ 0 & 0 \end{pmatrix}\begin{pmatrix} 0 \\ \psi_n \end{pmatrix} = \psi_p^\dagger \psi_n,$$

where ψ's in the middle and on the right-hand side stand for the spin-spatial parts of the nucleon wave functions.

We learn in Appendix A.12 about five classes of bilinear quantities in $\overline{\psi}$ and ψ. We can make those bilinear quantities by leptons and nucleons, because all four fermions, n, p, e^-, and $\overline{\nu}$ obey the Dirac equation. These are of the forms in Eq. (A.142), where $\overline{\psi}$ and ψ should be replaced by $\overline{\psi}_e$ and ψ_ν (or $\overline{\psi}_p$ and ψ_n), respectively.

There would be no *a priori* reason to assume a particular covariant form V_μ as the four-fermion coupling for beta decay. The most general Hamiltonian density that induces beta decay is, therefore, a linear combination of the following five covariants.

$$\begin{aligned} H_\beta = {} & C_S(\overline{\psi}_p \psi_n)(\overline{\psi}_e \psi_\nu) \\ & + C_V(\overline{\psi}_p \gamma_\mu \psi_n)(\overline{\psi}_e \gamma_\mu \psi_\nu) \\ & + C_T(\overline{\psi}_p i\gamma_\mu \gamma_\lambda \psi_n)(\overline{\psi}_e i\gamma_\mu \gamma_\lambda \psi_\nu) \\ & + C_A(\overline{\psi}_p i\gamma_\mu \gamma_5 \psi_n)(\overline{\psi}_e i\gamma_\mu \gamma_5 \psi_\nu) \\ & + C_P(\overline{\psi}_p \gamma_5 \psi_n)(\overline{\psi}_e \gamma_5 \psi_\nu) + \text{h.c.}, \qquad \mu < \lambda. \end{aligned} \tag{2.54}$$

(Strictly speaking, the interaction Hamiltonian density is expressed by the second quantized fields ψ. When we evaluate its matrix element for the beta decay of the neutron, we have the same expression as in Eq. (2.54) which is now given by the wave functions ψ.) Here we still assume the invariance of H_β under the Lorentz transformation, including the space inversion and time reversal. This requirement will be removed in the next chapter. If we use the matrices α and β instead of γ_μ in Eq. (2.49), we have

$$\begin{aligned} H_\beta = {} & C_S(\psi_p^\dagger \beta \psi_n)(\psi_e^\dagger \beta \psi_\nu) \\ & + C_V[(\psi_p^\dagger \psi_n)(\psi_e^\dagger \psi_\nu) - (\psi_p^\dagger \boldsymbol{\alpha} \psi_n)(\psi_e^\dagger \boldsymbol{\alpha} \psi_\nu)] \\ & + C_T[(\psi_p^\dagger \beta \boldsymbol{\sigma} \psi_n)(\psi_e^\dagger \beta \boldsymbol{\sigma} \psi_\nu) + (\psi_p^\dagger \beta \boldsymbol{\alpha} \psi_n)(\psi_e^\dagger \beta \boldsymbol{\alpha} \psi_\nu)] \\ & + C_A[(\psi_p^\dagger \boldsymbol{\sigma} \psi_n)(\psi_e^\dagger \boldsymbol{\sigma} \psi_\nu) - (\psi_p^\dagger \gamma_5 \psi_n)(\psi_e^\dagger \gamma_5 \psi_\nu)] \\ & + C_P(\psi_p^\dagger \beta \gamma_5 \psi_n)(\psi_e^\dagger \beta \gamma_5 \psi_\nu) + \text{h.c.}, \end{aligned} \tag{2.55}$$

with

$$\boldsymbol{\alpha} = \begin{pmatrix} 0 & \boldsymbol{\sigma} \\ \boldsymbol{\sigma} & 0 \end{pmatrix}, \quad \boldsymbol{\sigma} = \begin{pmatrix} \boldsymbol{\sigma} & 0 \\ 0 & \boldsymbol{\sigma} \end{pmatrix}, \quad \beta = \begin{pmatrix} 1 & 0 \\ 0 & -1 \end{pmatrix}, \quad \gamma_5 = \begin{pmatrix} 0 & 1 \\ 1 & 0 \end{pmatrix}. \tag{2.56}$$

Here the components of 4 x 4 matrices are given by 2 x 2 matrices. The magnitudes of the coupling constants C_i have to be determined in conformity with the experimental results in beta decay.

In Eq. (2.55), the four terms that do not involve $\boldsymbol{\alpha}$ or γ_5 have a larger contribution compared with the other four terms that do involve $\boldsymbol{\alpha}$ or γ_5. More precisely, the former can only induce the Fermi and Gamow–Teller transitions. Assuming the nucleons in the nucleus are moving with a velocity of about $0.1c$, we make the following nonrelativistic approximation (i.e., $|E| \approx Mc^2$).

$$\psi_{\mathrm{p}}^{\dagger} = \left(-\frac{\boldsymbol{\sigma} \cdot \mathbf{p}_{\mathrm{p}}}{2Mc} u_{\mathrm{p}}^{\dagger}, u_{\mathrm{p}}^{\dagger} \right) \quad \text{and} \quad \psi_{\mathrm{n}} = \begin{pmatrix} -\dfrac{\boldsymbol{\sigma} \cdot \mathbf{p}_{\mathrm{n}}}{2Mc} u_{\mathrm{n}} \\ u_{\mathrm{n}} \end{pmatrix}. \tag{2.57}$$

Here, $\mathbf{p}_{\mathrm{p}}$ and $\mathbf{p}_{\mathrm{n}}$ are momenta of the proton and neutron, and u's are two-component spinors. Since the nucleon momentum is small, the spinor u is larger compared with $-(\boldsymbol{\sigma} \cdot \mathbf{p}/2Mc)u$. The former refers to the large component and the latter to the small component. The four terms of Eq. (2.55) have odd operators, $\boldsymbol{\alpha}$ and γ_5, which connect the small component with the large one. Therefore, the contribution of these terms disappears in the limit as $(\mathbf{p}_{\mathrm{n}} - \mathbf{p}_{\mathrm{p}}) \to 0$. The other four terms have even operators β, $\boldsymbol{\sigma}$, 1, which connect the large component with the large one and the small component with the small one. In fact, the vector interaction gives the Fermi matrix

$$\int 1 = \int \psi_{\mathrm{p}}^{\dagger} \psi_{\mathrm{n}} \, d\mathbf{r} = \int u_{\mathrm{p}}^{\dagger} u_{\mathrm{n}} \, d\mathbf{r}. \tag{2.58}$$

On the other hand, the scalar interaction gives

$$\int \beta = \int \psi_{\mathrm{p}}^{\dagger} \beta \psi_{\mathrm{n}} \, d\mathbf{r} = -\int u_{\mathrm{p}}^{\dagger} u_{\mathrm{n}} \, d\mathbf{r}. \tag{2.59}$$

and

$$\int \beta = -\int 1. \tag{2.60}$$

This is because the matrix β has -1 for the large component. Therefore, both scalar and vector interactions can induce the Fermi transitions. A similar consideration leads us to conclude that the axial vector and tensor interactions induce the Gamow–Teller transitions. Furthermore, there is a relation

$$\int \beta \boldsymbol{\sigma} = -\int \boldsymbol{\sigma}. \tag{2.61}$$

It is noticed that we could not make the nonrelativistic approximations for the leptons, since the leptons are traveling at a near-light velocity.

2.6 BETA-RAY SPECTRUM

We are now ready to compute the beta-ray spectrum for the Fermi and Gamow–Teller transitions by using the relativistic interactions in Eq. (2.55).

We have to evaluate the square of the matrix element $H_{\mathrm{fi}} = \int H_\beta \, d\tau$,

$$|H_{\mathrm{fi}}|^2 = |\sum_k C_k(\psi_e^\dagger \Omega^k \psi_\nu) \int \Omega_k|^2 \qquad \text{with} \qquad \Omega^k = \Omega_k^\dagger. \tag{2.62}$$

Here the coupling constants and Dirac matrices are arranged as follows.

k	1	2	3	4	5	6	7	8
C_k	C_S	C_V	C_T	C_T	C_T	C_A	C_A	C_A
Ω_k	β	1	$\beta\sigma_0$	$\beta\sigma_1$	$\beta\sigma_{-1}$	σ_0	σ_1	σ_{-1} (2.63)

The subscript to σ refers to the magnetic quantum number

$$\sigma_0 = \sigma_z \qquad \text{and} \qquad \sigma_{\pm 1} = \mp(\tfrac{1}{2})^{1/2}(\sigma_x \pm i\sigma_y). \tag{2.64}$$

This leads to a relation

$$(\psi_\mathrm{f}^\dagger \sigma \psi_\mathrm{i})(\psi_\mathrm{e}^\dagger \sigma \psi_\nu) = \sum_M (-)^M (\psi_\mathrm{f}^\dagger \sigma_M \psi_\mathrm{i})(\psi_\mathrm{e}^\dagger \sigma_{-M} \psi_\nu). \tag{2.65}$$

Here the sum runs over $M = 0, \pm 1$. The neutron and proton wave functions are replaced by the initial and final wave functions. Nucleon operators $\Sigma_n \tau_+^{(n)} \sigma^{(n)}$ are abbreviated by $\boldsymbol{\sigma}$. This does not, however, change the following discussions. An abbreviation $(-)^M$ for $(-1)^M$ is also adopted throughout this book.

At first sight, there are 64 terms of the squares and interferences in Eq. (2.62). But most of the interferences disappear because of the rotational properties. In order to see this, we write the spin and magnetic quantum number of the nuclear states explicitly. Using the Wigner–Eckart theorem (Appendix A.4), we have

$$\int \psi_{J_\mathrm{f} M_\mathrm{f}}^\dagger (\Omega)_{LM} \psi_{J_\mathrm{i} M_\mathrm{i}} \, d\tau = (J_\mathrm{i} L M_\mathrm{i} M | J_\mathrm{f} M_\mathrm{f}) \langle J_\mathrm{f} \| \Omega \| J_\mathrm{i} \rangle \tag{2.66}$$

where $\langle J_\mathrm{f} \| \Omega \| J_\mathrm{i} \rangle$ is independent of the magnetic quantum numbers. L is the rank of the tensor for interactions. It is zero for β and 1, and one for $\beta\sigma$ and σ. $|H_{\mathrm{fi}}|^2$ is evaluated, as usual, by summing over the final states and averaging over the initial states. This gives the following factor for fixed values of M and M'.

$$\frac{1}{2J_\mathrm{i}+1} \sum_{M_\mathrm{i} M_\mathrm{f}} (J_\mathrm{i} L M_\mathrm{i} M | J_\mathrm{f} M_\mathrm{f})(J_\mathrm{i} L' M_\mathrm{i} M' | J_\mathrm{f} M_\mathrm{f}) = \frac{1}{2J_\mathrm{i}+1} \frac{2J_\mathrm{f}+1}{2L+1} \delta_{LL'} \delta_{MM'}. \tag{2.67}$$

From this, we have the following three rules.

Rule 1. There are no cross terms between different rank operators (or, more explicitly, no cross terms between Fermi and Gamow–Teller transitions).

Rule 2. There are no cross terms between operators with different M's.

Rule 3. The sum in Eq. (2.67) is independent of M.

These rules are valid only if the nuclei are not oriented in a particular direction. Calculation with nuclear orientation is discussed in the next chapter.

$|H_{\rm fi}|^2$ is evaluated with the trace method,

$$|H_{\rm fi}|^2 = \sum_{k,l=1}^{8} \sum_{\sigma_e, \sigma_\nu} C_k C_l^* (\psi_e^\dagger \Omega^k \psi_\nu)(\psi_e^\dagger \Omega^l \psi_\nu)^* \left(\int \Omega_k\right)\left(\int \Omega_l\right)^*. \tag{2.68}$$

Here the second summation refers to the sum over the electron and neutrino spins,

$$\begin{aligned}\sum_{\sigma_e,\sigma_\nu} (\psi_e^\dagger \Omega^k \psi_\nu)(\psi_e^\dagger \Omega^l \psi_\nu)^* &= \sum_{\sigma_e,\sigma_\nu} \sum_{\lambda,\mu,\rho,\delta} (\psi_\rho^{e\dagger} \Omega_{\rho\delta}^k \psi_\delta^\nu)(\psi_\lambda^{\nu\dagger} \Omega_{\lambda\mu}^{l\dagger} \psi_\mu^e) \\ &= \sum_{\sigma_e,\sigma_\nu} \sum_{\lambda,\mu,\rho,\delta} \Omega_{\rho\delta}^k \Omega_{\lambda\mu}^{l\dagger} \psi_\mu^e \psi_\rho^{e\dagger} \psi_\delta^\nu \psi_\lambda^{\nu\dagger} \\ &= \mathrm{Tr}(\Omega^k D_\nu \Omega^{l\dagger} D_e).\end{aligned} \tag{2.69}$$

In the first and second lines of Eq. (2.69), the particle names are set as superscripts, when the spinor components are expressed by the subscripts. D's are the projection operators, which are given by (A.95),

$$D_e = \tfrac{1}{2}\left[1 - \frac{c\boldsymbol{\alpha}\cdot\mathbf{p} + \beta mc^2}{E}\right] \quad \text{for the electron} \tag{2.70}$$

and

$$D_\nu = \tfrac{1}{2}\left[1 - \frac{c\boldsymbol{\alpha}\cdot\mathbf{q}}{E_\nu}\right] \quad \text{for the antineutrino} \tag{2.71}$$

where $\mathbf{q}$ is the momentum of the antineutrino, and the neutrino mass is assumed to vanish. We know that the matrices $\boldsymbol{\alpha}$, β, $\boldsymbol{\sigma}$ are traceless. Therefore, the nonzero values can be given only by the terms where the product of the even power of $\boldsymbol{\alpha}$, β, $\boldsymbol{\sigma}$, or γ_5 appears.

Example 1. Axial Vector

$M = 0$:

$$\begin{aligned}&\mathrm{Tr}\left(\sigma_z \tfrac{1}{2}\left[1 - \frac{c\boldsymbol{\alpha}\cdot\mathbf{q}}{E_\nu}\right]\sigma_z \tfrac{1}{2}\left[1 - \frac{c\boldsymbol{\alpha}\cdot\mathbf{p} + \beta mc^2}{E}\right]\right) \\ &\quad = 1 + \frac{-q_x p_x - q_y p_y + q_z p_z}{E_\nu E} c^2.\end{aligned} \tag{2.72}$$

$M = \pm 1$:

$$\begin{aligned}&\tfrac{1}{8}\,\mathrm{Tr}\left((\sigma_x \mp i\sigma_y)\left[1 - \frac{c\boldsymbol{\alpha}\cdot\mathbf{q}}{E_\nu}\right](\sigma_x \pm i\sigma_y)\left[1 - \frac{c\boldsymbol{\alpha}\cdot\mathbf{p} + \beta mc^2}{E}\right]\right) \\ &\quad = 1 - \frac{q_z p_z}{E_\nu E} c^2 \mp i\tfrac{1}{2}\,\mathrm{Tr}(\sigma_y D_\nu \sigma_x D_e) \pm i\tfrac{1}{2}\,\mathrm{Tr}(\sigma_x D_\nu \sigma_y D_e).\end{aligned} \tag{2.73}$$

Each trace should be multiplied by the nuclear matrix,

$$\frac{1}{2J_i + 1} \sum_{M_f, M_i} |\int \psi^{\dagger}_{J_f M_f} \sigma_M \psi_{J_i M_i} d\tau|^2 = \tfrac{1}{3} |\int \sigma|^2. \tag{2.74}$$

This coincides with the usual definition of $\int \sigma$, which absorbs the factor $(2J_f + 1)(2J_i + 1)^{-1}$ in Eq. (2.67). The factor $(2L + 1)^{-1}$, which is $\frac{1}{3}$, appears in Eq. (2.74) explicitly. Summing over M, we obtain

$$C_A^2 |\int \sigma|^2 \left(1 - \frac{\mathbf{q} \cdot \mathbf{p}}{3E_\nu E} c^2\right). \tag{2.75}$$

Multiplying Eq. (2.75) by the statistical shape (2.26), we obtain the beta-ray spectrum.

Example 2. Tensor

In the trace

$$\mathrm{Tr}(\Omega^k D_\nu \Omega^{l\dagger} D_e), \tag{2.76}$$

both Ω^k and $\Omega^{l\dagger}$ contain the matrix β. When β in $\Omega^{l\dagger}$ is moved from the right to the extreme left, $\boldsymbol{\alpha}$ changes sign, while $\boldsymbol{\sigma}$ and β commute. The final result is, therefore, different from that of axial vector only in the sign of $\mathbf{q} \cdot \mathbf{p}$,

$$C_T^2 |\int \beta\sigma|^2 \left(1 + \frac{\mathbf{q} \cdot \mathbf{p} c^2}{3E_\nu E}\right). \tag{2.77}$$

Example 3. Cross Term $\left(\int \sigma\right)\left(\int \beta\sigma\right)^*$ of Axial Vector and Tensor

$M = 0$:

$$\tfrac{1}{4}\,\mathrm{Tr}\left(\sigma_z \left[1 - \frac{c\boldsymbol{\alpha} \cdot \mathbf{q}}{E_\nu}\right] \beta\sigma_z \left[1 - \frac{c\boldsymbol{\alpha} \cdot \mathbf{p} + mc^2\beta}{E}\right]\right) = -\frac{mc^2}{E}. \tag{2.78}$$

$M = \pm 1$: These terms have the same result with Eq. (2.78). Hence we have

$$C_A C_T |\int \sigma|^2 \frac{mc^2}{E}, \tag{2.79}$$

where we have used a relation (2.61),

$$\int \beta\boldsymbol{\sigma} = -\int \boldsymbol{\sigma}. \tag{2.80}$$

The formula for the Fermi transition will be given in a similar way.

In Eqs. (2.75) and (2.77) the lepton momenta are explicitly given. The coefficient of the angular dependence $\mathbf{q} \cdot \mathbf{p} = qp \cos\theta$ changes with the coupling of the interaction, and is called the electron–neutrino angular correlation coefficient. This will be discussed in section 2.11. For the beta-ray spectrum, we have to integrate over the angular variables for leptons,

$$\left(\frac{1}{4\pi}\right)^2 \int d\Omega_e \int d\Omega_\nu, \tag{2.81}$$

and all $\mathbf{q}\cdot\mathbf{p}$ terms disappear. Here, $d\Omega$ is the surface element, which should not be confused with the transition operator Ω. Summarizing the result, we have the beta-ray spectrum

$$P(E)\,dE = \frac{1}{2\pi^3}\,PE(E_o - E)^2\,\xi\left[1 + \left(\frac{bmc^2}{E}\right)\right]dE \tag{2.82}$$

with

$$\xi = \left[(C_S^2 + C_V^2)|\textstyle\int 1|^2 + (C_T^2 + C_A^2)|\textstyle\int \sigma|^2\right], \tag{2.83}$$

$$\xi b = \pm 2\left[C_S C_V |\textstyle\int 1|^2 + C_A C_T |\textstyle\int \boldsymbol{\sigma}|^2\right] \qquad \text{for } \beta^{\mp} \text{ decay.} \tag{2.84}$$

Here

$$|\textstyle\int 1|^2 = \displaystyle\sum_{M_f} \left| \int \psi^\dagger_{J_f M_f} \sum_k \tau_\pm^{(k)} \psi_{J_i M_i} d\tau \right|^2 \tag{2.85}$$

and

$$|\textstyle\int \sigma|^2 = \displaystyle\sum_{M_f, M} \left| \int \psi^\dagger_{J_f M_f} \sum_k \tau_\pm^{(k)} \sigma_M^{(k)} \psi_{J_i M_i} d\tau \right|^2. \tag{2.86}$$

The relation of the reduced matrix elements in Eq. (2.66) to the matrices Eqs. (2.85) and (2.86) is

$$|\langle J_f \| \Omega \| J_i \rangle|^2 \frac{2J_f + 1}{2J_i + 1} = |\textstyle\int \Omega|^2. \tag{2.87}$$

For ξb in Eq. (2.84) the upper sign refers to the electron decay and the lower sign to the positron decay. This is seen from the projection operator (A.95), since the signs of $\mathbf{p}$, $\mathbf{q}$, and m are reversed in the positron decay. The effect of this modification appears only in the cross terms, called the Fierz interferences ([1]; see also section 2.8).

2.7 COULOMB CORRECTIONS

So far we have assumed that the electron wave functions as a plane wave. In reality, the Coulomb forces between the electron and the daughter nucleus are not negligible. The distortion from the plane wave due to the nuclear charges can be obtained by solving the Dirac equation with nuclear Coulomb potential. This will be done in the following. Readers who are not interested in this derivation can omit the material from here to Eq. (2.120).

In classical mechanics the equation of motion for a charged particle in a field is obtained from the free equations by replacing the energy $E \to E + e\phi$ and the momentum by $\mathbf{p} + (e/c)\mathbf{A}$. Here we are considering the case of the electron with charge $-e$ (e, positive). This will give the correct Lorentz force $-e(\mathbf{E} + \mathbf{v} \times \mathbf{H}/c)$. Substitutions

$$i\hbar \frac{\partial}{\partial t} \to i\hbar \frac{\partial}{\partial t} + e\phi, \qquad -i\hbar \nabla \to -i\hbar \nabla + \frac{e}{c} \mathbf{A}, \tag{2.88}$$

are made in the covariant free Dirac equations of motion and we have

$$\left(\frac{\partial}{\partial x_\mu} + \frac{ie}{\hbar c} A_\mu \right) \gamma_\mu \psi + \frac{mc}{\hbar} \psi = 0. \tag{2.89}$$

Let us consider a particle that is moving in an external static potential ϕ and write $V = -e\phi$. Then the wave equation for a stationary state with energy E is

$$E\psi = (-\boldsymbol{\alpha} \cdot \mathbf{p} - \beta + V(r))\psi. \tag{2.90}$$

Here we again use the atomic units $\hbar = m = c = 1$. The solutions were obtained by Rose in the case of the point nuclear charge. The outline of his method is as follows (for more detail, see [2]).

Let us start from the nonrelativistic description. As is well known, the spin of a particle in a central field gives rise to a spin-orbit coupling of the form

$$\phi(r)\mathbf{S} \cdot \mathbf{L}. \tag{2.91}$$

Assume $\phi(r)$ be a function that can be parameterized in the theory. The spin-orbit energy does not commute with the z component of the intrinsic spin S_z and orbital angular momentum L_z separately, but it commutes with the sum $J_z = S_z + L_z, \mathbf{J}^2$, $\mathbf{L}^2$, and $\mathbf{S}^2$. In this case we will introduce a spin angular function $\chi_{\kappa\mu}$ which diagonalizes the foregoing four operators,

$$\chi_{\kappa\mu} = \sum_m (l \tfrac{1}{2} \mu - m \;\; m | j\mu)\, Y_{l\mu-m}(\theta, \phi)\, \chi_m, \qquad m = \pm\tfrac{1}{2}, \tag{2.92}$$

with

$$Y_{lm}(\theta, \phi) = \frac{(-)^{l+m}}{2^l l!} \left[\frac{(2l+1)(l-m)!}{4\pi(l+m)!} \right]^{1/2} \sin^m\theta \left[\frac{\partial}{\partial \cos\theta} \right]^{l+m} \sin^{2l}\theta\, e^{im\phi}$$

and the intrinsic spin function χ_m defined in Eq. (A.73). The symbol κ specifies j and l simultaneously,

$$j = |\kappa| - \tfrac{1}{2}, \qquad l = \begin{cases} \kappa & \text{for } \kappa > 0, \\ -\kappa - 1 & \text{for } \kappa < 0. \end{cases} \tag{2.93}$$

Therefore $\kappa = -1, 1, -2, 2, \ldots$ refers to $s_{1/2}, p_{1/2}, p_{3/2}, d_{3/2}, \ldots$, respectively. We can observe from the definition the following properties of $\chi_{\kappa\mu}$.

$$
\begin{aligned}
\mathbf{J}^2\chi_{\kappa\mu} &= j(j+1)\chi_{\kappa\mu} = (|\kappa| - \tfrac{1}{2})(|\kappa| + \tfrac{1}{2})\chi_{\kappa\mu},\\
\mathbf{L}^2\chi_{\kappa\mu} &= l_\kappa(l_\kappa + 1)\chi_{\kappa\mu} = \kappa(\kappa+1)\chi_{\kappa\mu},\\
J_z\chi_{\kappa\mu} &= \mu\chi_{\kappa\mu},\\
\mathbf{S}^2\chi_{\kappa\mu} &= \tfrac{3}{4}\chi_{\kappa\mu}.
\end{aligned} \tag{2.94}
$$

The eigenvalue κ is given by the dynamic variable

$$-(1 + \mathbf{L}\cdot\boldsymbol{\sigma})\chi_{\kappa\mu} = \kappa\chi_{\kappa\mu}. \tag{2.95}$$

This is derived by using the relation

$$2\mathbf{S}\cdot\mathbf{L} = \mathbf{J}^2 - \mathbf{L}^2 - \mathbf{S}^2. \tag{2.96}$$

The nonrelativistic plane wave for an electron

$$\chi_m \exp\,[i(\mathbf{p}\cdot\mathbf{r} - Et)] \tag{2.97}$$

can be expressed as a superposition of the partial waves of the definite values of angular momenta, j, l, s, j_z. To do this we use the Rayleigh expansion

$$\exp\,(i\mathbf{p}\cdot\mathbf{r}) = 4\pi \sum_{l,m'} i^l j_l(pr) Y^*_{lm'}(\hat{\mathbf{p}}) Y_{lm'}(\hat{\mathbf{r}}) \tag{2.98}$$

and

$$\chi_m \exp\,(i\mathbf{p}\cdot\mathbf{r}) = 4\pi \sum_{\kappa\mu} i^l (l\tfrac{1}{2}\mu - m\; m|j\mu) Y^*_{l\mu-m}(\hat{\mathbf{p}})\chi_{\kappa\mu} j_l(pr). \tag{2.99}$$

For the relativistic descriptions, the plane wave solution is given in Eqs. (A.76), (A.80), and (A.81)

$$\psi(\mathbf{r}, t) = \left(\frac{E_+ + 1}{2E_+}\right)^{1/2} \begin{pmatrix} \dfrac{-c\boldsymbol{\sigma}\cdot\mathbf{p}}{E_+ + 1}\chi_m \\ \chi_m \end{pmatrix} \exp(i\mathbf{p}\cdot\mathbf{r} - Et). \tag{2.100}$$

From Eq. (2.99) we see easily that the large component (which is the lower one in our representation) can be decomposed into the spherical waves with j and l, while the corresponding small component (the upper one) has the same j, but a different l value, because of the $\boldsymbol{\sigma}\cdot\mathbf{p}$ operator. Therefore, the small component can be given by a multiple of $\chi_{\kappa\mu}$. As a consequence, we introduce a relativistic spherical wave in the form

$$\psi_{\kappa\mu}(\mathbf{r}) = \begin{pmatrix} -if_\kappa\chi_{-\kappa\mu} \\ g_\kappa\chi_{\kappa\mu} \end{pmatrix}, \tag{2.101}$$

where f_κ and g_κ are radial wave functions. $\psi_{\kappa\mu}$ satisfies

$$\begin{aligned} \mathbf{J}^2 \psi_{\kappa\mu} &= j(j+1)\psi_{\kappa\mu} \qquad \text{with} \qquad \mathbf{J} = \mathbf{L} + \tfrac{1}{2}\boldsymbol{\sigma}, \\ \mathbf{S}^2 \psi_{\kappa\mu} &= \tfrac{3}{4}\psi_{\kappa\mu}, \\ J_z \psi_{\kappa\mu} &= \mu\psi_{\kappa\mu}. \end{aligned} \tag{2.102}$$

However, $\psi_{\kappa\mu}$ is not an eigenfunction of $\mathbf{L}^2$ because

$$\mathbf{L}^2 \chi_{\mp\kappa\mu} = \mp\kappa(1 \mp \kappa)\chi_{\mp\kappa\mu}. \tag{2.103}$$

This means that $\mathbf{L}^2$ is not diagonal and the orbital angular momentum is not a constant of the motion.

Inserting Eq. (2.101) into the Dirac equation (2.89), we have a pair of coupled differential equations for radial wave functions,

$$\begin{aligned} \frac{df_\kappa}{dr} &= \frac{\kappa - 1}{r} f_\kappa - (E - 1 - V)g_\kappa, \\ \frac{dg_\kappa}{dr} &= (E - V + 1)f_\kappa - \frac{\kappa + 1}{r} g_\kappa. \end{aligned} \tag{2.104}$$

The solutions regular at $r = 0$ of these equations for the free equation $V = 0$ are

$$g_\kappa = \pi^{-1/2} p j_l(pr), \qquad f_\kappa = \frac{p^2}{E+1} \pi^{-1/2} S_\kappa j_{\bar{l}}(pr). \tag{2.105}$$

Here S_κ is the sign of κ

$$S_\kappa = \frac{\kappa}{|\kappa|} \tag{2.106}$$

and

$$\bar{l} = l - S_\kappa,$$

which refers to the orbital angular momentum for $\chi_{-\kappa\mu}$.

Distortion of the radial wave function due to the nuclear charge can be taken into account as follows. Most of the energetic electrons in beta decay have the de Broglie wavelength, which is long compared with nuclear dimensions. Therefore, it is usually sufficient to treat the nucleus as a point charge. The solutions of the equations (2.104), with

$$V = -\frac{Ze^2}{r} = -\frac{\alpha Z}{r} \tag{2.107}$$

where r is given in units of $\hbar/mc$, are given by Rose [2] as

$$g_\kappa = \left[\frac{p(E+1)}{\pi}\right]^{1/2} (Q + Q^*), \qquad f_\kappa = i\left[\frac{p(E-1)}{\pi}\right]^{1/2} (Q - Q^*), \tag{2.108}$$

where

$$Q = (\gamma + i\nu) \frac{|\Gamma(\gamma + i\nu)|}{\Gamma(2\gamma + 1)} (2pr)^{\gamma - 1}$$

$$\times \exp\left(\tfrac{1}{2}\pi\nu - ipr + i\eta\right) F(\gamma + 1 + i\nu, 2\gamma + 1; 2ipr). \tag{2.109}$$

The last factor F is the confluent hypergeometric function, which has the series representation

$$F(a, b; z) = \frac{\Gamma(b)}{\Gamma(a)} \sum_{n=0}^{\infty} \frac{\Gamma(a+n)}{\Gamma(b+n)} \frac{z^n}{n!}, \tag{2.110}$$

and Γ's are the gamma functions. The other symbols are

$$\gamma = (\kappa^2 - \alpha^2 Z^2)^{1/2} \qquad \text{and} \qquad \nu = \alpha Z E/p. \tag{2.111}$$

The phase factor is given by

$$\exp(2i\eta) = -\frac{(\kappa - i\alpha Z/p)}{(\gamma + i\nu)}. \tag{2.112}$$

The spherical wave $\psi_{\kappa\mu}$, normalized per unit energy, is thus

$$\int \psi^{\dagger}_{\kappa\mu}(E)\psi_{\kappa'\mu'}(E')\, d\mathbf{r} = \delta_{\kappa\kappa'}\delta_{\mu\mu'}\delta(E - E'). \tag{2.113}$$

Now we are ready to compute the beta-ray spectral shape by using the Coulomb field solutions for the electron and spherical wave solutions for the neutrino. Instead of the projection operator D, we introduce the matrix V,

$$V = \sum_{\kappa,\mu} \psi_{\kappa\mu}\psi^{\dagger}_{\kappa\mu}. \tag{2.114}$$

The trace in the case of the plane waves

$$\mathrm{Tr}(\Omega^k D_\nu \Omega^{l\dagger} D_e) \tag{2.115}$$

is now replaced by

$$\mathrm{Tr}(\Omega^k V_\nu \Omega^{l\dagger} V_e). \tag{2.116}$$

The range of the sum over κ is $\kappa = \pm 1$. In the allowed transitions, the orbital angular momentum of the lepton system is zero. Besides, the de Broglie wavelength is large compared with the nuclear dimension. Therefore, only the s waves of the electron and antineutrino have contributions. This restricts the range of the sum over κ. The following are the explicit forms of the relevant wave functions.

Electron:

$$\psi_{-1,1/2} = \begin{pmatrix} - \\ - \\ Y_{00}g_{-1} \\ - \end{pmatrix}, \qquad \psi_{-1,-1/2} = \begin{pmatrix} - \\ - \\ - \\ Y_{00}g_{-1} \end{pmatrix},$$

$$\psi_{1,1/2} = \begin{pmatrix} Y_{00}f_1 \\ - \\ - \\ - \end{pmatrix}, \qquad \psi_{1,-1/2} = \begin{pmatrix} - \\ Y_{00}f_1 \\ - \\ - \end{pmatrix}. \tag{2.117}$$

Neutrino (negative energy solution):

$$\psi_{-1,1/2} = \begin{pmatrix} - \\ Y_{00}g^{\nu}_{-1} \\ - \\ - \end{pmatrix}, \qquad \psi_{-1,-1/2} = \begin{pmatrix} Y_{00}g^{\nu}_{-1} \\ - \\ - \\ - \end{pmatrix},$$

$$\psi_{1,1/2} = \begin{pmatrix} - \\ - \\ - \\ Y_{00}f^{\nu}_1 \end{pmatrix}, \qquad \psi_{1,-1/2} = \begin{pmatrix} - \\ - \\ Y_{00}f^{\nu}_1 \\ - \end{pmatrix}. \tag{2.118}$$

The components denoted by a dash do not contribute to the beta decay considered here. These wave functions give

$$\begin{aligned} V_e &= \tfrac{1}{2} Y_{00}^2 [I(g_{-1}^2 + f_1^2) - \beta(g_{-1}^2 - f_1^2)], \\ V_\nu &= \tfrac{1}{2} Y_{00}^2 \{I[(g^{\nu}_{-1})^2 + (f^{\nu}_1)^2] + \beta[(g^{\nu}_{-1})^2 - (f^{\nu}_1)^2]\} = I Y_{00}^2 q^2, \end{aligned} \tag{2.119}$$

where I is the 4 x 4 unit matrix and β is the matrix in Eq. (2.56). For the spherical harmonics Y_{00} we have

$$Y_{00} = \left(\frac{1}{4\pi}\right)^{1/2}.$$

The radial wave functions g_{-1} and f_1 given by Eq. (2.108) are normalized in the energy scale. Now we use those normalized to one particle in a sphere of the unit radius. They are given by Eq. (2.108), by multiplying an additional factor $(\pi p/E)^{1/2}$. In the same normalization, we have $IY_{00}^2 q^2$ for V_ν in Eq. (2.119) from the radial wave functions of the neutrino (Eq. (2.105) with a vanishing rest mass). The beta-ray spectrum is given by Eqs. (2.14) and (2.116),

$$P(E)\,dE = 2\pi |H_{\mathrm{fi}}|^2 \frac{dn}{dE_o}$$

$$= 2\pi \left(\frac{1}{\pi}\frac{E}{p}\,dE \frac{1}{4\pi}\int d\Omega_{\mathrm{e}}\right)\left(\frac{1}{\pi}\frac{1}{4\pi}\int d\Omega_\nu\right) \mathrm{Tr}(\Omega^k V_\nu \Omega^{l\dagger} V_{\mathrm{e}}) \tag{2.120}$$

Here, the first parentheses stand for the number of the final state for the electron, and the second for the neutrino. The factors $(1/\pi)(E/p)\,dE$ and $(1/\pi)$ are the state densities obtained under the condition that the standing waves take place in the unit sphere. For the evaluation of the spatial integral, we again assume that the electron wavelength is large compared with the nuclear dimension and the electron wave functions are almost constant in the nucleus. The electron wave functions g and f are approximated with the values at the nuclear surface, $g(R)$ and $f(R)$, and we put them out of the nuclear matrix elements. As a consequence, the same definition of the Fermi and Gamow–Teller matrix elements holds as in section 2.6. Finally, the beta-ray spectrum becomes

$$P(E)\,dE = \frac{1}{2\pi^3}\,pE(E_o - E)^2 F(\pm Z, E)\xi\left(1 \pm 2\gamma\frac{b}{E}\right), \tag{2.121}$$

where $F(\pm Z, E)$ is usually divided into two parts L_0 and F_0,

$$F = F_0 L_0, \tag{2.122}$$

with

$$L_0 = \frac{(g_{-1}^2 + f_1^2)}{2p^2 F_0} \approx \frac{1+\gamma}{2}, \tag{2.123}$$

$$F_0 \equiv F_0(\pm Z, E) = 4(2pR)^{2\gamma-2}\exp(\pm\pi\nu)\frac{|\Gamma(\gamma \pm i\nu)|^2}{(\Gamma(2\gamma+1))^2}, \tag{2.124}$$

and

$$\gamma = (1 - \alpha^2 Z^2)^{1/2} \qquad \text{and} \qquad \nu = \alpha Z E/p. \tag{2.125}$$

We have also used a relation

$$g_{-1}^2 - f_1^2 \approx (\gamma/E)(g_{-1}^2 + f_1^2). \tag{2.126}$$

In the case of $(\alpha Z)^2 \ll 1$,

$$\gamma \to 1 \qquad \text{and} \qquad L_0 \to 1. \tag{2.127}$$

The deviation from the statistical shape $pE(E_o - E)^2$ due to the Coulomb distortion is, therefore, expressed by $F(Z, E)$, which is called the Fermi function. The effect of the Fermi function may be clearly seen in an example of Cu^{64}, which decays by emitting either an electron or a positron [3,4]. As is shown in Fig. 2.2, experimental data fit the spectra with Coulomb correction (2.121), and deviate substantially from the statistical spectra given by the solid curves. In general, the spectrum for positrons deviates to the higher

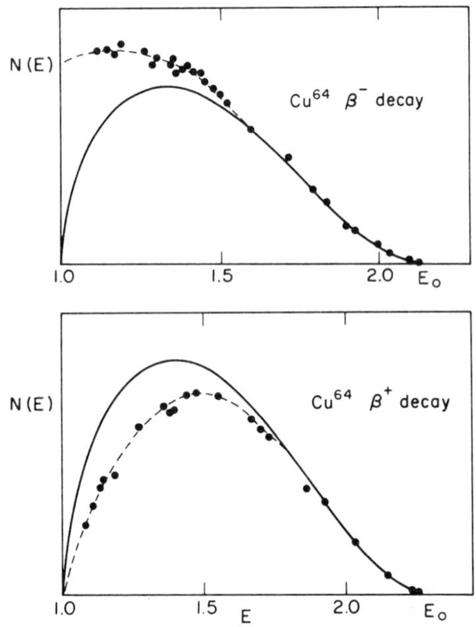

Fig. 2.2 Coulomb corrections for beta-ray spectra. The solid curves are statistical spectra, while the dashed curves are the spectra with Coulomb correction, Eq. (2.121). (These figures appear on page 11 of [3].) The experimental data $N(E)$, represented by dots, fit the dashed curves [4]. $N(E)$ is proportional to $P(E)$ and is given in arbitrary units.

energy side, since the nuclear Coulomb field accelerates the positrons, while the spectrum for electrons deviates to the lower energy side.

In the case of the nonrelativistic electron, the Fermi function can be approximated by

$$F(Z, E) = \frac{2\pi\nu}{1 - \exp(-2\pi\nu)}. \tag{2.128}$$

This can be proved by using a property of the gamma function

$$\Gamma(z)\Gamma(1 - z) = \frac{\pi}{\sin \pi z} \tag{2.129}$$

From Eqs. (2.121) and (2.124) we see that the energy spectrum for the electron has nonzero value at the lowest energy, $E = 1$, and the spectrum for the positron behaves like $\exp(-\text{constant}/p)$.

The Coulomb correction is included in the two parts, $F_0(Z, E)$ and L_0. For the former, the energy dependence is strong in the low energy region, while for the latter it is weak in every energy region. A numerical table for the Fermi function was made by Fano [5] in 1952. In order to take the Coulomb correction precisely, we have to examine the assumption for deriving Eq. (2.121) more carefully. The electron wave functions f_1 and g_{-1} originally in the integrand of nuclear matrix elements have a certain radial dependence. We first put them outside of the integral, and replace them by the values at the nuclear surface $r = R$. Strictly speaking, this procedure is valid only if

$$\int \psi_f^\dagger (r/R)^n \psi_i \, d\tau = \int \psi_f^\dagger \psi_i \, d\tau \qquad \text{for all } n. \tag{2.130}$$

Here n is a positive integer. In practice, this gives a good approximation in the case of allowed transitions. However, if we need more accurate values of F_0 and L_0, or if we study the forbidden transitions, we have to take into account the following three effects.

Finite de Broglie Wavelength Effect

In order to evaluate L_0 in Eq. (2.123), we adopted the values of g_{-1} and f_1 in Eq. (2.108) at $r = R$ and expanded in a power series of pR ($pR \ll 1$). The first term of this expansion gives Eq. (2.123). The next few terms are given by

$$\begin{aligned} L_0 &= \frac{(g_{-1}^2 + f_1^2)}{2p^2 F_0} \\ &= \frac{1}{2}(1 + \gamma) \mp \frac{5}{3}\alpha Z R E \mp \frac{\alpha Z R}{3E} - \frac{1}{3}p^2 R^2 + \cdots . \end{aligned} \tag{2.131}$$

The deviation from the approximation given in Eq. (2.123), that is, the contribution from terms other than the first term, is called the finite de Broglie wavelength effect. Numerical tables have been prepared for L_0 and the other spectral functions L_i, M_i, N_i (which will appear in the case of the forbidden transitions) in [6]. The finite de Broglie wavelength effect is negligibly small in the allowed transition, except for the high energy region.

Finite Nuclear Size Correction

If we take into account that the nuclear charge has a certain spatial distribution, the electron wave functions given by Eq. (2.108) should be modified. A correction due to this modification is called the finite nuclear size correction.

Rose and his collaborators made a calculation of the electron wave function, assuming the Coulomb field of the daughter nucleus to be

$$V = -\alpha Z/r \qquad \text{for} \quad r \geqslant R,$$
$$V = (-\alpha Z/2r)[3 - (r/R)^2] \qquad \text{for} \quad r < R, \tag{2.132}$$

with

$$R = 1.4A^{1/3} \text{ fm} \qquad (1 \text{ fm} = 10^{-13} \text{ cm}). \tag{2.133}$$

This potential corresponds to the uniform charge distribution in the nuclear volume. Rose and Holmes prepared numerical tables for the spectral functions L_i, M_i, and N_i [7]. Yamada also studied the finite nuclear size correction [8]. He noticed that the finite nuclear size correction becomes appreciable if the main part of the expression for the spectrum is dropped by accidental cancellation among nuclear matrix elements, e.g., in the case of the RaE beta-ray spectrum [9]. Extensive numerical studies of spectral shape were also performed by his group [10]. The same corrections in the beta-ray angular distributions were calculated by Dolginov and Toptigin [11] and by Matumoto [12]. (Here we mean the angular distributions as angular distributions with respect to the polarization axis of the nucleus or the direction of emission of the other particles.) Recently, Bühring investigated the beta-ray spectrum and longitudinal polarization in the first forbidden transitions in a similar way [13].

Generally speaking, the finite nuclear size correction is negligibly small for the beta-ray spectrum in the allowed transitions. The correction is, however, large, and sometimes reaches up to 50% for some forbidden spectral functions. Without this correction, we cannot study the beta-ray spectra and the problem of beta-ray angular distributions in the case of the cancellation among nuclear matrix elements.

Combination of the Two Effects

With the help of the electronic computer, we can take into account the finite nuclear size correction and the finite de Broglie wavelength effect simultaneously. Bhalla and Rose prepared numerical tables for the electron wave functions f_κ and g_κ [14], with the nuclear Coulomb potential (Eq. (2.132)), but with a more realistic nuclear radius

$$R = 1.2A^{1/3} \text{ fm}. \tag{2.134}$$

Their tables are sufficient to compute the beta-ray spectrum, longitudinal polarization of beta particles, beta–gamma angular correlations, etc. A correction to Bhalla and Rose's tables is reported by Bühring in the case of positron decay [15].

Until now, we have evaluated the lepton wave functions at the nuclear surface $r = R$. The combination, such as

$$L_0 = \left(\frac{g_{-1}^2 + f_1^2}{2p^2 F_0} \right)_{r=R}, \tag{2.135}$$

is expanded in the power series of pR and numerically evaluated. This is based on the assumption (2.130). If we omit this simplification, we have to introduce the matrix elements

$$\int r^2, \quad \int \sigma r^2, \quad \text{etc.} \tag{2.136}$$

Yamada devised a theory for taking account of this complication when evaluating the combined effect of finite nuclear size and the finite de Broglie wavelength [10]. Bühring investigated a systematic way of introducing the nuclear matrix elements for the same purpose [13]. A similar work was performed by Sodeman and Winther [16] for the beta decay of RaE.

Screening Effect

The nuclear Coulomb field is screened by the orbital electron cloud. The correction was estimated by Rose [17] and Longmire and Brown [18] using the modified WKB procedure, according to which the electron energy is shifted by an amount of the potential energy at the nucleus caused by the screening. This is because the negative charge distribution of the orbital electron accelerates the negative beta particles by the Coulombian repulsion, and changes the energy of the beta particle $E - V_o$ into E by the acceleration. Therefore the factor

$$pEF(\pm Z, E) \tag{2.137}$$

in Eq. (2.121) should be replaced by

$$[(E \mp V_o)^2 - 1]^{1/2} (E \mp V_o) F(\pm Z, E \mp V_o). \tag{2.138}$$

According to the Fermi-Thomas model,

$$V_o \approx 1.13 \alpha^2 Z^{4/3}. \tag{2.139}$$

Alternatively, numerically evaluated electron wave functions in a screening field are used. Such results have been obtained by Reitz for a few elements [19]. Screening corrections for electrons are seldom larger than 2% or 3% except for the heaviest elements. For positrons, corrections are appreciably larger and effects due to screening have been detected even for moderate Z. There is a marked discrepancy between the electron screening corrections by Rose, Longmire, and Brown and those by Reitz. Recently, Durand [20], Bühring [13], Yamada [21], and Bhalla [22] reexamined the screening effects by solving the Dirac equations numerically [21] or power series expansion in pr [13], and by using the Hulten model for the screened field with which the Schrödinger and Klein-Gordon equations are analytically solvable [20]. They reported that their results are in agreement with those obtained by Rose, Longmire, and Brown.

2.8 KURIE PLOT AND SPECTRAL SHAPE FACTOR

The energy spectrum of beta rays is given by Eq. (2.121). If we neglect a weak energy dependence of the type $1/E$, the energy spectrum is

$$P(E)\,dE = \text{const. } F(\pm Z, E)pE(E_o - E)^2\,dE. \tag{2.140}$$

The quantity $P(E)$ in formula (2.140), or a quantity proportional to it $N(E)$, is what we measure with a beta-ray spectrometer. In order to test the predic-

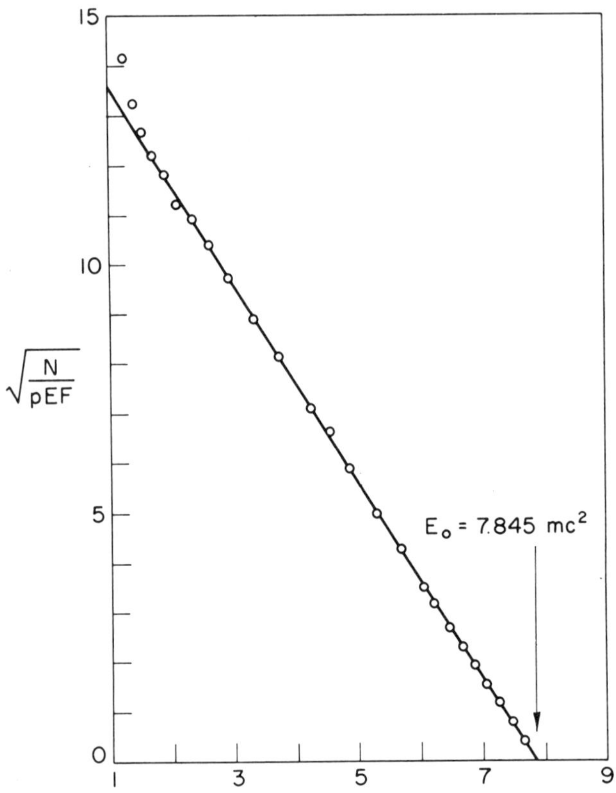

Fig. 2.3 Kurie plot of the He^6 beta-ray spectrum [24]. $K(E) = (N/pEF)^{1/2}$ is given in arbitrary units.

tion of the theory with experimental observation, we conventionally plot the quantity $K(E)$,

$$K(E) = \left[\frac{N(E)}{pEF(Z, E)}\right]^{1/2}, \tag{2.141}$$

versus the electron energy E. This is called Kurie plot [23]. If the theoretical spectrum Eq. (2.140) agrees with experimental data,

$$K(E) = \text{const.}\,(E_o - E) \tag{2.142}$$

and the Kurie plot is a straight line crossing the abscissa at the maximum energy E_o; see the Kurie plot of the He^6 beta-ray spectrum [24], Fig. 2.3. The linearity of the He^6 Kurie plot is in the energy range 0.25–3.50 MeV. Most of the precision experiments of the allowed beta-ray spectra confirm the allowed shape in Eq. (2.140). However, some allowed spectra of high Z or of large ft values could have a slight deviation from the straight Kurie plot at very low energies. (Here, the ft values will be given in section 2.10.) These effects could be understood in terms of several Coulombian effects in the last section and of contributions from nuclear matrix elements of forbidden transitions. If these effects are present, the beta-ray spectrum would have the form

$$P(E)\,dE = \frac{1}{2\pi^3}\,\xi F(\pm Z, E)pE(E_o - E)^2 C(E)\,dE, \tag{2.143}$$

where $C(E)$ represents the electron energy dependence other than the allowed shape. The $C(E)$, called the spectral shape factor, have the form

$$C(E) = 1 + aE + \frac{b}{E} + cE^2 + \cdots \tag{2.144}$$

for allowed transitions [25]. The magnitudes of the energy-independent coefficients a, b, and c are small compared with unity. These coefficients are dependent on more detailed consideration of the theory. For example, the effect of the beta decay of the meson cloud around nucleons gives the term aE, which is observable in the beta decays of B^{12} and N^{12} (see section 4.3).

If the spin and parity changes of the beta decay are different from those of the Fermi and Gamow–Teller transitions, they are called the forbidden transitions. In these cases $C(E)$ may have the same form as Eq. (2.144), but a, b, c may be quite different values from those in allowed transitions. For example, the Kurie plot for the beta decay of $Y^{91} \rightarrow Sr^{91} + e^- + \bar{\nu}$ is given in Fig. 2.4 [26], where Fig. 2.4a corresponds to the Kurie plot (2.141) and

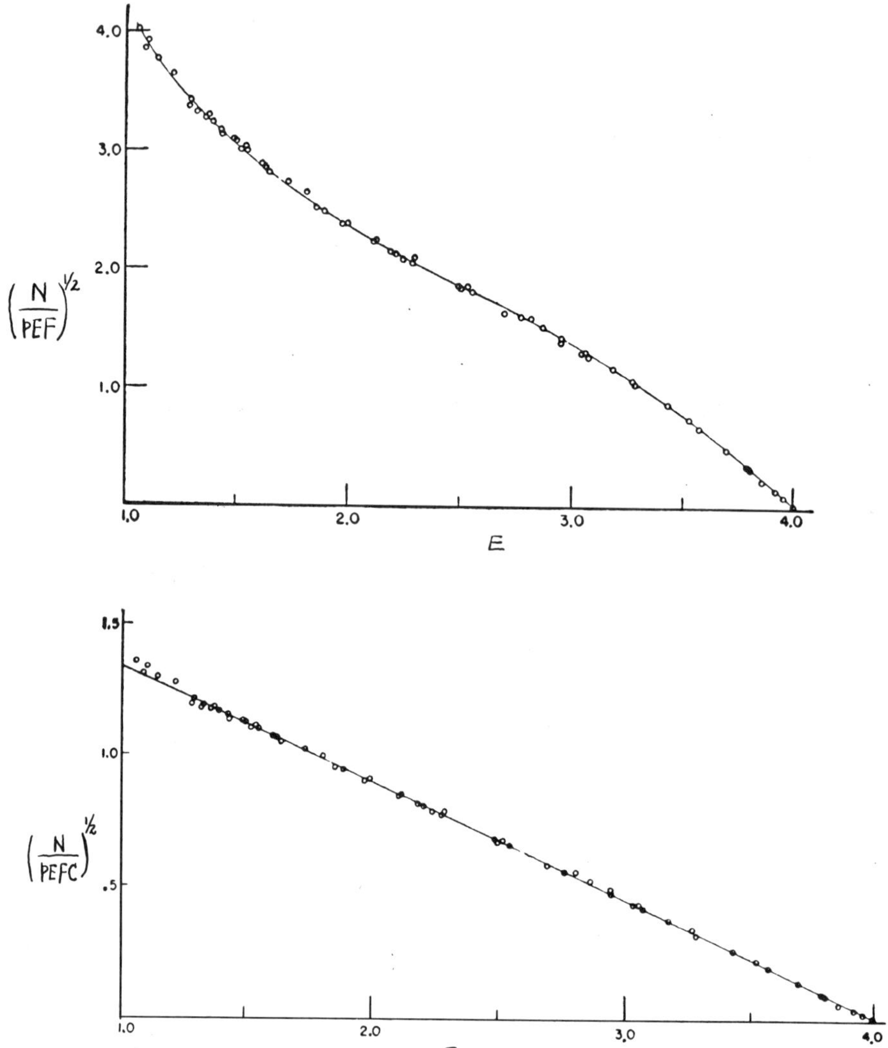

Fig. 2.4 Kurie plots of the Y^{91} beta-ray spectrum [26]. (a) Kurie plot, (2.141). (b) Kurie plot with the spectral shape factor, (2.146).

Fig. 2.4b corresponds to the modified Kurie plot

$$K(E) = \left[\frac{N(E)}{pEF(Z, E)C(E)} \right]^{1/2} \tag{2.145}$$

corrected by the spectral shape factor for the unique first forbidden transition

$$C(E) = p^2 + (E_o - E)^2. \tag{2.146}$$

Here the decay scheme of Y^{91} is

$$\tfrac{1}{2}^- \xrightarrow{\beta^-} \tfrac{5}{2}^+.$$

Fierz Interferences

As shown in Eq. (2.82), there is an interference term b/E in the spectrum. If this is the case, the Kurie plot will deviate slightly from the allowed shape, and the correction factor

$$C(E) = 1 + (b/E) \tag{2.147}$$

is required to make the Kurie plot straight. The Fierz term has been investigated for the determination of the relative strength C_S/C_V and C_T/C_A of the interactions. Since, however, the energy dependence $1/E$ is small, even the finest experiments can limit a relatively loose upper bound for b. On the other hand, the orbital electron capture by the nucleus, which is a transition analogous to the positron decay, is sensitive to the Fierz interference of the form b instead of b/E. Therefore, the ratio of the orbital electron capture to positron emission gives more strict limits for b:

$$b = -0.01 \pm 0.02 \qquad \text{for Gamow-Teller transitions [27]}, \tag{2.148}$$

and

$$b \lesssim 0.06 \qquad \text{for Fermi transitions [28]}. \tag{2.149}$$

Sherr investigated ft values that correspond to the integration of the spectrum (2.143) for several $0^+ \to 0^+$ transitions, and gave a limit

$$b = 0.00 \pm 0.10 \qquad \text{for Fermi transitions [29]}. \tag{2.150}$$

Since the discovery of parity nonconservation in beta decay, which is discussed in detail in Chapter 3, the Fierz interferences have had to be revised as follows.

$$b \propto \left[(C_V C_S + C_V' C_S')|\textstyle\int 1|^2 + (C_A C_T + C_A' C_T')|\textstyle\int \sigma|^2\right]. \tag{2.151}$$

Here, coupling constants C_i' also appear. Furthermore, the beta-ray angular distribution from polarized Co^{60} indicates that

$$C_T = -C_T' \qquad \text{and} \qquad C_A = C_A', \tag{2.152}$$

approximately. This gives a vanishing Fierz interference for Gamow-Teller interactions. We have to determine C_T/C_A by some other experiments which involve measurements of nuclear recoils. This is also the case for Fermi interactions.

2.9 NEUTRINO MASS AND SPECTRAL SHAPE NEAR END POINT ENERGY E_o

The statistical spectrum in Eq. (2.30) is derived under the assumption of zero neutrino mass, $m_\nu = 0$. The beta-ray spectrum with $m_\nu \neq 0$ is given by

$$P(E)\,dE = \frac{1}{4\pi^3} F(\pm Z, E) pE(E_o - E + m_\nu)[(E_o - E + m_\nu)^2 - m_\nu^2]^{1/2} \times \left[(C_V^2 + C_V'^2) + (C_V'^2 - C_V^2)\frac{m_\nu m}{E_\nu E} \right] dE. \tag{2.153}$$

The formula is given only for the vector interaction, for simplicity. Since the units $\hbar = m = c = 1$ are adopted, m in the last term is unity. It is, however, kept for convenience. The $m_\nu m/E_\nu E$ term is very small and vanishes in the limit $C_V' = C_V$. From Eq. (2.153) the derivative of the beta-ray spectrum with respect to the electron energy is

$$\begin{aligned} \frac{\partial P(E)}{\partial E} &= 0 \qquad \text{for } m_\nu = 0, \\ &= -\infty \qquad \text{for } m_\nu \neq 0. \end{aligned} \tag{2.154}$$

Therefore, the upper limit of the neutrino mass can be studied by precise measurement of the beta-ray spectrum near the end point energy. This is done by Langer and Moffat [30] in the case

$$\begin{aligned} &H^3 \to He^3 + \beta^- + \bar{\nu}. \\ &\tfrac{1}{2}^+ \to \tfrac{1}{2}^+ \end{aligned} \tag{2.155}$$

They obtained the upper limit for the beta neutrino mass

$$m_{\nu_\beta} c^2 < 250 \text{ eV}, \tag{2.156}$$

which is less than 1/2000 times the electron mass, Fig. 2.5. A lower value of 60 eV for Eq. (2.156) was reported recently by Bergkvist [31]. (There are two kinds of neutrinos in nature; see section 5.8. One is the beta neutrino accompanying the beta particle, and the other is the muon neutrino accompanying the muon.)

There is no such precision experiment for the muon neutrino mass. The upper limit is only given by the energy-momentum conservation in the $\pi \to \mu + \nu$ decay [32], which gives

$$m_{\nu_\mu} c^2 < 1.5 \text{ MeV}. \tag{2.157}$$

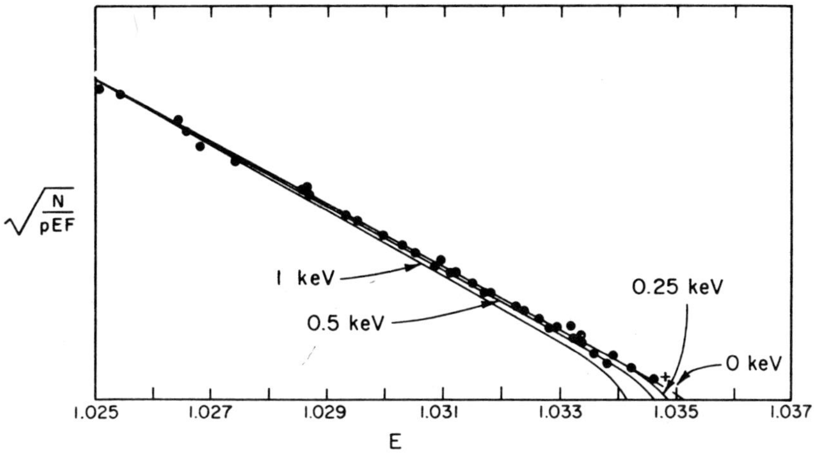

Fig. 2.5 Kurie plot of the H^3 beta-ray spectrum [30], with different values of neutrino mass, which are shown in units of kiloelectron volts (keV).

2.10 $ft_{1/2}$ VALUES AND COUPLING CONSTANTS

We derived the probability of the emission of beta particles with energy E to $E + dE$ per second in sections 2.6 and 2.7. Now, we investigate the total probability of the beta decay per second. This is obtained by integrating Eq. (2.121) over E. We define the integrated Fermi function f as

$$f \equiv f(\pm Z, E_o) = \int_1^{E_o} F(\pm Z, E)(E^2 - 1)^{1/2} E(E_o - E)^2 \, dE. \tag{2.158}$$

The probability of beta decay per second is given by

$$W = \frac{mc^2}{\hbar} \frac{1}{2\pi^3} \left(C_F^2 |\textstyle\int 1|^2 + C_{GT}^2 |\textstyle\int \sigma|^2 \right) f, \tag{2.159}$$

with

$$C_F^2 = C_S^2 + C_V^2 \qquad \text{and} \qquad C_{GT}^2 = C_T^2 + C_A^2. \tag{2.160}$$

The Fierz interference terms are set to be zero for the reason described in section 2.8. We will also conclude, in section 2.11, that

$$C_F^2 = C_V^2, \qquad C_S = 0, \qquad C_{GT}^2 = C_A^2, \qquad C_T = 0. \tag{2.161}$$

We notice that all the quantities on the right-hand side of Eq. (2.159) are given in units of $\hbar = m = c = 1$, whereas the left-hand side is given in inverse seconds. The mean life τ and half life $t_{1/2}$ are given by

$$W = \frac{1}{\tau} = \frac{\log 2}{t_{1/2}}. \tag{2.162}$$

The half life of beta decay depends on E_o and Z through the function f. If the Coulomb effect is omitted, the dependence on E_o is analytically seen as follows. That is, in the case of

$$F(Z, E) = 1 \qquad \text{for } Z = 0, \tag{2.163}$$

the function f is given by

$$\begin{aligned} f_{Z=0} &= (E_o^2 - 1)^{1/2}\left(\frac{E_o^4}{30} - \frac{3E_o^2}{20} - \frac{2}{15}\right) + \frac{E_o}{4}\log(E_o + (E_o^2 - 1)^{1/2}) \\ &\approx 0.216(E_o - 1)^{7/2} \qquad \text{if} \quad E_o - 1 \ll 1, \\ &\approx \frac{1}{30} E_o^5 \qquad \text{if} \quad E_o \gg 1. \end{aligned} \tag{2.164}$$

However, an exact value of f is required, and it is obtained by numerical integration of the Fermi function, or more quickly from the monograph prepared by Moszkowsi and Jantzen [33].

The half life is also dependent on the magnitude of $|\int 1|^2$ and $|\int \sigma|^2$. The square of the matrix is inversely proportional to a quantity called the $ft_{1/2}$ value,

$$ft_{1/2} = \frac{\hbar}{mc^2}\frac{2\pi^3 \log 2}{C_F^2 |M|^2} = \frac{B}{|M|^2} \tag{2.165}$$

with

$$|M|^2 = |\textstyle\int 1|^2 + \left(\dfrac{C_{\mathrm{GT}}}{C_{\mathrm{F}}}\right)^2 |\textstyle\int \sigma|^2.$$

In this formula, f and $t_{1/2}$ can be obtained experimentally, while B is a universal constant. (More precisely, f can be calculated with an experimentally known value of E_o.) This means that from experiments we find the magnitude of $|\int 1|^2$ and $|\int \sigma|^2$ if the constant B and the ratio $C_{\mathrm{GT}}/C_{\mathrm{F}}$ are known. Conversely, if $|M|^2$ is known theoretically, we find the magnitude of the coupling constants C_{GT}^2 and C_{F}^2. Let us study the latter case first.

Coupling Constants of Beta Decay

The $0^+ \to 0^+$ transition depends only on the Fermi matrix element. An example of this is the positron decay of Al^{26m} (see Fig. 2.6). It is the transition among the members of an isotriplet. Therefore, the Fermi matrix is given by (see section 7.1)

$$|\int 1|^2 = 2. \tag{2.166}$$

Freeman *et al.* obtained experimental results on the half life and end point energy [34]

$$t_{1/2} = 6.346 \pm 0.005 \text{ sec}, \qquad E_{\max} = E_o - mc^2 = 3208.6 \pm 1.3 \text{ keV}. \tag{2.167}$$

The $ft_{1/2}$ value can be calculated by using the integrated Fermi function in Eq. (2.158) with the screening correction, finite nuclear size correction, and radiative correction. The first two are about 0.13% and 0.20% [22]. The last one, which we discuss in the next subsection, is about 1.12% [35]. In this way, Blin-Stoyle and Freeman [35] found the $ft_{1/2}$ value

$$ft_{1/2} = 3073 \pm 5 \text{ sec}. \tag{2.168}$$

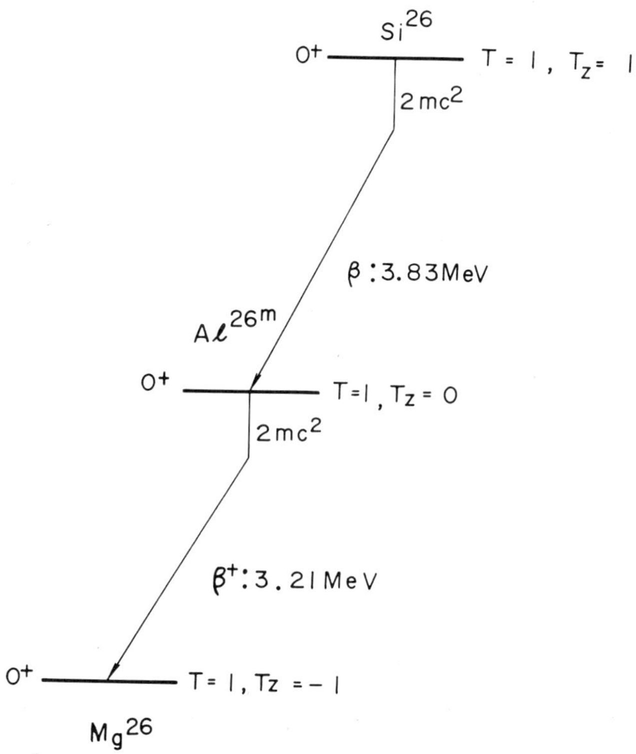

Fig. 2.6 Decay scheme of Al^{26m}.

Substituting Eqs. (2.166) and (2.168) into Eq. (2.165), we have

$$B = 6146 \pm 10 \text{ sec} \tag{2.169}$$

and

$$C_{\text{F}} = (3.001 \pm 0.002) \times 10^{-12} \frac{\hbar^3}{m^2 c}$$
$$= (1.4150 \pm 0.0011) \times 10^{-49} \text{ erg cm}^3. \tag{2.170}$$

The coupling constant C_{F} is extremely small compared with the electron charge in the following sense. The fine structure constant α of the electromagnetic interactions is nondimensional

$$\alpha = \frac{e^2}{\hbar c} \sim 10^{-2}, \tag{2.171}$$

while

$$\left(C_{\text{F}} \frac{m^2 c}{\hbar^3}\right)^2 \sim 10^{-23} \tag{2.172}$$

for the beta decay. This is the reason why the beta-decay interaction is called a weak interaction. The numerical constant in Eq. (2.172) depends on the choice of the mass. If we adopt the nucleon mass as the dimensional unit, we have

$$\left(C_{\text{F}} \frac{M^2 c}{\hbar^3}\right)^2 \sim 10^{-10}. \tag{2.173}$$

The coupling constant C_{GT} of the Gamow–Teller transition can be derived from the neutron beta decay $\text{n} \to \text{p} + \text{e}^- + \bar{\nu}$ (see Fig. 2.7). Since the neutron and proton have the same wave function except for the isospin, the matrices are

$$|\textstyle\int 1|^2 = 1 \qquad \text{and} \qquad |\textstyle\int \sigma|^2 = 3. \tag{2.174}$$

The half life of the neutron beta decay was measured by Sosnovski *et al.* [36].

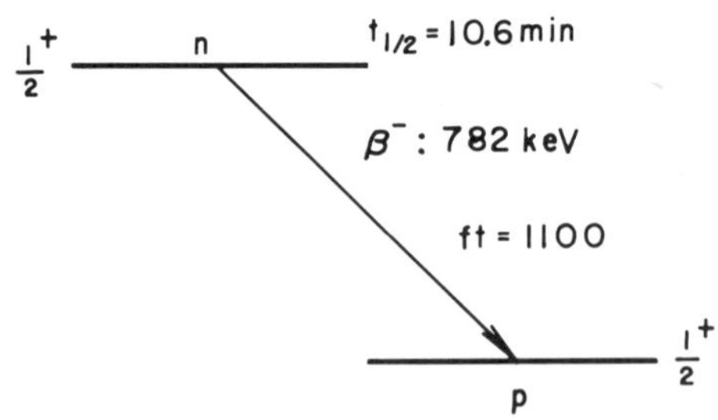

Fig. 2.7 Decay scheme of a neutron.

Recently, Christensen *et al.* performed a refined experiment and they gave a value [37]

$$t_{1/2} = 10.61 \pm 0.16 \text{ min.} \tag{2.175}$$

Using this value and a calculated Fermi function by Bhalla [38], we have

$$ft_{1/2} = 1100 \pm 17 \text{ sec.} \tag{2.176}$$

The magnitude of C_{GT}/C_F can be derived from Eq. (2.165) with Eqs. (2.169), (2.174), and (2.176). The sign of C_{GT}/C_F was also found, in the measurement of the beta-ray angular distribution in neutron decay [39], to be negative. Finally, Christensen *et al.* [37] obtained a ratio

$$C_{GT}/C_F = -1.239 \pm 0.011. \tag{2.177}$$

More Detail on Coupling Constants and Radiative Correction

Freeman's group in Harwell is studying the $0^+ \to 0^+$ transitions extensively. Their experimental results on the end point energies and half lives are summarized in Table 2.1. Since these transitions are all positron decays among isotriplet members, they have the Fermi matrices given in Eq. (2.166). As a result, calculated $ft_{1/2}$ values should, in principle, coincide with each other. This was confirmed in theoretical works performed by Freeman's group (see the references in Table 2.1), Yamada and Kodama [40], Bhalla [22], Behrens and Bühring [41], and Damgaard [42], except in the case of Al^{26m}. We should

Table 2.1 Half Life and End Point Energy for the $0^+ \to 0^+$ Positron Decays

Decay	End Point Energy (keV)	Half Life (msec)	Reference
$C^{10} \to B^{10**}$	888.1 ± 1.8	$1275 \begin{smallmatrix} +18 \\ -15 \end{smallmatrix}$ sec	*a*
$O^{14} \to N^{14*}$	1809.1 ± 1.5	71056 ± 36	*b*
$Al^{26m} \to Mg^{26}$	3208.6 ± 1.3	6346 ± 5	*c*
$Cl^{34} \to S^{34}$	4459.7 ± 4.0	1565 ± 7	*d*
$Sc^{42} \to Ca^{42}$	5409.0 ± 2.3	683·0 ± 1.5	*d*
$V^{46} \to Ti^{46}$	6032.1 ± 2.2	425.9 ± 0.8	*d*
$Mn^{50} \to Cr^{50}$	6609.0 ± 2.6	285.7 ± 0.6	*d*
$Co^{54} \to Fe^{54}$	7227.7 ± 3.8	193.7 ± 1.0	*d*

[a] J. M. Freeman, J. G. Jenkin, and G. Murrary, *Nucl. Phys.* **A124**, 393 (1969).

[b] J. M. Freeman, J. G. Jenkin, D. C. Robinson, G. Murray, and W. E. Burcham, *Phys. Letters* **27B**, 156 (1968); G. J. Clark, J. M. Freeman, D. C. Robinson, J. S. Ryder, W. E. Burcham, and G. T. A. Squire, *Phys. Letters* **35B**, 503 (1971).

[c] J. M. Freeman, J. G. Jenkin, G. Murray, D. C. Robinson, and W. E. Burcham, *Nucl. Phys.* **A132**, 593 (1969).

[d] J. M. Freeman, J. H. Montague, G. Murray, R. E. White, and W. E. Burcham, *Phys. Letters* **8**, 115 (1964); J. M. Freeman, G. Murray, and W. E. Burcham, *Phys. Letters* **17**, 317 (1965).

notice here that the refined half-life measurement in O^{14} by Freeman's group gave a slightly different value from the old data [43]. As a consequence, the new $ft_{1/2}$ value of O^{14} is [44]

$$ft_{1/2} = 3081 \pm 11 \text{ sec.} \tag{2.178}$$

This is, in fact, in good agreement with that of Al^{26m} (see Eq. (2.168)). A

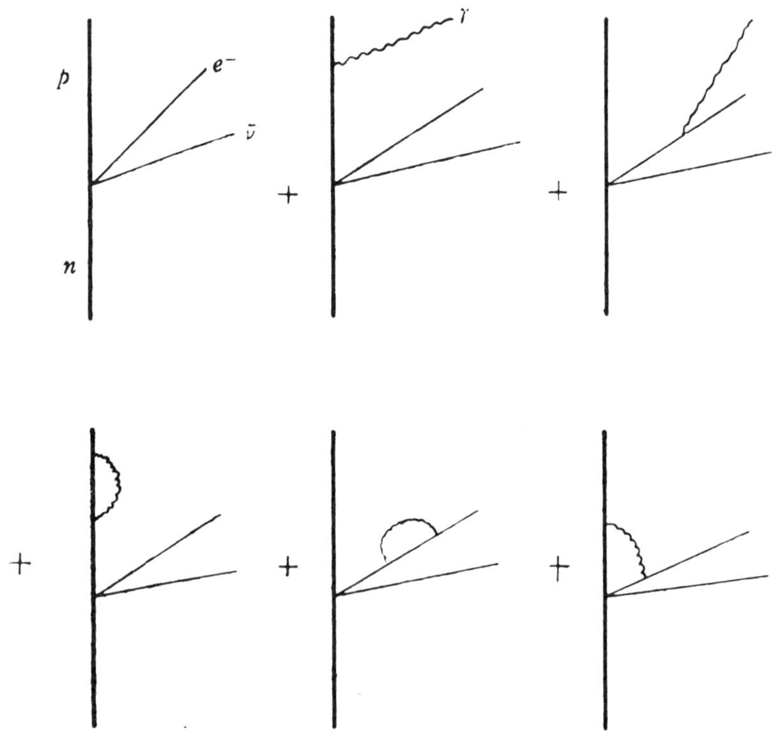

Fig. 2.8 Feynman diagrams for beta decay with radiative corrections. In nuclear beta decay, the neutron and proton lines should read as the nuclear lines (A, Z) and $(A, Z \pm 1)$, respectively. The photon line comes from either of these two nuclear lines in the second, fourth, and sixth diagrams. Furthermore, a diagram where the photon line starts from a nuclear line and ends at the other one should be included.

slight discrepancy in the $ft_{1/2}$ values of the other $0^+ \to 0^+$ transitions may be removed if the refined half-life measurement is adopted.

Next, let us discuss the radiative corrections. As is shown in Fig. 2.8, beta decay is accompanied by various radiative processes, that is, the emission and absorption of both real and virtual photons. These photons emitted during beta decay are called the internal bremsstrahlung. Kinoshita and Sirlin, and in essentially the same way Berman and Sirlin, made calculations of the radiative correction in a covariant way, so that all the diagrams in Fig. 2.8 contribute up to the first order of the fine structure constant [45]. A well-known logarithmic divergence appears in the radiative correction to the $ft_{1/2}$ value,

$$\frac{\Delta ft_{1/2}}{ft_{1/2}} = \frac{\alpha}{2\pi}\left(6 \log \frac{\Lambda}{M_{\rm p}} + 3 \log \frac{M_{\rm p}}{2E_o} - 2.85\right). \tag{2.179}$$

Here Λ is the cutoff parameter. In the numerical calculation, we avoid this divergence by using a cutoff momentum equal to the nucleon mass. The numerical value of Eq. (2.179) varies from 1.6% (for O^{14}) to 1.1% (for Co^{54}). Källen considered the structure of the vertex functions and the possibility of nonlocal space-time [46]. Källen's correction is about 0.4% higher than that of Kinoshita and Sirlin. The radiative correction is slightly modified if the weak interaction takes place with an intermediate vector boson [47]. The importance of this effect is noted by Blin-Stoyle and Freeman [35].

Further investigation of the experimental and theoretical works on the $ft_{1/2}$ values for neutron decay and the $0^+ \to 0^+$ transitions is being performed in several laboratories, since it should provide more precise values of the coupling constants $C_{\rm F}$ and $C_{\rm GT}$, which are the fundamental quantities in the weak interactions, as the electron charge is in the electromagnetic interactions.

2.11 ELECTRON-NEUTRINO ANGULAR CORRELATION

Experimental determination of the specific form of the interactions for nuclear beta decays has remained a central problem since the original formation of the theory by Fermi. Though Fermi's original theory was based on the assumption of vector interaction, five relativistic invariants are available (see Eq. (2.55)). There is, however, practically no contribution of P to the allowed transitions. The electron-neutrino angular correlation has different forms for different interactions, as shown in Eqs. (2.180)-(2.183), and we can decide which interaction is correct for the Gamow-Teller transition, T or A, and similarly for the Fermi transition, S or V. Derivation of the formula was shown in Eqs. (2.72)-(2.80). Performing the computation for all possible interactions, we have [48, 49]

$$W(\theta) = 1 + \frac{b}{E} + a\frac{p}{E}\cos\theta, \tag{2.180}$$

with

$$\begin{aligned}\xi = {} & (|C_{\rm S}|^2 + |C_{\rm V}|^2 + |C'_{\rm S}|^2 + |C'_{\rm V}|^2)|\textstyle\int 1|^2 \\ & + (|C_{\rm T}|^2 + |C_{\rm A}|^2 + |C'_{\rm T}|^2 + |C'_{\rm A}|^2)|\textstyle\int \sigma|^2,\end{aligned} \tag{2.181}$$

$$a\xi = -(|C_S|^2 - |C_V|^2 + |C_S'|^2 - |C_V'|^2)|\textstyle\int 1|^2$$
$$+\tfrac{1}{3}(|C_T|^2 - |C_A|^2 + |C_T'|^2 - |C_A'|^2)|\textstyle\int \boldsymbol{\sigma}|^2$$
$$\pm 2[\mathrm{Im}(C_S^* C_V + C_S'^* C_V')|\textstyle\int 1|^2 - \tfrac{1}{3}\,\mathrm{Im}(C_T^* C_A + C_T'^* C_A')|\textstyle\int \boldsymbol{\sigma}|^2]\left(\dfrac{\alpha Z}{p}\right), \tag{2.182}$$

$$b\xi = \pm 2[1-(\alpha Z)^2]^{1/2}[\mathrm{Re}(C_S^* C_V + C_S'^* C_V')|\textstyle\int 1|^2$$
$$+ \mathrm{Re}(C_T^* C_A + C_T'^* C_A')|\textstyle\int \boldsymbol{\sigma}|^2] \qquad \text{for} \quad \beta^{\mp}. \tag{2.183}$$

Here θ is the angle between the momenta of the electron (positron) and antineutrino (neutrino). (The original calculation in [48] assumed parity conservation, $C_i' = 0$, but included the imaginary part. The C_i' terms are added in [49], after the discovery of the parity nonconservation; see Chapter 3.) For convenience later, we have derived the formulas for the most general case where the Coulomb correction, parity nonconservation, and possible time-reversal noninvariance (complex coupling constants) are taken into consideration.

In this section, we will discuss Eq. (2.180) by omitting all terms of C_i' and of $\alpha Z/p$. The coefficients a are -1 for S, $+1$ for V, $+\frac{1}{3}$ for T, and $-\frac{1}{3}$ for A, if a single interaction induces the beta decay. The measurements of a are provided by those of the recoiled nucleus, since the neutrino is hard to detect. The experiments have been performed since the late 1940s and are reviewed excellently by Allen [50] and Kofoed-Hansen [51].

The recent data for the Gamow–Teller transitions are

He^6 $(0^+ \xrightarrow{\beta^-} 1^+)$

$$\begin{aligned} a &= -0.39 \pm 0.05 && [52] \\ &= -0.353 \begin{smallmatrix} +0.033 \\ -0.053 \end{smallmatrix} && [53] \\ &= -0.319 \pm 0.028 && [54] \\ &= -0.3343 \pm 0.0030, && [55] \end{aligned} \tag{2.184}$$

Ne^{23} $\left(\frac{5}{2}^+ \xrightarrow{\beta^-} \frac{3}{2}^+ \text{ and } \frac{5}{2}^+ \xrightarrow{\beta^-} \frac{5}{2}^+\right.$, mostly Gamow–Teller from isospin selection rule$\left.\right)$

$$\begin{aligned} a &= -0.37 \pm 0.04 && [54] \\ &= -0.33 \pm 0.03 && [56]. \end{aligned} \tag{2.185}$$

These data indicate the dominance of A. A possible admixture of T is very small, and its limit is given, for example, by [55]

$$\frac{(|C_T|^2 + |C_T'|^2)}{(|C_A|^2 + |C_A'|^2)} \leqslant 0.004. \tag{2.186}$$

For the Fermi interaction, we cannot use the $0^+ \rightarrow 0^+$ transitions, because of the source problem. There is, however, a suitable isotope A^{35}, which is known mostly in the Fermi transition. The Gamow–Teller matrix element in the positron decay of A^{35} $\left(\frac{3}{2}^+ \xrightarrow{\beta^+} \frac{3}{2}^+\right)$ is very small from the analysis of the

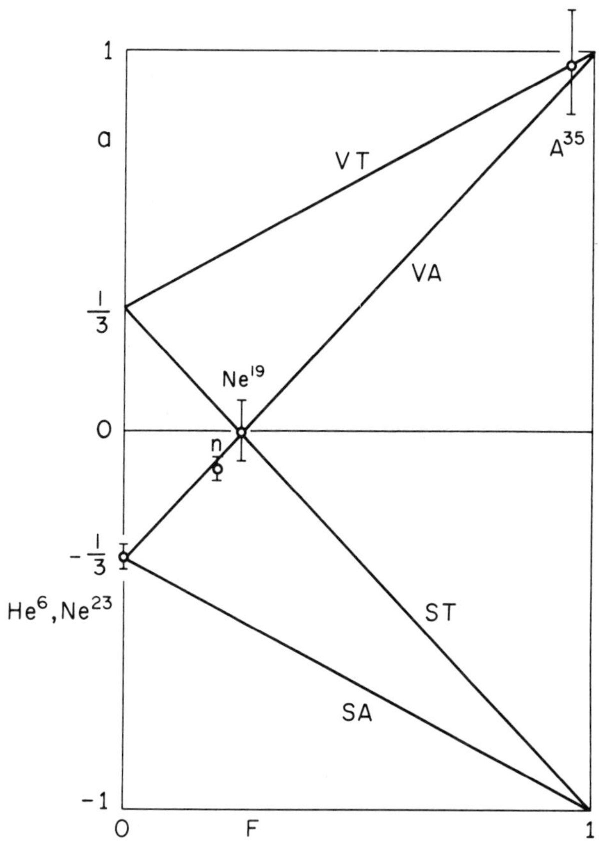

Fig. 2.9 Electron–neutrino angular correlation coefficients plotted against the fraction *F* of the decays that are allowed by Fermi selection rules. See Eq. (2.189).

measured $ft_{1/2}$ value [57], although the initial and final states have the same spin and parity. Allen's group successfully measured the coefficient a

$$a = 0.97 \pm 0.14 \qquad \text{for } A^{35}, \tag{2.187}$$

which indicates V dominance in the Fermi transition [58]. This result became an important support for the coming VA theory.

The latest data on He^6, Ne^{23}, A^{35} [55, 56, 58] and those of the mixed transitions,

$$\begin{aligned} a &= 0.00 \pm 0.08 \qquad \text{for } Ne^{19} \qquad [52] \\ a &= -0.12 \pm 0.04 \qquad \text{for neutrons} \qquad [59], \end{aligned} \tag{2.188}$$

are plotted against the fraction F,

$$F = \frac{C_F^2 |\int 1|^2}{C_F^2|\int 1|^2 + C_{GT}^2|\int \sigma|^2}, \tag{2.189}$$

in Fig. 2.9, where all the data are consistent with a combination *VA*.

In earlier days, difficulties in experimental technique once introduced a positive sign for a in He^6. If this is the case, the Gamow–Teller transition is *T*. Combining it with the data on Ne^{19}, a combination *ST* was concluded to exist (see the line labeled *ST* in Fig. 2.9). This combination was accepted until Allen's group made a measurement on A^{35} and showed that the *V* is dominant in the Fermi interaction.

Finally, we should mention the experiments on beta-neutrino-alpha directional correlation in the beta decay of Li^8. Li^8 emits a beta ray and decays into the Be^8 nucleus, which instantly splits into two alpha particles. The neutrino recoil experiments can be done in terms of the directional correlation of a beta and two alpha particles. A possibility of this experiment was discussed by the present author [60], and it was performed by Barnes *et al.* [61]. They concluded that the beta decay of Li^8 is mostly due to the Gamow–Teller part, that the Fermi part is less than 10%, and that the Gamow–Teller part is mostly axial vector with less than 10% of a possible admixture of tensor. An independent measurement on Li^8 made by Lauterjung *et al.* [62] resulted in agreement with the axial vector.

REFERENCES

1 M. Fierz, *Z. Phys.* **104,** 553 (1937).
2 M. E. Rose, *Phys. Rev.* **51,** 484 (1937); *Relativistic Electron Theory,* Wiley, New York, 1960.
3 E. J. Konopinski, *The Theory of Beta Radioactivity,* Clarendon Press, Oxford, 1966.
4 L. M. Langer, R. D. Moffat, and H. C. Price, *Phys. Rev.* **76,** 1725 (1949).
5 U. Fano, *Tables for the Analysis of Beta Spectra,* Nat. Bur. of Standards Appl. Math. Ser. 13 (U. S. Govt. Printing Office, Washington, D. C., 1952).
6 M. E. Rose, C. L. Perry, and N. M. Dismuke, *ORNL* 1459 (1953); M. E. Rose and C. L. Perry, *Phys. Rev.* **90,** 479 (1953).
7 M. E. Rose and D. K. Holmes, *Phys. Rev.* **83,** 190 (1951); *ORNL* 1022 (1951).
8 M. Yamada, *Progr. Theoret. Phys.* **10,** 241 (1953).
9 M. Yamada, *Progr. Theoret. Phys.* **10,** 245, 252 (1953).
10 Z. Matumoto and M. Yamada, *Progr. Theoret. Phys.* **19,** 285 (1958).
11 A. Z. Dolginov and I. N. Toptigin, *Nucl. Phys.* **2,** 147 (1956–1957).
12 Z. Matumoto, *Progr. Theoret. Phys.* **23,** 531 (1960).

13 W. Bühring, *Nucl. Phys.* **40,** 472 (1963); **49,** 190 (1963); **61,** 110 (1965). W. Bühring and L. Schülke, *Nucl. Phys.* **65,** 369 (1965).
14 C. P. Bhalla and M. E. Rose, *Phys. Rev.* **128,** 1774 (1962); *ORNL* 3207 (1962). C. P. Bhalla, Nat. Bur. of Standards Monograph 81 (1964).
15 W. Bühring, A report presented at Kernforschungszentrum Karlsruhe, *KFK* 559 (1967).
16 J. Sodeman and A. Winther, *Nucl. Phys.* **69,** 369 (1965).
17 M. E. Rose, *Phys. Rev.* **49,** 727 (1936).
18 C. Longmire and H. Brown, *Phys. Rev.* **75,** 1102 (1949).
19 J. R. Reitz, *Phys. Rev.* **77,** 10 (1950).
20 L. Durand, *Phys. Rev.* **135,** B310 (1964).
21 M. Yamada, *Sci. Eng. Inst. Rept. Waseda Univ.* **36,** 67 (1966).
22 C. P. Bhalla, *J. Phys. Soc. Japan Suppl.* **24,** 730 (1968).
23 F. N. D. Kurie, J. R. Richardson, and H. C. Paxton, *Phys. Rev.* **49,** 368 (1936).
24 A. Schwarzschild, Ph.D. thesis, Columbia Univ., New York, 1957.
25 M. Morita, *Progr. Theoret. Phys. Suppl.* **26,** 1 (1963).
26 L. M. Langer and H. C. Price, Jr., *Phys. Rev.* **76,** 641 (1949).
27 R. Sherr and R. H. Miller, *Phys. Rev.* **93,** 1076 (1954).
28 J. Scobie and G. M. Lewis, *Phil. Mag.* **2,** 1089 (1957).
29 R. Sherr and J. B. Gerhart, *Bull. Amer. Phys. Soc.* **1,** 219 (1956).
30 L. M. Langer and R. J. D. Moffat, *Phys. Rev.* **88,** 689 (1952).
31 K. E. Bergkvist, *Topical Conf. on Weak Interaction* (CERN 69-7), p. 91 (1969).
32 W. H. Barkas, W. Birnbaum, and F. M. Smith, *Phys. Rev.* **101,** 778 (1956); J. People, *Nevis Rept.* 147, Columbia Univ. (1965).
33 S. A. Moszkowski and D. Jantzen, *UCLA Tech. Rept.* **10,** 26 (1955).
34 J. M. Freeman, J. G. Jenkin, G. Murray, D. C. Robinson, and W. E. Burcham, *Nucl. Phys.* **A132,** 593 (1969).
35 R. J. Blin-Stoyle and J. M. Freeman, *Nucl. Phys.* **A150,** 369 (1970).
36 A. N. Sosnovski, P. E. Spivak, Yu. A. Prokofiev, I. E. Kutinov, and Yu. P. Dobrynin, *Soviet Phys. JETP* **8,** 739 (1959).
37 C. J. Christensen, A. Nielsen, A. Bahnsen, W. K. Brown, and B. M. Rustad, *Phys. Rev.* **D5,** 1628 (1972); *Phys. Letters* **26B,** 11 (1967).
38 C. P. Bhalla, *Phys. Letters* **19,** 691 (1966).
39 M. T. Burgy, V. E. Krohn, T. B. Novey, G. R. Ringo, and V. L. Telegdi, *Phys. Rev.* **110,** 1214 (1958).
40 M. Yamada and T. Kodama, *J. Phys. Soc. Japan Suppl.* **24,** 729 (1968).
41 H. Behrens and W. Bühring, *Nucl. Phys.* **A106,** 433 (1967).
42 J. Damgaard, *Nucl. Phys.* **A130,** 233 (1969).
43 R. K. Bardin, C. A. Barnes, W. A. Fowler, and P. A. Seeger, *Phys. Rev.* **127,** 583 (1962).
44 G. J. Clark, J. M. Freeman, D. C. Robinson, J. S. Ryder, W. E. Burcham, and G. T. A. Squire, *Phys. Letters* **35B,** 503 (1971).
45 T. Kinoshita and A. Sirlin, *Phys. Rev.* **113,** 1652 (1959); S. M. Berman and A. Sirlin, *Ann. Phys.* **20,** 20 (1962).
46 G. Källen, *Nucl. Phys.* **B1,** 225 (1967).
47 A. Sirlin, *Phys. Rev.* **164,** 1767 (1967).
48 M. Morita, *Progr. Theoret. Phys.* **10,** 364 (1953).
49 M. Morita and R. S. Morita, *Phys. Rev.* **107,** 139 (1957).
50 J. S. Allen, *The Neutrino,* Princeton Univ. Press, Princeton, N. J., 1958.
51 O. Kofoed-Hansen, *α, β, γ Ray Spectroscopy* (K. Siegbahn, ed.), p. 1397. North-Holland, Amsterdam, 1965.

52 J. S. Allen, R. L. Burman, W. B. Herrmannsfeldt, P. Stähelin, and T. H. Braid, *Phys. Rev.* **116,** 134 (1959).

53 B. W. Ridley, *Nucl. Phys.* **25,** 483 (1961).

54 J. B. Vise and B. M. Rustad, *Phys. Rev.* **132,** 2573 (1963).

55 C. H. Johnson, F. Pleasonton, and T. A. Carlson, *Phys. Rev.* **132,** 1149 (1963).

56 T. A. Carlson, *Phys. Rev.* **132,** 2239 (1963).

57 O. C. Kistner, A. Schwarzschild, and B. M. Rustad, *Phys. Rev.* **104,** 154 (1956).

58 W. B. Herrmannsfeldt, D. R. Maxson, P. Stähelin, and J. S. Allen, *Phys. Rev.* **107,** 641 (1957).

59 V. V. Vladimirski, V. K. Gregorev, V. A. Ergakov, D. P. Sjarkov, and Y. V. Trebukhovski, *Izv. Akad. Nauk USSR Fys.* **25,** 1121 (1961).

60 M. Morita, *Phys. Rev. Letters* **1,** 112 (1958).

61 T. Lauritsen, C. A. Barnes, W. A. Fowler, and C. C. Lauritsen, *Phys. Rev. Letters* **1,** 326 (1958). C. A. Barnes, W. A. Fowler, H. B. Greenstein, C. C. Lauritsen, and M. E. Nordberg, *Phys. Rev. Letters* **1,** 328 (1958).

62 K. H. Lauterjung, B. Schimmer, and H. Maier-Leibnitz, *Z. Phys.* **150,** 657 (1958).

Chapter Three

DISCOVERY OF PARITY NONCONSERVATION IN BETA DECAY

3.1 INTRODUCTION

Fermi originally introduced direct four-fermion coupling of the vector type to explain nuclear beta decay. This interaction was generalized to the linear combination of five different covariants, Eq. (2.54), under the assumption of Lorentz invariance. That is, the interaction Hamiltonian (or Lagrangian) density is invariant under the proper Lorentz transformation, as well as under the space inversion $\mathbf{r} \to -\mathbf{r}$, $t \to t$ and time reversal $\mathbf{r} \to \mathbf{r}$, $t \to -t$. In 1956, Lee and Yang [1] introduced a more general form for H_β,

$$H_\beta = (\tfrac{1}{2})^{1/2} \Big[\sum_i C_i(\bar{\psi}_p O_i \tau_+ \psi_n)(\bar{\psi}_e O_i \psi_\nu) + \sum_i C_i'(\bar{\psi}_p O_i \tau_+ \psi_n)(\bar{\psi}_e O_i \gamma_5 \psi_\nu) + h.c.\Big]. \tag{3.1}$$

Here the summation i runs over the interactions S, V, T, A, and P. The Dirac matrices O_i for these interactions are, respectively, 1, γ_μ, $i\gamma_\mu\gamma_\nu$ $(\mu < \nu)$, $i\gamma_\mu\gamma_5$, and γ_5. The first five terms are identical with Eq. (2.54), while the second five terms have an extra γ_5 in the lepton covariants. The former are even under the space inversion, and the latter odd. (Here we assume the same intrinsic parity for all four fermions. The factor $(\frac{1}{2})^{1/2}$ is adopted to keep a relation $[|C_i|^2 + |C_i'|^2]_{\text{new}} = [2|C_i|^2]_{\text{old}}$. Since this factor changes only the denominator of the beta-decay probability (2.161) from $2\pi^3$ to $4\pi^3$, but does not change angular correlations, we will drop this factor in the following discussions, except for the absolute decay rate.) That is, Eq. (3.1) is no longer invariant under the space inversion. But H as a whole is invariant under the *proper* Lorentz transformation. The interaction (3.1) induces parity-nonconserving phenomena in beta decay. The first successful verification of parity nonconservation was obtained by Wu and her collaborators by measuring the angular distribution of beta rays from polarized Co^{60} [6]. Parity non-conservation was also found in the other weak processes, such as $\pi \to \mu + \nu$,

$\mu \to e + \nu + \bar{\nu}$, $\Lambda \to p + \pi$, and these will be given in Chapter 5. The discovery of parity nonconservation had a tremendous impact on the development of these fields. Information about the exact form of beta-decay interactions, coupling constants, neutrino helicity, etc., was obtained from parity-nonconserving phenomena. In this chapter, we first describe why Lee and Yang introduced the interaction (3.1). Second, we study the transformation properties of the Dirac wave functions, and beta-decay interactions under space inversion, time reversal, particle-antiparticle (or charge) conjugation, and their combinations. Third, we derive the formulas for parity-nonconserving phenomena and discuss the experimental results.

3.2 THE θ–τ PUZZLE

Until 1956, the fundamental laws for physics were supposed to be invariant under the space inversion P,

$$\mathbf{r} \to -\mathbf{r}, \qquad t \to t.$$

For example, Newton's equations of motion and Maxwell's equations of the classical electromagnetic theory are invariantly constructed under the transformation P. This holds also in quantum mechanics, which gives the classical theory in the limiting case as the correspondence principle requires.

The eigenfunction of the Hamiltonian of the system goes under P,

$$P\psi(\mathbf{r}, t) = \psi(-\mathbf{r}, t), \tag{3.2}$$

and it becomes the original state and P is applied twice. Therefore, the eigenvalue of P is $+1$ or -1. The quantum number that is the eigenvalue of P is called the parity, and we have the even parity state ψ_+ or the odd parity state ψ_-, according to the equation

$$P\psi_\pm(\mathbf{r}, t) = \pm\psi_\pm(\mathbf{r}, t). \tag{3.3}$$

In experiments relating to photons, atoms, molecules, and nuclei, we can assign the parity of the states with extremely good precision. Therefore, it was customarily required to make the Hamiltonian of the system be invariant under space inversion P.

The concept of the invariance of physical laws under space inversion was generally accepted, and consequently parity conservation became subject to question when the properties of strange particles were studied. Among these strange particles is a group, called K mesons, that have very short lives. They have masses about 1000 times heavier than the electron has. They are nicknamed according to their decay processes; that is, the particles that decay into two pions are called theta mesons and those that decay into three pions are called tau mesons:

$$\theta^{\pm} \to \pi^{\pm} + \pi^{0}, \qquad \tau^{\pm} \to \pi^{\pm} + \pi^{+} + \pi^{-} \tag{3.4}$$

where the superscripts refer to the electric charge of the relevant particles. The pions are known to be bosons with zero spin and odd parity. If we assume the conservation of angular momentum and that of parity in the system, then the possibilities of spins and parities for θ and τ are

$$\theta^{\pm}: 0^{+}, 1^{-}, 2^{+}, \ldots; \qquad \tau^{\pm}: 0^{-}, 1^{\pm}, 2^{\pm}, \ldots. \tag{3.5}$$

In the beginning, experiments which then used cosmic rays were less accurate, and θ and τ were tentatively classed as different particles. With the use of high-energy accelerators, such as the Bevatron at Berkeley and Cosmotron at Brookhaven, experiments became increasingly accurate. Detailed analysis of the data finally led to the conclusion that θ and τ have zero spin and that their mean lives and masses coincide within the experimental errors

$$m_{\theta^+} = (968 \pm 5)m, \qquad m_{\tau^+} = (966.1 \pm 0.7)m,$$
$$\tau_{\theta^+} = (1.21 \pm 0.04) \times 10^{-8} \text{ sec}, \qquad \tau_{\tau^+} = (1.19 \pm 0.05) \times 10^{-8} \text{ sec}. \tag{3.6}$$

It is, therefore, natural to assume that θ^+ and τ^+ are identical, and the K^+ particle decays into two different modes (2π and 3π modes). Most physicists, however, were reluctant to accept this, because parity conservation requires the plus parity for K^+ from the θ mode and simultaneously the minus parity from the τ mode. This difficulty was called the θ–τ puzzle, and a number of ideas were proposed to explain the spin and parity of θ and τ, and the approximate equality of their lifetimes and masses. None of them succeeded in solving the θ–τ puzzle in a consistent way. Finally, Lee and Yang decided to abandon the assumption of parity conservation [1]. In their theory, θ and τ are identical and are K^+ particles of spin and parity 0^-. Lee and Yang assumed that parity is not conserved in the decay process, so that the K^+ particle can decay into either 2π or 3π with an appropriate branching ratio. They proposed the interaction Eq. (3.1) for the beta decays and suggested possible experiments on parity nonconservation in beta decay, $\pi \to \mu$, $\mu \to e$ decays, etc. The proposal was immediately tested by experimental groups at Columbia University, and the great discovery of parity nonconservation was made in January 1957.

3.3 TRANSFORMATION PROPERTIES OF THE DIRAC WAVE FUNCTIONS UNDER *P, T, C*

Before discussing the experimental implications introduced by the interaction Hamiltonian, Eq. (3.1), we have to study its transformation properties under the space inversion P, time reversal T, and particle–antiparticle (or charge) conjugation C, and their combinations. To do this, we first have to study those for Dirac wave functions.

Space Inversion P

The Dirac wave function $\psi(x)$ satisfies the Dirac equation

$$\left(\gamma_\mu \frac{\partial}{\partial x_\mu} + \frac{mc}{\hbar}\right) \psi(\mathbf{r}, t) = 0. \tag{3.7}$$

Invariance of the Dirac equation under space inversion means that the transformed wave function $\psi'(x')$ under P

$$\mathbf{r}' = -\mathbf{r}, \qquad t' = t \tag{3.8}$$

satisfies the transformed equation which has the same form as Eq. (3.7),

$$\left(\gamma_\mu \frac{\partial}{\partial x'_\mu} + \frac{mc}{\hbar}\right)\psi'(x') = 0, \tag{3.9}$$

with

$$\psi'(x') = P\psi(-\mathbf{r}, t) = \eta_P U_P \psi(\mathbf{r}, t). \tag{3.10}$$

Here, P is the operator of the space inversion, U_P and η_P are a unitary matrix and a phase factor, and x is the abbreviation of four variables $\mathbf{r}$, ict. Putting Eq. (3.10) into Eq. (3.9) and multiplying U_P^{-1} from the left, we obtain

$$U_P^{-1}\left(\gamma_4 \frac{\partial}{\partial x_4} - \gamma_k \frac{\partial}{\partial x_k} + \frac{mc}{\hbar}\right) U_P \psi(\mathbf{r}, t) = 0. \tag{3.11}$$

Equation (3.11) reduces to Eq. (3.7) if U_P satisfies the relations

$$U_P^{-1}\gamma_4 U_P = \gamma_4, \qquad U_P^{-1}\gamma_k U_P = -\gamma_k, \qquad k = 1, 2, 3. \tag{3.12}$$

From this we find

$$U_P = \gamma_4. \tag{3.13}$$

Applying the operation P twice on ψ, we have ψ. Therefore,

$$\psi'' = \eta_P U_P \psi' = \eta_P^2 \psi = \psi$$

and

$$\eta_P = \pm 1. \tag{3.14}$$

Consequently,

$$P\psi(\mathbf{r}, t) = \pm\gamma_4 \psi(-\mathbf{r}, t). \tag{3.15}$$

The adjoint wave function $\bar{\psi}$ transforms by

$$\overline{[P\psi(\mathbf{r}, t)]} = \pm\bar{\psi}(-\mathbf{r}, t)\gamma_4, \tag{3.16}$$

which makes the adjoint equation invariant.

Time Reversal T

In general, the wave function of a system satisfies the time-dependent Schrödinger equation

$$H(t)\psi(t) = i\hbar \frac{\partial \psi(t)}{\partial t}. \tag{3.17}$$

The time-reversed wave function cannot be obtained by a simple replacement $t \to -t$, since this would change the sign of the energy. Wigner introduced the antilinear operator for the time reversal, and the transformed wave function is given by

$$\psi'(t') = T\psi(t') = \psi^*(t), \tag{3.18}$$

where $t' = -t$. The complex conjugate of Eq. (3.18) is

$$H^*(t)\psi^*(t) = -i\hbar \frac{\partial \psi^*(t)}{\partial t}. \tag{3.19}$$

From Eqs. (3.18) and (3.19) we have

$$H(t')\psi'(t') = i\hbar \frac{\partial \psi'(t')}{\partial t'} \tag{3.20}$$

if

$$H(t') = H^*(t). \tag{3.21}$$

Since Eq. (3.20) has the same form as Eq. (3.17), the physical law is the same in the original system and the time-reversed system. The condition of this invariance is given by Eq. (3.21).

Now we study the Dirac equation, (3.7). The invariance of this equation under the time reversal T

$$\mathbf{r}' = \mathbf{r}, \qquad t' = -t \tag{3.22}$$

means that the transformed wave function $\psi'(x')$ satisfies

$$\left(\gamma_\mu \frac{\partial}{\partial x'_\mu} + \frac{mc}{\hbar}\right)\psi'(x') = 0. \tag{3.23}$$

Here $\psi'(x')$ is given by the antilinear transformation

$$\psi'(x') = T\psi(\mathbf{r}, -t) = \eta_T U_T \psi^*(\mathbf{r}, t). \tag{3.24}$$

Substituting Eq. (3.24) in Eq. (3.23) and multiplying U_T^{-1} from the left, we obtain

$$U_T^{-1}\left(-\gamma_4 \frac{\partial}{\partial x_4} + \gamma_k \frac{\partial}{\partial x_k} + \frac{mc}{\hbar}\right) U_T \psi^*(\mathbf{r}, t) = 0, \tag{3.25}$$

while the complex conjugate of Eq. (3.7) is

$$\left(-\gamma_4^* \frac{\partial}{\partial x_4} + \gamma_k^* \frac{\partial}{\partial x_k} + \frac{mc}{\hbar}\right)\psi^*(\mathbf{r}, t) = 0. \tag{3.26}$$

That is, Eq. (3.25) reduces to Eq. (3.26) if

$$U_T^{-1}\gamma_\mu U_T = \gamma_\mu^*. \tag{3.27}$$

A solution of U_T can be given as follows. Since we are adopting a particular system of Dirac matrices where γ_2 and γ_4 are real and γ_1 and γ_3 are pure imaginary, Eq. (3.27) is written as

$$U_T^{-1}\gamma_\mu U_T = \pm\gamma_\mu, \qquad \text{plus sign for } \mu = 2, 4; \quad \text{minus sign for } \mu = 1, 3. \tag{3.28}$$

From this, it is obvious that

$$U_T = \gamma_3\gamma_1 = i\sigma_2. \tag{3.29}$$

Applying the time reversal twice on ψ, we obtain

$$\psi'' = \eta_T U_T \eta_T^* U_T^* \psi = -|\eta_T|^2 \psi. \tag{3.30}$$

The phase factor should therefore be arbitrary with modulus unity

$$\eta_T = \exp(i\phi_T). \tag{3.31}$$

Consequently,

$$T\psi(\mathbf{r}, t) = \eta_T \gamma_3 \gamma_1 \psi^*(\mathbf{r}, -t) = \eta_T \gamma_3 \gamma_1 \gamma_4 \bar{\psi}^T(\mathbf{r}, -t) \tag{3.32}$$

and

$$\overline{[T\psi(\mathbf{r}, t)]} = \eta_T^* \bar{\psi}(\mathbf{r}, -t)^* \gamma_1 \gamma_3 = \eta_T^* \psi^T(\mathbf{r}, -t) \gamma_4 \gamma_1 \gamma_3.$$

Here ψ^T refers to the transpose of ψ.

Charge Conjugation C

This is the relation of the wave functions for a normal particle and its antiparticle. We will start from the Dirac equation in the presence of an electromagnetic field

$$\left[\gamma_\mu \left(\frac{\partial}{\partial x_\mu} + \frac{ie}{\hbar c} A_\mu\right) + \frac{mc}{\hbar}\right] \psi(x) = 0 \qquad \text{for an electron.} \tag{3.33}$$

The antiparticle is the positron, and its wave equation is identical to the foregoing except for the sign of the charge, since the electron and positron are assumed here to have the same dynamic variables.

$$\left[\gamma_\mu \left(\frac{\partial}{\partial x_\mu} - \frac{ie}{\hbar c} A_\mu\right) + \frac{mc}{\hbar}\right] \psi'(x) = 0. \tag{3.34}$$

Here, the positron wave function ψ' is obtained from the electron's by the charge conjugation operation C

$$\psi' = C\psi = \eta_C U_C \psi^*. \tag{3.35}$$

Substituting Eq. (3.35) into Eq. (3.34), and multiplying Eq. (3.34) by U_C^{-1} from the left gives

$$U_C^{-1}\left[\gamma_4 \left(\frac{\partial}{\partial x_4} - \frac{ie}{\hbar c} A_4\right) + \gamma_k \left(\frac{\partial}{\partial x_k} - \frac{ie}{\hbar c} A_k\right) + \frac{mc}{\hbar}\right] U_C \psi^* = 0, \tag{3.36}$$

while the complex conjugate of Eq. (3.33) is

$$\left[-\gamma_4^* \left(\frac{\partial}{\partial x_4} - \frac{ie}{\hbar c} A_4\right) + \gamma_k^* \left(\frac{\partial}{\partial x_k} - \frac{ie}{\hbar c} A_k\right) + \frac{mc}{\hbar}\right] \psi^* = 0. \tag{3.37}$$

Equation (3.36) reduces to Eq. (3.37) if

$$U_C^{-1} \gamma_k U_C = \gamma_k^*, \qquad U_C^{-1} \gamma_4 U_C = -\gamma_4^*. \tag{3.38}$$

The solution of U_C is immediately given by

$$U_C = \gamma_2. \tag{3.39}$$

The second application of charge conjugation on ψ is

$$\psi'' = |\eta_C|^2 \psi,$$

and the η_C is an arbitrary phase factor

$$\eta_C = \exp(i\phi_C). \tag{3.40}$$

Finally

$$C\psi = \eta_C \gamma_2 \psi^* \qquad \text{and} \qquad \overline{C\psi} = \eta_C^* \psi^T \gamma_2 \gamma_4. \tag{3.41}$$

CPT Operation

Summarizing P, T, C operations, we have

$$P\psi(\mathbf{r}, t) = \eta_P \gamma_4 \psi(-\mathbf{r}, t),$$

$$T\psi(\mathbf{r}, t) = \eta_T \gamma_3 \gamma_1 \psi^*(\mathbf{r}, -t), \qquad \gamma_3\gamma_1 = i\sigma_2,$$

$$C\psi(\mathbf{r}, t) = \eta_C \gamma_2 \psi^*(\mathbf{r}, t).$$

A successive application of T, P, C gives

$$CPT\psi(\mathbf{r}, t) = \eta_C \eta_P^* \eta_T^* \gamma_5 \psi(-\mathbf{r}, -t). \tag{3.42}$$

Therefore, the combined CPT operation or its permutations always transform the Dirac wave functions

$$\psi(x) \to \gamma_5 \psi(-x), \tag{3.42$'$}$$

except for a phase factor which is a modulus of unity.

3.4 TRANSFORMATION PROPERTIES OF BILINEAR QUANTITIES

In order to see the transformation properties of the interaction Hamiltonian density H_β, it is necessary to know those for bilinear products of the Dirac wave functions, which are covariant in the proper Lorentz transformation. For simplicity, we set the phase factors

$$\eta_P = \eta_T = \eta_C = 1$$

for all the particles we are concerned with. This is because the products of these phase factors appearing for a particular operator are common for the first ten terms in Eq. (3.1), as they are for the ten terms in the hermitian conjugate.

Now let us define the phase factors for bilinear quantities for the particular transformations that follow.

Space Inversion P

$$(\bar{\psi}_a O_i \psi_b) \to (\bar{\psi}'_a(x')O_i\psi'_b(x')) = (\bar{\psi}_a(-\mathbf{r}, t)\gamma_4 O_i \gamma_4 \psi_b(-\mathbf{r}, t))$$
$$= \eta_i^P (\bar{\psi}_a(-\mathbf{r}, t)O_i\psi_b(-\mathbf{r}, t)). \tag{3.43}$$

Time Reversal T

$$(\bar{\psi}_a O_i \psi_b) \to (\bar{\psi}'_a(x')O_i\psi'_b(x')) = (\psi_a^T(\mathbf{r}, -t)\gamma_4\gamma_1\gamma_3 O_i \gamma_3\gamma_1\gamma_4 \bar{\psi}_b^T(\mathbf{r}, -t))$$
$$= \eta_i^T (\bar{\psi}_b(\mathbf{r}, -t)O_i\psi_a(\mathbf{r}, -t)). \tag{3.44}$$

Charge Conjugation C

$$(\bar{\psi}_a O_i \psi_b) \to (\bar{\psi}_{\bar{a}} O_i \psi_{\bar{b}}) = -(\bar{\psi}_b(\gamma_2\gamma_4 O_i \gamma_4\gamma_2)^T \psi_a)$$
$$= \eta_i^C (\bar{\psi}_b O_i \psi_a). \tag{3.45}$$

(We are dealing with the Dirac wave functions $\psi(x)$. The phase factors are, therefore, in conformity with those given by de Groot and Tolhoek [2]. If $\psi(x)$'s are the second quantized fields, we have to antisymmetrize each covariant in a form

$$\bar{\psi}_a(x)O_i\psi_b(x) \to \tfrac{1}{2}(\bar{\psi}_a(x)O_i\psi_b(x) - \psi_b(x)O_i\bar{\psi}_a(x)),$$

so that the ambiguity due to the δ function in the equal-time commutator will disappear. Furthermore, all signs for η_i^C, η_i^{PC}, and η_i^{TPC} in Table 3.1 will be reversed. See, also the remark below Eq. (2.54).)

PC Operation

$$(\bar{\psi}_a O_i \psi_b) \to \eta_i^C (\bar{\psi}'_b(x')O_i\psi'_a(x'))$$
$$= \eta_i^C \eta_i^P (\bar{\psi}_b(-\mathbf{r}, t)O_i\psi_a(-\mathbf{r}, t)), \tag{3.46}$$

and we define

$$\eta_i^{PC} = \eta_i^C \eta_i^P. \tag{3.47}$$

TPC Operation

$$(\bar{\psi}_a O_i \psi_b) \to (\eta_i^C \eta_i^P)^* (\bar{\psi}'_b(-\mathbf{r}, t')O_i\psi'_a(-\mathbf{r}, t'))$$
$$= (\eta_i^C \eta_i^P)^* \eta_i^T (\bar{\psi}_a(-x)O_i\psi_b(-x)), \tag{3.48}$$

and we define

$$\eta_i^{TPC} = \eta_i^{C*}\eta_i^{P*}\eta_i^T. \tag{3.49}$$

In our representation, all Dirac matrices O_i are chosen to be hermitian. The phase factors η_i are easily obtained from the second and third terms of each transformation, and they are listed in Table 3.1. The phase factors for the combined operations *TC, CT, PT, TP, CP,* and permutations of *TCP* are not listed there. They are, however, easily obtained from similar calculations. The results in Table 3.1 give us several important consequences.

Table 3.1 Phase Factors of the Bilinear Quantities under P, C, T^a

Covariant	O_i	η_i^P	η_i^C	η_i^T	η_i^{PC}	η_i^{TPC}
S	1	1	-1	1	-1	-1
V	γ_μ	ϵ_μ	1	ϵ_μ	ϵ_μ	1
T	$\frac{1}{2i}[\gamma_\mu, \gamma_\nu]$	$\epsilon_\mu \epsilon_\nu$	1	$-\epsilon_\mu \epsilon_\nu$	$\epsilon_\mu \epsilon_\nu$	-1
A	$i\gamma_\mu \gamma_5$	$-\epsilon_\mu$	-1	ϵ_μ	ϵ_μ	1
P	γ_5	-1	-1	-1	1	-1

[a] Here $\epsilon_\mu = 1$ for $\mu = 4$ and -1 for $\mu = 1, 2, 3$.

CTP Theorem

Lüders and Pauli proved a theorem of outstanding importance that connects P, T, C transformations and gives a relation between these transformations and the proper Lorentz transformations [3, 4]. The theorem claims that if a theory of interacting fields is invariant under the proper Lorentz transformations, then it will be automatically invariant under the combination of the successive application of space inversion, time reflection, and particle-antiparticle conjugation. This is seen by considering that the interaction Hamiltonian density for beta decays (3.1) is a linear combination of the products of the lepton and nucleon bilinear quantities,

$$H_\beta = (\tfrac{1}{2})^{1/2} \sum_i [C_i(\bar{\psi}_p O_i \psi_n)(\bar{\psi}_e O_i \psi_\nu) + C_i'(\bar{\psi}_p O_i \psi_n)(\bar{\psi}_e O_i \gamma_5 \psi_\nu)] + \text{h.c.} \tag{3.50}$$

This is, of course, invariant under the proper Lorentz transformation, and it is a local interaction, since the space and time coordinates are commonly chosen for all four wave functions. Under the *TPC* operation, the phase factor η_i^{TPC} is -1 for even rank tensors (S, T, P) and $+1$ for odd rank tensors (V, A). Therefore, Eq. (3.50) is immediately seen to be invariant. This holds also for the combined transformations of the permutation of *TPC*. τ_+ in H_β is again dropped in this section throughout, for the same reason as given after Eq. (2.53).

Parity Nonconservation and Charge Conjugation Noninvariance

If C_i and C_i' are nonvanishing, H_β is noninvariant under P or C separately [1]. For simplicity, we discuss the scalar interaction with real coupling constants

$$H_\beta = C_S(\bar{\psi}_p \psi_n)(\bar{\psi}_e \psi_\nu) + C_S'(\bar{\psi}_p \psi_n)(\bar{\psi}_e \gamma_5 \psi_\nu) + C_S(\bar{\psi}_n \psi_p)(\bar{\psi}_\nu \psi_e) - C_S'(\bar{\psi}_n \psi_p)(\bar{\psi}_\nu \gamma_5 \psi_e). \quad (3.51)$$

η_i^P is +1 for scalar and −1 for pseudoscalar. Therefore, C_S terms remain unchanged, while C_S' terms change their sign. Thus, H_β is noninvariant under P.

Under the charge conjugation C, the first and third terms are mutually interchanged, as are the second and fourth terms. The phase factor η_i^C is −1 for both scalar and pseudoscalar. Consequently, H_β becomes

$$H_\beta = C_S(\bar{\psi}_n \psi_p)(\bar{\psi}_\nu \psi_e) + C_S'(\bar{\psi}_n \psi_p)(\bar{\psi}_\nu \gamma_5 \psi_e) + C_S(\bar{\psi}_p \psi_n)(\bar{\psi}_e \psi_\nu) - C_S'(\bar{\psi}_p \psi_n)(\bar{\psi}_e \gamma_5 \psi_\nu). \quad (3.52)$$

Comparing Eqs. (3.51) and (3.52), we notice that the signs for the interactions of the electron and positron decays are interchanged in the C_S' couplings. This proves that the interaction (3.51) is noninvariant under C. On the other hand, H_β remains invariant under the successive operation of C and P. This can be seen from the phase factor η_i^{PC} in Table 3.1.

Condition for T Invariance

Let us now assume complex coupling constants

$$H_\beta = C_S(\bar{\psi}_p \psi_n)(\bar{\psi}_e \psi_\nu) + C_S'(\bar{\psi}_p \psi_n)(\bar{\psi}_e \gamma_5 \psi_\nu) + C_S^*(\bar{\psi}_n \psi_p)(\bar{\psi}_\nu \psi_e) - C_S'^*(\bar{\psi}_n \psi_p)(\bar{\psi}_\nu \gamma_5 \psi_e). \quad (3.53)$$

We emphasize that the phases in the coupling constants are meaningful only if more than two kinds of interactions are present. Under time reversal, annihilation and creation of the particles are interchanged. The phase factors η_i^T for this transformation are 1 for scalar and −1 for pseudoscalar. Consequently, H_β becomes

$$H_\beta = C_S(\bar{\psi}_n \psi_p)(\bar{\psi}_\nu \psi_e) - C_S'(\bar{\psi}_n \psi_p)(\bar{\psi}_\nu \gamma_5 \psi_e) + C_S^*(\bar{\psi}_p \psi_n)(\bar{\psi}_e \psi_\nu) + C_S'^*(\bar{\psi}_p \psi_n)(\bar{\psi}_e \gamma_5 \psi_\nu). \quad (3.54)$$

Therefore, H_β in Eq. (3.53) remains invariant under T if

$$C_S'/C_S = C_S'^*/C_S^*, \quad (3.55)$$

since a common phase factor for the transformed interaction is physically irrelevant. That is, the coupling constants should be mutually real for the

T-invariant interactions. The condition (3.55) is referred to hereafter simply as real coupling constants.

If C_S and C'_S are real, then H_β is invariant under T as well as under CP and PC, while it is noninvariant under P and C, separately. It is noticed here that T and CP are, in general, different operators, although the invariance properties of H_β are the same under T and CP. As we remember, single transformations P, T, C are accompanied by the phase factors η_P, η_T, η_C for individual Dirac wave functions, and they are tentatively set equal to unity.

If C_S is real but C'_S is purely imaginary, then H_β in Eq. (3.53) is noninvariant under T or P, but is invariant under C.

Transformation of Formulas for Electrons to Those for Positrons

In Chapter 2 we devised a formula for the beta-ray spectrum which depends on the sign of the electric charge of the beta particles. We will also derive formulas for various phenomena in beta decays, such as angular distributions and longitudinal polarization. Explicit calculations of these formulas for the positron are not required if the formulas for the electron are already given. The former are derived from the latter by using the following substitution rules [5].

(a) Nuclear matrix element for beta decay:

$$\int \psi_p^\dagger \Omega \psi_n d\mathbf{r} \rightarrow \int \psi_n^\dagger \Omega \psi_p d\mathbf{r}, \tag{3.56}$$

(b) Atomic number of the daughter nucleus:

$$Z \rightarrow -Z, \tag{3.57}$$

(c) Coupling constants:

$$C_i \rightarrow \mp C_i^*, \qquad C'_i \rightarrow \pm C_i'^*, \tag{3.58}$$

where the upper sign holds for $i = S, A, P$; the lower sign for $i = V, T$. The last rule can be seen from the sign of η_i^C and the relative signs among the C'_i terms and their hermitian conjugate in Eq. (3.50).

Five Different Classes of Transformation Properties of Interaction

Summarizing the foregoing discussion, we have five cases of possible noninvariant interaction Hamiltonian density under the discrete transformations P, T, C and their combinations. These classifications are listed in Table 3.2, with the following interaction of scalar and pseudoscalar,

$$\begin{aligned} H_\beta = {} & C_S(\bar{\psi}_p\psi_n)(\bar{\psi}_e\psi_\nu) + C'_S(\bar{\psi}_p\psi_n)(\bar{\psi}_e\gamma_5\psi_\nu) \\ & + C_P(\bar{\psi}_p\gamma_5\psi_n)(\bar{\psi}_e\gamma_5\psi_\nu) + C'_P(\bar{\psi}_p\gamma_5\psi_n)(\bar{\psi}_e\psi_\nu) + \text{h.c.} \end{aligned} \tag{3.59}$$

Table 3.2 Transformation Properties of Possible Noninvariant Interactions[a]

Coupling Constants	Noninvariant under	Invariant under
C_S = real, C_P = real, $C'_S = C'_P = 0$	None	P, T, C, PCT
C_S = real, C'_S = real, $C_P = C'_P = 0$	P, C	T, PC, CP, PCT
C_S = real, C'_S = imag., $C_P = C'_P = 0$	P, T	C, PT, TP, PCT
C_S = real, C_P = complex, $C'_S = C'_P = 0$	C, T	P, CT, TC, PCT
C_S = real, C'_S = complex, $C_P = C'_P = 0$	P, C, T	PCT

[a] The invariance under the proper Lorentz transformation and the locality of interaction are assumed.

3.5 PSEUDOSCALAR QUANTITIES

In order to detect parity-nonconserving phenomena, Lee and Yang proposed to measure a pseudoscalar formed by the experimentally measurable quantities. Under space inversion

$$\mathbf{r} \to -\mathbf{r}, \qquad t \to t,$$

the linear and angular momenta transform as vector and axial vector, respectively,

$$\mathbf{p} \to -\mathbf{p}, \qquad \mathbf{J} \to \mathbf{J},$$

since $\mathbf{p}$ is $(\hbar/i)\,\partial/\partial\mathbf{x}$ and $\mathbf{J}$ transforms like $\mathbf{r} \times \mathbf{p}$. Pseudoscalar terms have, therefore, the forms of $\mathbf{J} \cdot \mathbf{p}$ and $\mathbf{p}_1 \cdot (\mathbf{p}_2 \times \mathbf{p}_3)$. Here $\mathbf{J}$ is the spin of the nucleus or particle and $\mathbf{p}_i$'s are measurable linear momenta.

There are essentially three possible ways of measuring pseudoscalar quantities.

Beta Asymmetry from Polarized Nuclei

One is to measure the angular distribution of the electron coming from beta decays of polarized nuclei. Here, the pseudoscalar is $\mathbf{J} \cdot \mathbf{p}$, $\mathbf{J}$ and $\mathbf{p}$ being the nuclear spin and the momentum of the beta ray, respectively. The angular distribution has an asymmetry for parity nonconservation.

Beta-Gamma Circular Polarization Correlation

This method involves measuring the angular correlation of the beta ray and circularly polarized gamma ray in beta-gamma cascade transitions, or in other words measuring the beta-ray angular distribution with respect to the photon spin direction, which is parallel or antiparallel to the direction of emission of gamma ray. This angular correlation contains the odd powers of $\cos\theta$, where θ is the relative angle of beta and gamma rays, for parity nonconservation. The pseudoscalar is $\boldsymbol{\sigma}_\gamma \cdot \mathbf{p}$. It should be noticed here that the observation of the circular polarization is unavoidable. Otherwise the photon spin direction is averaged, and the pseudoscalar vanishes.

Longitudinal Polarization of Electron

The pseudoscalar is $\boldsymbol{\sigma}\cdot\mathbf{p}$ where $\boldsymbol{\sigma}$ and $\mathbf{p}$ are the spin and momentum of the beta particle, respectively. Nonzero value of $\boldsymbol{\sigma}\cdot\mathbf{p}$ means that electrons from the beta decay of unoriented nuclei show longitudinal polarization in the case of parity nonconservation.

3.6 ANGULAR DISTRIBUTION OF BETA RAYS FROM POLARIZED NUCLEI

The angular distribution of beta rays from polarized nuclei is obtained by evaluating the transition probability by leaving the angular variables for beta particles unintegrated. The relative population of the initial magnetic substate is not constant, but it depends on the magnetic quantum number. This fact should be taken into account. The computation is easier in the spherical representation for the nuclear states as well as for the interaction Hamiltonian. We need a bit of Racah algebra (Appendix A.5), too.

As an example, we will demonstrate the calculation for the Gamow-Teller transition of the electron decay for which H_β becomes

$$H_\beta = \sum_m (\psi_{\mathrm{p}}^\dagger \tau_+ \sigma_m \psi_{\mathrm{n}})(\psi_{\mathrm{e}}^\dagger \sigma_m^\dagger (C_{\mathrm{A}} + C'_{\mathrm{A}}\gamma_5)\psi_\nu)$$

$$+ \sum_m (\psi_{\mathrm{p}}^\dagger \tau_+ \beta\sigma_m \psi_{\mathrm{n}})(\psi_{\mathrm{e}}^\dagger \beta\sigma_m^\dagger (C_T + C'_T\gamma_5)\psi_\nu), \tag{3.60}$$

with

$$\sigma_0 = \sigma_z \qquad \text{and} \qquad \sigma_{\pm 1} = \mp(\tfrac{1}{2})^{1/2}(\sigma_x \pm i\sigma_y).$$

In Chapter 2 the reduced nuclear matrix element $\langle j_1 \| \sigma \| j\rangle$ for the transition from the state j, m to the state j_1, m_1 was introduced by Eq. (2.66)

$$\int \psi_{j_1 m_1}^\dagger \sum_k \tau_+^{(k)} \sigma_M^{(k)} \psi_{jm}\, d\mathbf{r}_1\, d\mathbf{r}_2 \cdots d\mathbf{r}_A = (j1mM \mid j_1 m_1)\langle j_1 \| \sigma \| j\rangle.$$

As the Wigner-Eckart theorem states, the reduced nuclear matrix element is independent of magnetic quantum numbers, while they are completely absorbed in the Clebsch-Gordan coefficients $(j1mM \mid j_1 m_1)$. Furthermore, a relation of the reduced nuclear matrix element to the Gamow–Teller matrix element was introduced earlier as Eq. (2.87)

$$\frac{2j_1+1}{2j+1}\,|\langle j_1\,\|\,\sigma\,\|\,j\rangle|^2 = |\textstyle\int\sigma|^2.$$

The amplitude of subtransition $j,\ m \to j_1,\ m_1$ is given by

$$(j1mM\,|\,j_1m_1)\,\langle j_1\,\|\,\sigma\,\|\,j\rangle\,(\psi_e^\dagger\sigma_M^\dagger(C_A + C_A'\gamma_5)\psi_\nu) \tag{3.61}$$

for the axial vector. We introduce the function

$$F_{LL'}^M(\theta) = \sum_n\,(-)^M(L'L - MM\,|\,n0)b_{LL'}^{(n)}P_n(\cos\theta), \tag{3.62}$$

which is proportional to the square of the lepton covariants. The function describes the angular distribution of the partial wave (of the electron-neutrino system) with quantum numbers L, L', and $-M$. (We emphasize here that the emitted electron-neutrino should have the magnetic quantum number $-M$. This is obvious in the Clebsch-Gordan coefficient in Eq. (3.61).) In our case, $L = L' = 1$ and $M = 0, \pm 1$. Therefore the order of the Legendre polynomial $P_n(\cos\theta)$ becomes $n = 0, 1, 2$. Here, $b_{LL'}^{(n)}$'s are the parameters depending on the interactions and the electron energy. More explicitly, the $F_{11}^1(\theta)$ for the axial vector is defined by

$$\begin{aligned} F_{11}^1(\theta) &= \frac{1}{4\pi}\int d\Omega_\nu \sum_{\sigma_e,\sigma_\nu} |(\psi_e^\dagger\sigma_1^\dagger(C_A + C_A'\gamma_5)\psi_\nu)|^2\,|\langle j_1\,\|\,\sigma\,\|\,j\rangle|^2 \\ &= \frac{1}{4\pi}\int d\Omega_\nu\,\mathrm{Tr}(D_e\sigma_1^\dagger(C_A + C_A'\gamma_5)D_\nu(C_A^* + C_A'^*\gamma_5)\sigma_1)|\langle j_1\,\|\sigma\|j\rangle|^2 \end{aligned} \tag{3.63}$$

with $\sigma_1 = -(\frac{1}{2})^{1/2}(\sigma_x + i\sigma_y)$. Here the trace method of section 2.6 is adopted. From this, we can easily obtain the result

$$F_{11}^1(\theta) = |\langle j_1\,\|\,\sigma\,\|\,j\rangle|^2\,[|C_A|^2 + |C_A'|^2 + (C_A C_A'^* + \text{c.c.})(p/E)\cos\theta], \tag{3.64}$$

where θ is the angle between the spin of the nuclear polarization and the direction of the momentum of the emitted electron. On the other hand, Eq. (3.62) becomes, for $L = L' = M = 1$,

$$F_{11}^1(\theta) = -(\tfrac{1}{3})^{1/2}b_{11}^{(0)}P_0 + (\tfrac{1}{2})^{1/2}b_{11}^{(1)}P_1 - (\tfrac{1}{6})^{1/2}b_{11}^{(2)}P_2. \tag{3.65}$$

For the angular correlation, the common factor for all terms is irrelevant.

Therefore, we obtain the following parameters $b_{LL'}^{(n)}$ from Eqs. (2.87), (3.64), and (3.65).

$$b_{11}^{(0)} = -3^{1/2} [|C_A|^2 + |C'_A|^2] |\textstyle\int \sigma|^2, \tag{3.66}$$

$$b_{11}^{(1)} = 2^{1/2} 2 \operatorname{Re}(C_A^* C'_A) \frac{p}{E} |\textstyle\int \sigma|^2, \tag{3.67}$$

$$b_{11}^{(2)} = 0. \tag{3.68}$$

The angular distribution of beta rays is now obtained by taking the weighted sum of the absolute squares of Eq. (3.61), and substituting Eq. (3.62) in it,

$$W(\theta) = \sum_{m, m_1} a_m (jLmM | j_1 m_1)(jL'mM | j_1 m_1) \times (-)^M (L'L - MM | n0) b_{LL'}^{(n)} P_n(\cos\theta). \tag{3.69}$$

After performing the sum over the magnetic quantum number m_1, we have

$$W(\theta) = \sum_n \hat{f}_n(j)(-)^{j_1 - j - n + L + L'} W(jjLL', nj_1) b_{LL'}^{(n)} P_n(\cos\theta) \tag{3.70}$$

with

$$\hat{f}_n(j) = \sum_m a_m (-)^{j-m} (jjm - m | n0).$$

The sum over m is still kept in $\hat{f}_n(j)$, since the relative populations of the magnetic substates a_m are not specified.

Explicit forms for the Racah and Clebsch-Gordan coefficients are simple in the cases of $n = 0$ and 1. We substitute these into Eq. (3.70). Consequently, the angular distribution of the beta ray from the polarized nuclei is

$$W(\theta) = 1 \mp \frac{\langle j_z \rangle}{j} \frac{2 \operatorname{Re}(C_A^* C'_A)}{|C_A|^2 + |C'_A|^2} \frac{p}{E} \cos\theta \qquad \text{for} \quad j_1 = j - 1 \tag{3.71}$$

where the degree of the nuclear polarization is

$$\frac{\langle j_z \rangle}{j} = \frac{\sum_m a_m m}{j}. \tag{3.72}$$

Again, the upper sign refers to the electron decay and the lower sign to the positron decay. If these are both axial vector and tensor interactions, the angular distribution in the same decay scheme becomes

$$W(\theta) = 1 \mp \frac{\langle j_z \rangle}{j} \times \frac{2 \operatorname{Re}(C_A^* C'_A - C_T^* C'_T)}{[|C_A|^2 + |C'_A|^2 + |C_T|^2 + |C'_T|^2 + (2/E) \operatorname{Re}(C_T^* C_A + C_T'^* C'_A)]} \times \frac{p}{E} \cos\theta. \tag{3.73}$$

The Polarized Co^{60} Experiment

The first experiment that provided evidence of parity nonconservation in beta decay was performed by Wu and her collaborators [6]. A source of Co^{60} that decays by a pure Gamow–Teller transition ($5^+ \to 4^+$; see Fig. 3.1) was selected for this test, since the asymmetry of the beta distribution due to parity nonconservation was expected to be smaller in the case of $\Delta J = 0$, from Eq. (3.78), together with the isospin selection rule. The radioactive nuclei were introduced into the crystal [$2\ Ce\,(NO_3)_3 \cdot 3\ Mg\,(NO_3)_2 \cdot 24H_2O$], which

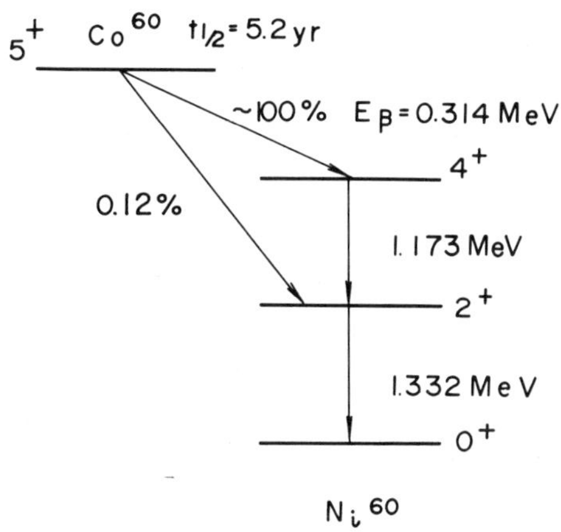

Fig. 3.1 Decay scheme of Co^{60}. Here, E_β is the maximum kinetic energy of the beta particle.

was cooled by an adiabatic demagnetization method down to an absolute temperature of 0.01°K. The degree of nuclear polarization was determined by measuring simultaneously the anisotropy of its gamma radiation

$$\epsilon = \frac{W(\pi/2) - W(0)}{W(\pi/2)}. \tag{3.74}$$

Here, $W(\pi/2)$ and $W(0)$ are the relative intensities of the emitted gamma ray at $\theta = \pi/2$ and 0 with respect to the nuclear polarization axis. A large beta asymmetry was observed, as shown in Fig. 3.2. The actually observed asymmetry, which is the coefficient of the cos θ term in Eq. (3.73), is about -25% at the nuclear polarization $\langle j_z \rangle / j = 0.65$, the electron velocity $v/c = p/E = 0.6$, and the attenuation factor $\frac{2}{3}$. This means that the coefficient A is nearly equal to -1,

$$A = \frac{-2\,\mathrm{Re}\,(C_A^* C_A' - C_T^* C_T')}{|C_A|^2 + |C_A'|^2 + |C_T|^2 + |C_T'|^2 + 2\,\mathrm{Re}\,(C_T^* C_A + C_T'^* C_A')}$$

$$= \frac{-0.25}{0.65 \times 0.60} \times \frac{3}{2} \approx -1. \tag{3.75}$$

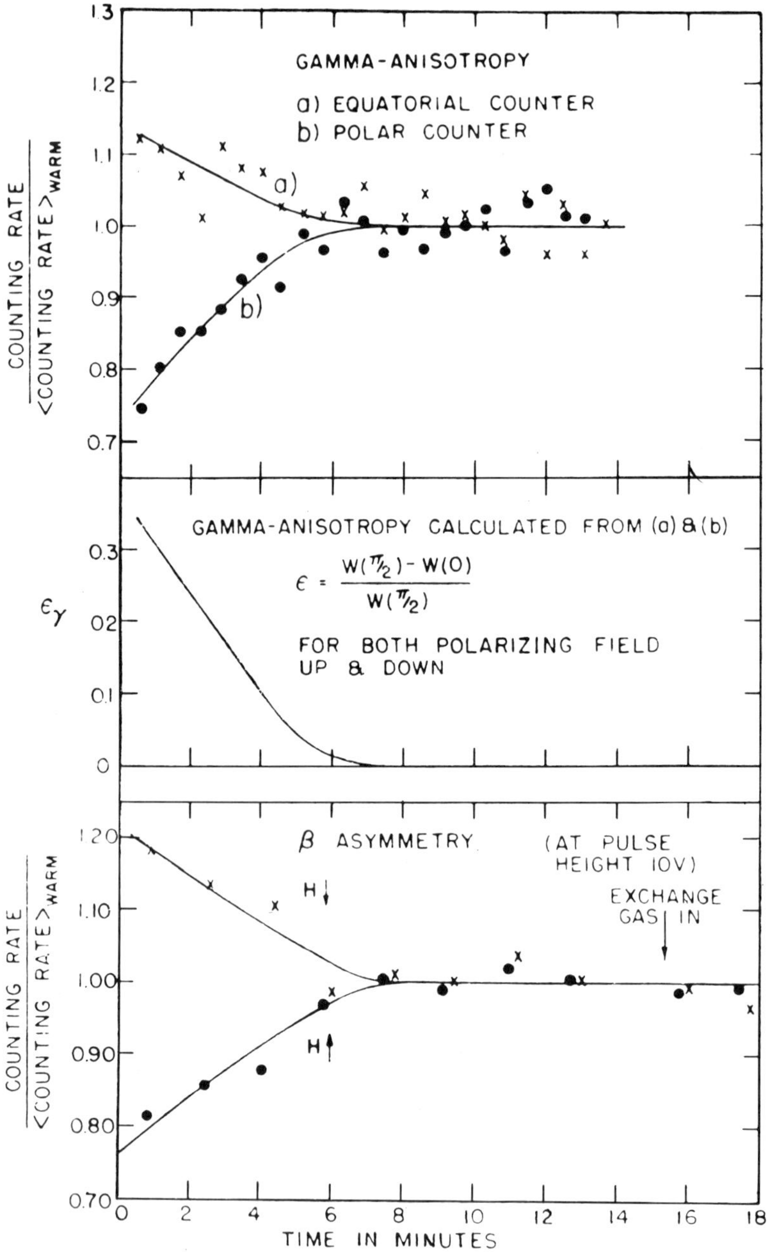

Fig. 3.2 Results of beta asymmetry and gamma anisotropy from polarized Co^{60} experiment. The disappearance of the beta asymmetry coincides exactly with the time of disappearance of gamma anisotropy. The measured asymmetry indicates that the emission of electrons is preferred in the direction opposite to that of the nuclear spin. From [6].

This result clearly demonstrates the violation of parity conservation. Equation (3.75) implies also the noninvariance under the charge conjugation, since C_i and C_i' have simultaneously nonzero values. Furthermore, the time-reversal invariance in beta decay holds approximately, since the asymmetry coefficient A is almost unity, and consequently the coupling constants are approximately real numbers,

$$C_{\mathrm{A}} \approx C'_{\mathrm{A}} \qquad \text{and/or} \qquad C_{\mathrm{T}} \approx -C'_{\mathrm{T}}. \tag{3.76}$$

As we discussed in section 2.11, it was shown later that $C_{\mathrm{T}}/C_{\mathrm{A}} \approx 0$.

General Formula for Beta-Ray Angular Distributions

In Eq. (3.73) we gave the formula for the beta-ray angular distribution in the case of Gamow–Teller transitions with no Coulomb correction. The formula can be generalized easily in the case of no-spin change $\Delta j = 0$ where the interference of the Gamow–Teller and Fermi interactions appears, with the Coulomb corrections. For this purpose, we have to compute the particle parameters $b^{(n)}_{LL'}$ defined in Eq. (3.62), which are obtained in the same manner as described in section 2.6. We cannot, however, apply the three rules which are given under Eq. (2.67), since the initial nuclear state is oriented. We simply perform calculations of all interferences. In fact, such calculations were done, to different extents, by several groups, including those of Morita; Curtis and Lewis; Jackson, Treiman, and Wyld; Feld; Alder, Stech, and Winther; Ebel and Feldman. (Works of these authors are summarized in [6a].) The following formula, which was given by the present author and was published in a paper by Wu *et al.* [7], is derived from Eq. (3.70), using the parameters $b^{(n)}_{LL'}$ given in the Appendix to [8] with explicit forms of the Racah coefficients.

$$W(\theta) = 1 + \mathrm{A}\,\frac{\langle j_z \rangle}{j}\,\frac{p}{E}\,\frac{1}{\xi(1 + b/E)}\cos\theta \tag{3.77}$$

with

$$\begin{aligned} A\xi = {} & \pm\lambda_{jj_1} 2\,\mathrm{Re}\,[(C_{\mathrm{T}}C_{\mathrm{T}}'^{*} - C_{\mathrm{A}}C_{\mathrm{A}}'^{*}) \pm i\frac{\alpha Z}{p}(C_{\mathrm{A}}C_{\mathrm{T}}'^{*} + C_{\mathrm{A}}'C_{\mathrm{T}}^{*})](\textstyle\int\sigma)^2 \\ & + \delta_{jj_1} 2\left(\frac{j}{j+1}\right)^{1/2} \mathrm{Re}\,[(C_{\mathrm{T}}'C_{\mathrm{S}}^{*} + C_{\mathrm{T}}C_{\mathrm{S}}'^{*} - C_{\mathrm{A}}'C_{\mathrm{V}}^{*} - C_{\mathrm{A}}C_{\mathrm{V}}'^{*}) \\ & \pm i\frac{\alpha Z}{p}(C_{\mathrm{A}}'C_{\mathrm{S}}^{*} + C_{\mathrm{A}}C_{\mathrm{S}}'^{*} - C_{\mathrm{T}}'C_{\mathrm{V}}^{*} - C_{\mathrm{T}}C_{\mathrm{V}}'^{*})](\textstyle\int\sigma)(\textstyle\int 1), \end{aligned} \tag{3.78}$$

and

$$\lambda_{jj_1} = 1, \qquad j \to j_1 = j - 1,$$
$$= \frac{1}{j+1}, \qquad j \to j_1 = j,$$
$$= -\frac{j}{j+1}, \qquad j \to j_1 = j + 1.$$

Here, ξ and $b\xi$ are as given in Eqs. (2.181) and (2.183), and δ_{jj_1} is the Kronecker delta.

3.7 CONSERVATION OF THE MAGNETIC QUANTUM NUMBERS AND LONGITUDINAL POLARIZATION OF LEPTONS

From the conservation law of the magnetic quantum numbers we find that the leptons from polarized nuclei in the Co^{60} experiment are longitudinally polarized (see Fig. 3.3). For simplicity, we assume that the initial state of

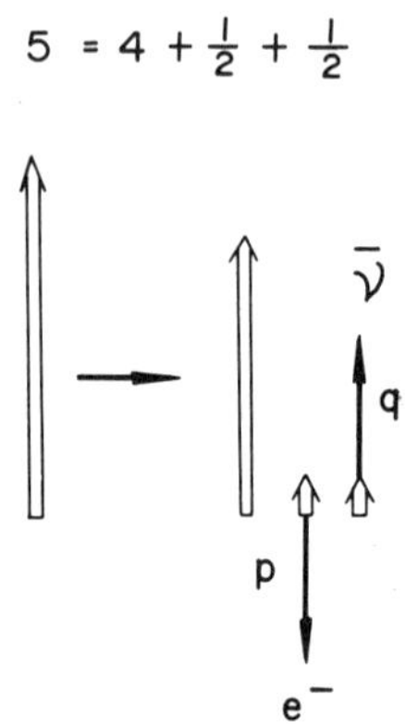

Fig. 3.3 Conservation of magnetic quantum numbers in the polarized Co^{60} experiment.

Co^{60} is completely polarized and the magnetic quantum number of the initial system is 5. This will split into three parts, 4 for Ni^{60}, $\frac{1}{2}$ for the electron, and $\frac{1}{2}$ for the antineutrino. The sum of the magnetic quantum numbers is 5 and it is conserved in the final system. This means that spins of leptons are in the same direction as the nuclear polarization. On the other hand, the electron is emitted in the direction opposite to the nuclear polarization of Co^{60}. The spin and momentum of the electron are, thus, antiparallel, and the electron has left-handed polarization. For the antineutrino, there were two possibilities at the time of the Co^{60} experiment. First, consider the case of the axial vector interaction, for which the electron and antineutrino are going out in opposite directions (see Eq. (2.75)). Therefore, the spin and momentum of antineutrino are parallel, and the antineutrino has right-handed polarization (Fig. 3.3). If the interaction is a tensor, the sign of the electron–antineutrino angular correlation is reversed (see Eq. (2.77)), so that the antineutrino has left-handed polarization. But this second possibility was disproved by later experiments, as we discussed in section 2.11.

3.8 THE TWO-COMPONENT THEORY OF THE NEUTRINO

If the relations in Eq. (3.76) are taken strictly, the neutrino wave function appears in a combination of either

$$\tfrac{1}{2}(1+\gamma_5)\psi_\nu \qquad \text{or} \qquad \tfrac{1}{2}(1-\gamma_5)\psi_\nu. \tag{3.79}$$

We will see that the operators $\frac{1}{2}(1 \pm \gamma_5)$ are the projection operators for the longitudinally polarized state. From Eq. (A.86) and Table A.2, we have as the Dirac spinors for the massless neutrino $E_\nu > 0$

$$\psi_\nu^\uparrow = \frac{1}{\sqrt{2}}\begin{pmatrix}-1\\0\\1\\0\end{pmatrix} \qquad \text{and} \qquad \psi_\nu^\downarrow = \frac{1}{\sqrt{2}}\begin{pmatrix}0\\1\\0\\1\end{pmatrix}, \tag{3.80}$$

where the direction of the momentum is chosen to be the z axis and the arrow indicates the spin direction. If we define the helicity operator as

$$\boldsymbol{\sigma}\cdot\hat{\mathbf{p}} \qquad \text{with} \qquad \hat{\mathbf{p}} = \mathbf{p}/|\mathbf{p}| \tag{3.81}$$

where $\boldsymbol{\sigma}$ is the spin operator and $\hat{\mathbf{p}}$ the unit vector of the neutrino momentum, it is obvious that $\psi_\nu^\uparrow$ and $\psi_\nu^\downarrow$ are eigenstates for the helicity operator with eigenvalues +1 and −1, respectively. From Eq. (3.80) we have

$$\begin{aligned} &\tfrac{1}{2}(1+\gamma_5)\psi_\nu^\uparrow = 0, \qquad &&\tfrac{1}{2}(1-\gamma_5)\psi_\nu^\uparrow = \psi_\nu^\uparrow,\\ &\tfrac{1}{2}(1+\gamma_5)\psi_\nu^\downarrow = \psi_\nu^\downarrow, \qquad &&\tfrac{1}{2}(1-\gamma_5)\psi_\nu^\downarrow = 0. \end{aligned} \tag{3.82}$$

Therefore, the combinations

$$\tfrac{1}{2}(1 \pm \gamma_5)\psi_\nu = \begin{cases}\psi_\nu^L\\ \psi_\nu^R\end{cases} \tag{3.83}$$

make the left-handed polarized neutrino (helicity −1) ψ_ν^L and the right-handed polarized neutrino (helicity +1) ψ_ν^R.

The two-component theory of the neutrino was proposed, immediately after the discovery of parity nonconservation, by Landau [9], Salam [10], and Lee and Yang [11]. In this theory, there are only two states instead of four states (see Table A.2). One is the completely polarized particle state (neutrino) and the other, also completely polarized (but different in sign), is the antiparticle state (antineutrino). In other words, if we take the left-handed neutrino ψ_ν^L, which has the helicity -1, the wave function is the first one in Eq. (3.80). Its antiparticle, which has the helicity $+1$, is given by

$$\psi_{\bar{\nu}}^L = (\tfrac{1}{2})^{1/2} \begin{pmatrix} 1 \\ 0 \\ -1 \\ 0 \end{pmatrix} \tag{3.84}$$

from the third line of Table A.2, where we have changed $\mathbf{p} \to -\mathbf{p}$ so that $\mathbf{p}$ refers to the antineutrino momentum. Furthermore, we see from Eqs. (3.80) and (3.84) that the lower two components are repetitions of the upper two components (except for a sign), and that therefore the two components out of four are redundant.

γ_5 Transformation and Mass Zero Neutrino

If the neutrino mass is zero, the Dirac equation will be

$$E\psi_\nu = -c\boldsymbol{\alpha}\cdot\mathbf{p}\psi_\nu. \tag{3.85}$$

The solutions of this equation satisfy the chirality equation

$$\gamma_5\,\psi_\nu = \mp\psi_\nu. \tag{3.86}$$

Substituting Eq. (3.86) into Eq. (3.85), we have

$$E\psi_\nu = \pm c\boldsymbol{\sigma}\cdot\mathbf{p}\psi_\nu. \tag{3.87}$$

It is interesting to note that the mathematical possibility of this sort of two-component relativistic particle with spin $\frac{1}{2}$ and mass zero was proposed by Weyl in 1929 [12].

Conversely, the Dirac equation, Eq. (3.7), is invariant under the transformation

$$\psi_\nu \to \pm\gamma_5\,\psi_\nu \tag{3.88}$$

if the neutrino mass vanishes. The sign for Eq. (3.88) should be determined in conformity with the type of interaction in Eq. (3.76).

3.9 LONGITUDINAL POLARIZATION OF LEPTONS FROM BETA DECAY OF UNORIENTED NUCLEI

Here we study that the electron and antineutrino from beta decay of unoriented nuclei are longitudinally polarized, as are those in the beta decay of polarized Co^{60}, by the following calculations. The longitudinal polarization of the electron is given by the expectation value of the helicity operator

$$P_L = \langle \sigma \cdot \hat{\mathbf{p}} \rangle \tag{3.89}$$

with the electron wave function evaluated at $r = \infty$. This wave function ψ_e is, of course, the perturbed wave function by the interaction of the beta decay, Eq. (3.1), and it is the superposition of the form

$$\psi_e = \sum_f \psi_{\mathbf{p},f} \langle f | \Omega | i \rangle \tag{3.90}$$

where $\langle f | \Omega | i \rangle$ is the transition amplitude to the final state where the electron is created in the state $\psi_{\mathbf{p},f}$, subscripts $\mathbf{p}$ and f being the electron momentum and the set of the quantum numbers, respectively. In the case of the plane wave, by neglecting the nuclear Coulomb field, the $\psi_{\mathbf{p},f}$ are those in Table A.2; therefore f corresponds to spin up and down. The longitudinal polarization of the electron, Eq. (3.89), is obtained by taking the expectation value of $\sigma \cdot \hat{\mathbf{p}}$ with the electron wave function ψ_e in Eq. (3.89), and by normalizing with its absolute square,

$$\langle \sigma \cdot \hat{\mathbf{p}} \rangle = \frac{\sum_{f,f'} \langle i | \Omega^\dagger | f \rangle \langle f | \sigma \cdot \hat{\mathbf{p}} | f' \rangle \langle f' | \Omega | i \rangle}{\sum_f |\langle f | \Omega | i \rangle|^2} \tag{3.91}$$

where Ω represents the operator part of the interaction in Eq. (3.1), and i and f are initial neutron state and the final proton–electron–antineutrino states. The summation f signifies the sum over the quantum numbers (for the electron as well as the antineutrino) and averaging over the angular variables of the antineutrino momentum. The denominator of Eq. (3.91) is exactly the same as the spectral shape factor in section 2.6. On the other hand, the numerator includes the following trace of the lepton covariants,

$$\begin{aligned} &\sum (\psi_\nu^\dagger \Omega^{l\dagger} \psi_e)(\psi_e^\dagger \sigma \cdot \hat{\mathbf{p}} \psi_e')(\psi_e'^\dagger \Omega^k \psi_\nu) \\ &= \frac{1}{4\pi} \int d\Omega_\nu \, \mathrm{Tr}\,(\Omega^k D_\nu \Omega^{l\dagger} D_e \sigma \cdot \hat{\mathbf{p}} D_e). \end{aligned} \tag{3.92}$$

Here Σ means the spin sum and the integral over the unobserved angular variables. As an example, we give the foregoing trace for the $C_V^* C_V'$ term,

$$\begin{aligned} &C_V^* C_V' \frac{1}{8} \frac{1}{4\pi} \int d\Omega_\nu \, \mathrm{Tr} \left[\gamma_5 \left(1 - \frac{\alpha \cdot \mathbf{q}}{E_\nu} \right) 1 \left(1 - \frac{\alpha \cdot \mathbf{p} + \beta}{E} \right) \right. \\ &\times (\sigma \cdot \hat{\mathbf{p}}) \left. \left(1 - \frac{\alpha \cdot \mathbf{p} + \beta}{E} \right) \right] |\textstyle\int 1|^2 = -\frac{p}{E} C_V^* C_V' |\textstyle\int 1|^2. \end{aligned} \tag{3.93}$$

We notice here that the trace (3.92) vanishes for all $C_i^* C_j$ and $C_i'^* C_j'$ including $i = j$ and the nonzero contribution comes only from $C_i^* C_i'$ terms and their complex conjugate. Consequently, the longitudinal polarization of the electron is generally given by

$$P_L = \frac{p}{E}\frac{N}{\xi(1+b/E)}, \tag{3.94}$$

with

$$N = \pm 2\,\mathrm{Re}\left[(C_S^* C_S' - C_V^* C_V')\,|\textstyle\int 1|^2 + (C_T^* C_T' - C_A^* C_A')\,|\int \sigma|^2\right], \tag{3.95}$$

$$\xi = (|C_S|^2 + |C_S'|^2 + |C_V|^2 + |C_V'|^2)\,|\textstyle\int 1|^2$$
$$+ (|C_T|^2 + |C_T'|^2 + |C_A|^2 + |C_A'|^2)\,|\textstyle\int \sigma|^2, \tag{3.96}$$

$$b\xi = \pm 2\,\mathrm{Re}\left[(C_S^* C_V + C_S'^* C_V')\,|\textstyle\int 1|^2 + (C_T^* C_A + C_T'^* C_A')\,|\int \sigma|^2\right]. \tag{3.97}$$

The formulas above are given in the $Z = 0$ approximation. There is, however, no essential change introduced by the Coulomb corrections. Again, the upper sign refers to the electron and the lower sign to the positron. If the coupling constants obey the relations

$$C_i = C_i' \qquad \text{for} \quad i = V \text{ and } A, \tag{3.98a}$$

$$C_j = -C_j' \qquad \text{for} \quad j = S \text{ and } T, \tag{3.98b}$$

the longitudinal polarization Eq. (3.94) becomes

$$P_L = \mp \frac{p}{E} \qquad \text{for } \beta^{\mp}.$$

It should be noticed that (3.98a) or (3.98b) corresponds to the two-component theory of the neutrino. In either case of Eqs. (3.98), or in the case where both Eq. (3.98a) and Eq. (3.98b) are present, the Fierz interference (3.97) vanishes identically.

A similar calculation can be performed for the longitudinal polarization of the emitted antineutrino (for electron decay) or neutrino (for positron decay). The expectation value of the helicity operator is evaluated with the neutrino wave function, which has a form similar to that of Eq. (3.89). Consequently, Eq. (3.92) should read

$$\sum (\psi_\nu^\dagger \Omega^{l\dagger} \psi_e)(\psi_\nu'^\dagger (\sigma\cdot\hat{q})\,\psi_\nu)(\psi_e^\dagger \Omega^k \psi_\nu')$$
$$= \frac{1}{4\pi}\int d\Omega_e\, \mathrm{Tr}\,(\Omega^{l\dagger} D_e \Omega^k D_\nu(-\sigma\cdot\hat{q})D_\nu), \tag{3.99}$$

with the projection operator for the antineutrino in Eq. (2.71)

$$\frac{1}{2}\left(1 - \frac{\alpha\cdot q}{E_\nu}\right)$$

where $\mathbf{q}$ is the antineutrino momentum. The minus sign for $\sigma\cdot\hat{q}$ in the second line of Eq. (3.99) is the consequence of the transformation from the energy-negative neutrino state to the energy-positive antineutrino state. The longitudinal polarization is now given by

$$P_L = \frac{N_\nu}{\xi(1 + b/E)}, \tag{3.100}$$

with

$$N_\nu = \pm 2\,\mathrm{Re}\Big[(C_S^* C_S' + C_V^* C_V')|\textstyle\int 1|^2 + (C_T^* C_T' + C_A^* C_A')|\textstyle\int \sigma|^2\Big], \tag{3.101}$$

and ξ and $b\xi$ the same as in Eqs. (3.96) and (3.97). The upper sign now refers to the antineutrino for electron decay, and the lower sign to the neutrino for positron decay. The coupling constants Eq. (3.98a) give

$$P_L = \begin{cases} +1 & \text{for } \bar{\nu}, \\ -1 & \text{for } \nu, \end{cases} \tag{3.102}$$

while Eq. (3.98b) gives

$$P_L = \begin{cases} -1 & \text{for } \bar{\nu}, \\ +1 & \text{for } \nu. \end{cases} \tag{3.103}$$

These confirm that the interacting two-component neutrinos have 100% polarization, as have the free neutrinos. The latter was seen from Eq. (3.87) immediately. On the other hand, coexistence of the coupling constants, Eqs. (3.98a) and (3.98b), does not introduce a complete longitudinal polarization of the neutrino.

Experiments on the Longitudinal Polarization of the Electron

The first experiment for detecting longitudinal polarization that proved parity nonconservation was performed by Frauenfelder's group [13]. They obtained $P_L \sim -p/E$ for the electron from unoriented Co^{60}. They used the Mott scattering method, the basic idea behind which is illustrated in Fig. 3.4. In the rest system of the electron, where the nucleus moves around the electron, the nucleus produces a magnetic field at the electron. The attractive potential increases when the magnetic moment and magnetic field are parallel. The existence of this sort of spin-orbit interaction causes electrons with spin up that approach a nucleus of positive charge to be scattered predominantly to the right. Since this is a suitable method for detecting transverse polarization, later experiments have been improved: in these more recent efforts, the longitudinal electrons are transformed to transverse ones by using a cylindrical condenser, as proposed by Tolhoek [14].

Whereas Mott scattering is useful only for electrons, electron–electron scattering is useful for both charges of beta particles. Both electron–electron (Møller) and electron–positron (Bhabha) scatterings are spin dependent. Therefore, beta particles are differently scattered when they hit a magnetized ferromagnetic foil, and we can use this characteristic to measure the longitudinal polarization of beta particles.

Longitudinally polarized electrons produce circularly polarized bremsstrahlung. The degree of circular polarization is determined by scattering or transmission, using magnetized iron. From this we can also determine the longitudinal polarization of beta particles.

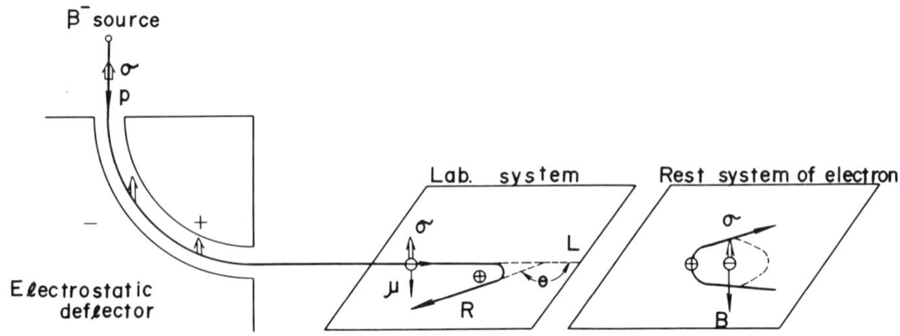

Fig. 3.4 Mott scattering. The longitudinal polarization of the beta particle β^- is converted into transverse polarization by the electrostatic deflector. Therefore, in the laboratory system the incident electron with spin σ perpendicular to the plane of scattering is deflected by an angle θ to the right. In the rest system of the electron, the nucleus orbits partway around the electron and creates a magnetic field **B**. Consequently, the magnetic field and the spin are antiparallel, and the magnetic moment μ and the field **B** are parallel. The spin-orbit interaction acts in the same sense as the Coulomb attraction and the scattering to the right is enhanced [15].

Table 3.3 gives data on the helicities of the beta particles collected by Frauenfelder and Steffen [15]. In the allowed transitions, the magnitude of the helicity is always nearly equal to p/E, and its sign is minus for the electron and plus for the positron. This evidence supports again the coupling constants in Eqs. (3.98a) and (3.98b), proving parity nonconservation and charge conjugation noninvariance in beta decay.

All allowed and most forbidden decays show, within rather wide limits of error, a polarization of $P_L = \pm p/E$. A large deviation from this rule was found first in Bi^{210}, for which P_L is less than p/E [16–19]. This can be explained by mutual cancellation effects of the several first forbidden matrix elements, section 7.9.

Table 3.3 Helicities of Beta Particles†

Decay	Character	E_{max} (MeV)	Observed Kinetic Energy (MeV)	$\frac{P_L}{p/E}$	Method†	Reference
$B^{12} \xrightarrow{\beta^-} C^{12}$	$1^+ \to 0^+$ Allowed, G–T	13.4	7–13	-0.98 ± 0.06	EB	*a*
$O^{14} \xrightarrow{\beta^+} N^{14}$	$0^+ \to 0^+$ Allowed, F	1.81	1.0	$+0.97 \pm 0.19$	e–e	*b*
$P^{32} \xrightarrow{\beta^-} S^{32}$	$1^+ \to 0^+$ Allowed, G–T	1.71	Upper end	-1.05 ± 0.04	EB	*c*
			0.66–0.99	-1.00 ± 0.02	e–e	*d*
			0.616	-0.99 ± 0.01	M	*e*
$Co^{60} \xrightarrow{\beta^-} Ni^{60}$	$5^+ \to 4^+$ Allowed, G–T	0.31	0.166	-0.96 ± 0.06	M	*f*
			0.194	-0.99 ± 0.06	M	*g*
$Ga^{68} \xrightarrow{\beta^+} Zn^{68}$	$1^+ \to 0^+$ Allowed, G–T	1.9	1.03–1.3	$+0.99 \pm 0.09$	e–e	*d*
$Au^{198} \xrightarrow{\beta^-} Hg^{198}$	$2^- \to 2^+$ First forbidden	0.96	0.46–0.81	-0.98 ± 0.03	e–e	*d*

†Data from [15]. Key: e–e, electron–electron scattering; M, Mott scattering; EB, external bremsstrahlung.

[a] P. Lipnik, J. P, Deutsch, L. Grenacs, and P. Macq, *Nucl. Phys.* **30**, 312 (1962).
[b] J. C. Hopkins, J. B. Gerhart, F. H. Schmidt, and J. E. Stroth, *Phys. Rev.* **121**, 1185 (1961).
[c] A. Bisi, A. Fasana, and L. Zappa, *Nuovo Cimento* **16**, 1374 (1960); *Nucl. Phys.* **45**, 405 (1963).
[d] J. D. Ullman, H. Frauenfelder, H. J. Lipkin, and A. Rossi, *Phys. Rev.* **122**, 536 (1961).
[e] A. R. Brosi, A. I. Galonsky, B. H. Ketelle, and H. B. Willard, *Nucl. Phys.* **33**, 353 (1962).
[f] H. Bienlein, R. Fleischmann, and H. Wegener, *Z. Phys.* **150**, 80 (1958).
[g] J. S. Greenberg, D. P. Malone, R. L. Gluckstern, and V. W. Hughes, *Phys. Rev.* **120**, 1393 (1960).

Measurements of the Helicity of the Neutrino

A measurement of the neutrino was performed by Goldhaber, Grodzins, and Sunyar [20] in the cascade transitions of K capture and gamma decay in Eu^{152} (Fig. 3.5). The idea of this experiment is summarized in the following three steps.

First, in order to relate the spin directions of the neutrino and the gamma ray, let us consider the conservation of magnetic quantum numbers. Choosing the direction of the emission of the gamma ray to be the z axis, we have two polarizations for it, $m_\gamma = \pm 1$. The K orbital electron is of course the s electron, and its magnetic quantum number is $\frac{1}{2}$ or $-\frac{1}{2}$. This is also the case for the s neutrino, since the K capture in this case is the allowed transition. The final state consists of three particles; the neutrino, the gamma ray, and the nucleus of Sm^{152}, which is the ground state of spin zero. Therefore, the requirement for conservation of magnetic quantum numbers is represented by $m_e = m_\gamma + m_\nu$. This means that m_ν should be $-\frac{1}{2}$ (or $\frac{1}{2}$) for $m_\gamma = 1$ (or -1). As a result, the spin direction of the gamma ray is always opposite to that of the neutrino in this particular decay scheme.

Second, we utilize the phenomenon of nuclear fluorescence resonance to measure the direction of the neutrino momentum. When the state of the excitation energy E emits the gamma ray, E is divided into two parts: the gamma energy E_γ and the nuclear recoil energy E_R ($E_R = E_\gamma/2Mc^2$, M being the mass of the nucleus). When the ground state of the same target nucleus is excited to the same state as before, by absorbing the gamma ray, there is also a recoil energy loss E_R. Therefore, the gamma ray emitted from the state of E cannot be resonantly absorbed unless the energy $2E_R$ is compensated by

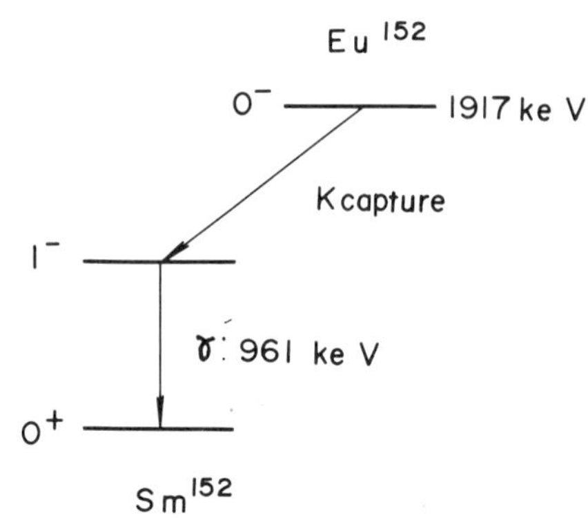

Fig. 3.5 Decay scheme of Eu^{152}.

some other sources. This will be done if the nuclear recoil due to the preceding neutrino emission is properly taken into account. For example, E_γ increases due to the Doppler shift when the gamma ray is emitted in a direction opposite to the momentum of the neutrino. In fact, in the geometry above for Eu^{152} the compensation is achieved. (The rigorous condition is given by $E_\nu \cos\theta = E$, θ being the angle between the directions of recoil and the gamma ray. As a matter of fact, $E_\nu < E$ in the case of Eu^{152}. A small overlapping is, however, expected by considering the line breadth.)

Third, the resonant condition is fulfilled only for those gamma rays that are emitted opposite to the neutrino. The detection of resonance scattering is equivalent to the detection of neutrino momentum. Furthermore, the gamma ray and neutrino have opposite spin directions. Consequently, the helicities of these two particles are the same in sign, and it is possible to measure the helicity of the neutrino by measuring the circular polarization of those gamma rays which cause resonance scattering.

In this experiment [20], Goldhaber *et al.* measured the circular polarization of those gamma rays. It was $-(67 \pm 15)\%$, compared with the theoretically expected 84%. The experimental data proved that the neutrino from the K capture is left-handed and the Gamow-Teller interaction is predominantly an axial vector.

The possibility of measuring the helicity of the antineutrino (neutrino) in electron (positron) decay was theoretically studied, and various kinds of experiments were proposed by Treiman [21], Morita *et al.* [22, 23], Lewis and Curtis [24], and Bincer [25]. Formulation is a little more complicated than that for K capture, since the electron, antineutrino, and gamma ray are emitted.

Detailed considerations show that the beta-gamma directional correlation where the gamma ray is resonantly absorbed supplies information on the helicity of the antineutrino [22-24]. An experiment was done in Birmingham on the beta-gamma cascade of Ne^{23} [26]. The gamma-(resonantly absorbed) gamma directional correlation also gives the helicity of the antineutrino [23], and this experiment was done by Burgov and Terekhov using the beta-gamma-gamma cascade decay of Na^{24} [27]. In these two cases, the beta decays are mixed Fermi and Gamow-Teller transitions, but the Fermi transitions are hindered due to the isospin selection rule. The experimental data in both cases indicated that the antineutrino in beta decay has a positive helicity and that the Gamow-Teller transitions occur mainly via axial vector interaction.

3.10 BETA-GAMMA CIRCULAR POLARIZATION CORRELATION

The beta-gamma angular correlation is expressed in the form

$$W(\theta) = 1 + \tau A \frac{p}{E} \cos\theta. \tag{3.104}$$

$\tau = +1$ for right-handed circular polarization and -1 for left-handed circular polarization where θ is the angle between the directions of the emission of beta and gamma rays for the allowed beta transition in cascade decays (Fig. 3.6). As we studied in the last section, the beta particle is longitudinally polarized, and the intermediate state j_1 is also polarized with respect to the direction

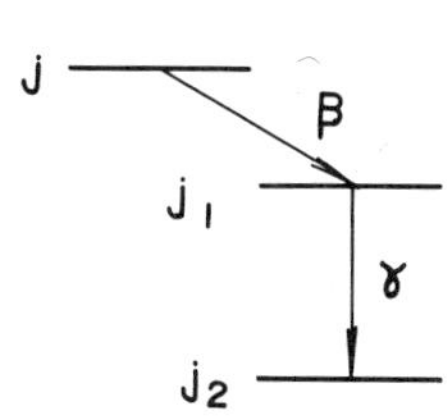

Fig. 3.6 Beta-gamma cascade transition.

of the emission of the beta ray. (Remember the conservation of the magnetic quantum number.) The subsequently emitted gamma ray in the transition $j_1 \rightarrow j_2$ is, therefore, circularly polarized, and is distributed asymmetrically with respect to the polarization axis. Consequently, the beta-gamma angular correlation has a $\cos\theta$ term, if the circular polarization of the gamma ray is observed; otherwise, it is isotropic. In this discussion, a short lifetime is assumed for the intermediate state to preserve polarization of the nuclear spin before the emission of the gamma ray.

The general form of the beta-gamma angular correlation is given in Chapter 6, and its derivation is discussed in Appendix A.7. Here, we show a formula for the *VA* theory in the simplest decay scheme

$$j \xrightarrow{\beta} j_1 \xrightarrow{\gamma} j_2$$

with $j_2 = j_1 - L$, and beta decay is the allowed transition. The symbol L refers to the multipolarity of the gamma ray.

$$A = \frac{2}{\xi(1+\mathrm{L})}\left\{\mp\lambda'_{jj_1}\,\mathrm{Re}\,(C_A^* C_A')\,|\textstyle\int\sigma|^2 - \delta_{jj_1}\left(\dfrac{j+1}{j}\right)^{1/2} \right.$$
$$\left. \times\,\mathrm{Re}[(C_A^* C_V' + C_A'^* C_V)(\textstyle\int\sigma)^*(\int 1)]\right\} \qquad \text{for } \beta^{\mp}, \tag{3.105}$$

with

$$\xi = (|C_V|^2 + |C_V'|^2)|\textstyle\int 1|^2 + (|C_A|^2 + |C_A'|^2)|\int\sigma|^2, \tag{3.106}$$

and

$$\lambda'_{jj_1} = \begin{cases} 1 & \text{for } j_1 = j - 1, \\ -\dfrac{1}{j} & \text{for } j_1 = j, \\ -\dfrac{j+2}{j+1} & \text{for } j_1 = j + 1. \end{cases} \tag{3.107}$$

Here $\int 1$ and $\int\sigma$ are generally complex. But they are real by assuming the strong interactions to be invariant under time reversal.

For pure Gamow-Teller transitions, we have $A = -\frac{1}{3}$ in the decay of Co^{60}

$$\left(5^+ \xrightarrow{\beta^-} 4^+ \xrightarrow{\gamma} 2^+ \xrightarrow{\gamma} 0^+\right)$$

and $A = \frac{1}{3}$ in the decay of Na^{22}

$$\left(3^+ \xrightarrow{\beta^+} 2^+ \xrightarrow{\gamma} 0^+\right).$$

For mixed Fermi and Gamow-Teller transitions $j \xrightarrow{\beta} j$, we can determine the ratio $\int 1/\int\sigma$ by measuring the asymmetry coefficient A. This ratio gives

Table 3.4 Beta–Gamma Circular Polarization Data on Allowed Gamow–Teller Beta Transitions[a]

β Emitter	β–γ Cascade	Asymmetry Coefficient A	Reference
Na^{22}	$3 \xrightarrow{\beta^+} 2 \xrightarrow{\gamma} 0$	$+0.39 \pm 0.08$	H. Schopper, *Phil. Mag.* **2,** 710 (1957)
		$+0.295 \pm 0.054$	H. Appel *et al., Phys. Rev.* **109,** 2211 (1958)
		$+0.35 \pm 0.02$	R. M. Steffen, *Phys. Rev.* **115,** 980 (1959)
		$+0.36 \pm 0.08$	H. Appel, *Z. Phys.* **155,** 580 (1959)
		$+0.336 \pm 0.017$	H. Behrens, *KFK* No. 434 (1966)
Co^{60}	$5 \xrightarrow{\beta^-} 4 \xrightarrow{\gamma} 2 \xrightarrow{\gamma} 0$	-0.41 ± 0.07	H. Schopper, *Phil. Mag.* **2,** 40 (1957)
		-0.40 ± 0.09	F. Boehm and A. H. Wapstra, *Phys. Rev.* **106,** 1364 (1957)
		-0.38 ± 0.04	R. M. Steffen, *Proc. Rehovoth Conf.* (Amsterdam, 1958) p. 419
		-0.35 ± 0.05	H. Appel and H. Schopper, *Z. Phys.* **149,** 103 (1957)
		-0.344 ± 0.09	P. Debrunner and W. Kündig, *Helv. Phys. Acta* **30,** 261 (1957)
		-0.32 ± 0.07	A. Lundby *et al., Nuovo Cimento* **6,** 745 (1957)
		-0.335 ± 0.018	H. Appel *et al., Phys. Rev.* **109,** 2211 (1958)
		-0.41 ± 0.08	F. Boehm and A. H. Wapstra, *Phys. Rev.* **109,** 456 (1958)
		-0.345 ± 0.019	R. M. Steffen, *Phys. Rev.* **115,** 980 (1959)
		-0.335 ± 0.018	H. Appel, *Z. Phys.* **155,** 580 (1959)
		-0.203 ± 0.138	P. Jager, *Z. Phys.* **158,** 214 (1960)
		-0.361 ± 0.082	P. Jager, *Z. Phys.* **158,** 214 (1960)
		-0.320 ± 0.034	J. Berthier, Unpublished thesis, 1962
		-0.33 ± 0.02	V. M. Lobashov and V. A. Nazarenko, *JETP* **15,** 247 (1962)
		-0.322 ± 0.014	H. Behrens, *KFK* No. 434 (1966)

[a] From Frauenfelder and Steffen [28].

Table 3.5 Beta-Gamma Circular Polarization Data on Allowed $\Delta j = 0$ Beta Transitions[a]

β Emitter	β-γ Cascade	Asymmetry Coefficient A	$\frac{C_V \int 1}{C_A \int \sigma}$	Reference
Na^{24}	$4 \xrightarrow{\beta^-} 4 \xrightarrow{\gamma} 2 \xrightarrow{\gamma} 0$	-0.063 ± 0.047	$+0.21 \pm 0.07$	H. Appel and H. Schopper, *Z. Phys.* **149,** 103 (1957)
		$+0.05 \pm 0.04$	$+0.04 \pm 0.05$	R. M. Steffen, *Proc. Rehovoth Conf.* (Amsterdam, 1958), p. 419
		$+0.07 \pm 0.04$	$+0.02 \pm 0.05$	F. Boehm and A. H. Wapstra, *Phys. Rev.* **109,** 456 (1958)
		$+0.07 \pm 0.03$	$+0.02 \pm 0.04$	R. M. Steffen, *Phys. Rev.* **115,** 980 (1959)
		$+0.12 \pm 0.03$	-0.05 ± 0.04	Th. Mayer-Kuckuk and R. Nierhaus, *Z. Phys.* **154,** 383 (1959)
		$+0.085 \pm 0.028$	0.00 ± 0.02	S. D. Bloom *et al., Phys. Rev.* **125,** 2021 (1962)
		$+0.104 \pm 0.026$	-0.028 ± 0.035	E. L. Haase *et al. Phys. Letters* **4,** 338 (1963)
		$+0.102 \pm 0.006$	-0.026 ± 0.008	H. Behrens, *KFK* No. 434 (1966)
Sc^{46}	$4 \xrightarrow{\beta^-} 4 \xrightarrow{\gamma} 2 \xrightarrow{\gamma} 0$	$+0.33 \pm 0.04$	-0.39 ± 0.07	F. Boehm and A. H. Wapstra, *Phys. Rev.* **107,** 1202 (1957); **109,** 456 (1958)
		$+0.23 \pm 0.06$	-0.20 ± 0.12	R. M. Steffen, *Proc. Rehovoth Conf.* (Amsterdam, 1958), p. 419
		$+0.29 \pm 0.11$	$-0.33^{+0.31}_{-0.19}$	A. Lundby *et al., Nuovo Cimento* **7,** 891 (1958)
		$+0.24 \pm 0.04$	-0.22 ± 0.08	W. Jungst and H. Schopper, *Z. Naturforsch.* **13a,** 505 (1958)
		$+0.24 \pm 0.02$	-0.22 ± 0.04	R. M. Steffen, *Phys. Rev.* **115,** 980 (1959); **118,** 1667(E) (1960)

		+ 0.085 ± 0.017	0.00 ± 0.02	S. D. Bloom *et al.*, *Phys. Rev. Letters* **5**, 326 (1960)
		+ 0.10 ± 0.02	−0.02 ± 0.02	H. Daniel and M. Kuntze, *Z. Phys.* **162**, 229 (1961)
		+ 0.075 ± 0.018	+ 0.020 ± 0.025	S. D. Bloom *et al.*, *Phys. Rev.* **125**, 2021 (1962)
		+ 0.215 ± 0.019	−0.19 ± 0.03	F. Boehm and J. Rogers, *Nucl. Phys.* **33**, 118 (1962)
		+ 0.19 ± 0.03	−0.15 ± 0.05	R. M. Singru and R. M. Steffen, *Nucl. Phys.* **43**, 537 (1963)
		+ 0.113 ± 0.008	−0.040 ± 0.011	H. Daniel *et al.*, *Nucl. Phys.* **45**, 529 (1963)
		+ 0.103 ± 0.004	−0.027 ± 0.005	H. Behrens, *KFK* No. 434 (1966)
Mn^{52}	$6 \xrightarrow{\beta^+} 6 \xrightarrow{\gamma} 4 \xrightarrow{\gamma} 2$; $\xrightarrow{\gamma} 0$	−0.16 ± 0.05	+ 0.15 ± 0.06	F. Boehm, *Phys. Rev.* **109**, 1018 (1958)
		−0.089 ± 0.028	+ 0.04 ± 0.04	S. D. Bloom *et al.*, *Phys. Rev.* **125**, 2021 (1962)
		0.00 ± 0.02	−0.08 ± 0.03	H. Daniel *et al.*, *Phys. Rev.* **128**, 261 (1962)
		−0.019 ± 0.010	−0.051 ± 0.014	H. Behrens, *KFK* No. 434 (1966)
Co^{56}	$4 \xrightarrow{\beta^+} 4 \xrightarrow{\gamma} 2 \xrightarrow{\gamma} 0$	−0.02 ± 0.06	−0.08 ± 0.08	H. Daniel *et al.*, *Z. Naturforsch.* **16a**, 1118 (1961)
		0.00 ± 0.03	−0.11 ± 0.04	L. G. Mann *et al.*, *Phys. Rev.* **127**, 2134 (1962)
		+ 0.002 ± 0.010	−0.108 ± 0.013	H. Behrens *KFK* No. 434 (1966)

[a] From Frauenfelder and Steffer [28].

us invaluable information on nuclear structure. In this case, A is simply given by

$$A = \frac{1}{(1 + \mathrm{L})(1 + R^2)} \left[\pm\frac{1}{j} - 2\left(\frac{j+1}{j}\right)^{1/2} R \right], \tag{3.108}$$

with $R = (C_V \int 1)/(C_A \int \sigma)$ and the coupling constants in Eq. (3.98a).

Experimentally, A is obtained by the coincidence technique of detecting beta and gamma rays, combined with the simultaneous measurement of the circular polarization of the gamma ray. This is done, for example, by the transmission method. After passing through the magnetized iron, the circularly polarized gamma radiation differs in intensity according to the orientation of the electron spin and polarization of the gamma ray. (For details, see [28].) The first experiment of this sort was done by Schopper on the beta decay of Co^{60} [29]. Here we give only a few examples from the data on A that were collected by Steffen and Frauenfelder [28] and are presented in Tables 3.4 and 3.5. (For all cases in these tables, the gamma radiation is quadrupole, $L = 2$.)

In Table 3.4, the latest experimental values converge to the theoretically expected values $A = \pm\frac{1}{3}$ for the pure Gamow–Teller transitions. In the case of the mixed transition, the Fermi matrix vanishes if the isospin selection rule is strictly valid. If it is,

$$A = \pm \frac{1}{(1 + L)j}, \tag{3.109}$$

which is

$$A \approx \pm 0.08 \qquad \text{for} \quad j = 4 \text{ and } L = 2. \tag{3.110}$$

The data on Al^{24}, Mn^{52}, and Co^{58} agree with this assumption. The latest data on Na^{24} and Sc^{46} show the hindrance of the Fermi matrix elements. As a whole, data on the asymmetry coefficient A strongly support parity nonconservation and charge conjugation noninvariance in beta decay.

3.11 TIME-REVERSAL INVARIANCE IN BETA DECAY

The discovery of parity nonconservation in beta decay and the related weak processes also stimulated experimental tests of the validity of time-reversal invariance. Such attempts are summarized in this section. As a matter of fact, the possibility of testing time-reversal invariance was studied to some extent even before the discovery of parity nonconservation. In the early 1950s, it was known that the invariance under the pure time reversal (T, the Wigner-type time reversal) results in the real coupling constants, while the invariance under the product of the charge conjugation and time reversal (CT, the Pauli-type time reversal) determines the anticommutability of the fermion fields. In order to test whether coupling constants are real or complex, it was proposed to measure the energy dependence of the electron–neutrino angular correlation [30].

As in the parity experiments, time-reversal invariance is tested by measuring the pseudoscalar quantities under the time reversal

$$\mathbf{r} \to \mathbf{r}, \qquad t \to -t.$$

The momentum and angular momentum transform

$$\mathbf{p} \to -\mathbf{p}, \qquad \mathbf{J} \to -\mathbf{J}.$$

From this, we see that the general form of the pseudoscalar is expressed by

$$(\mathbf{p}_1 \cdot \mathbf{p}_2 \times \mathbf{p}_3)^a (\mathbf{p}_1 \cdot \mathbf{p}_2)^b (\mathbf{p}_2 \cdot \mathbf{p}_3)^c (\mathbf{p}_3 \cdot \mathbf{p}_1)^d \tag{3.111}$$

where the powers b, c, d are zero or integral and a is odd. (A general discussion of these powers a, b, c, d is given in [31].) The $\mathbf{p}_i$'s are either the linear or angular momenta.

The other form of the pseudoscalars is αZ times Eq. (3.111) with zero or even integral for a. The factor αZ, the product of the fine structure constant and the atomic number of the daughter nucleus, which comes from the Coulomb phase shift for the electron wave functions, changes its sign under time reversal, since the incoming and outgoing waves are interchanged. For example, the electron–neutrino angular correlations have ([30]; see also Eq. (2.182))

$$\mathrm{Im}(C_\mathrm{A}^* C_\mathrm{T} + C_\mathrm{A}'^* C_\mathrm{T}') \frac{\mathbf{p} \cdot \mathbf{q}}{E E_\nu} \frac{\alpha Z}{p}. \tag{3.112}$$

This kind of pseudoscalar gives a smaller observable effect, because of the factor αZ. Furthermore, Eq. (3.112) vanishes for the two-component neutrino. Therefore, the T experiments are usually performed by detecting pseudoscalars in Eq. (3.111).

A measurement of pseudoscalars

$$\mathrm{Im}(C_\mathrm{V}'^* C_\mathrm{A} + C_\mathrm{V}^* C_\mathrm{A}')(\textstyle\int 1)(\int \sigma)(\mathbf{j} \cdot \mathbf{p} \times \mathbf{k})(\mathbf{j} \cdot \mathbf{k})^n, \qquad n \text{ odd} \tag{3.113}$$

was proposed by Morita and Morita [31, 32] and Curtis and Lewis [33] in 1957. Here $\mathbf{j}$, $\mathbf{p}$, and $\mathbf{k}$ are the nuclear spin vector, electron momentum, and photon momentum, respectively. This proposal is to measure the beta-gamma directional correlation in the cascade transition of oriented nuclei. The first experiment of this sort was performed by Ambler *et al.* [34] for Co^{58} nuclei, which have the decay scheme

$$2^+ \xrightarrow{\beta^+} 2^+ \xrightarrow{\gamma} 0^+.$$

Co^{58} nuclei were polarized along the z axis by application of an 800-gauss magnetic field. Coincidences were recorded between beta particles emitted along the x axis with $p/E > 0.5$ and each of two gamma-ray channels. Each channel consisted of two detectors connected in parallel and located diametrically opposite one another in the y-z plane. These detector pairs were positioned symmetrically with respect to, and made an angle of 37° on either side of, the polarization axis. The observed correlation was given by

$$\frac{W(\mathbf{j}, \mathbf{p}, \mathbf{k}) - W(\mathbf{j}, -\mathbf{p}, \mathbf{k})}{W(\mathbf{j}, \mathbf{p}, \mathbf{k}) + W(\mathbf{j}, -\mathbf{p}, \mathbf{k})} = -0.014 \pm 0.034. \tag{3.114}$$

This experimental result is consistent with the assumption of time reversal invariance of the beta interaction, Eq. (3.1), while the theoretical formula [32, 33] involves the terms of $n = 1$ and 3 in Eq. (3.113).

The most general form of the beta–gamma angular correlation function $W(\mathbf{j}, \mathbf{p}, \mathbf{k})$ in the decay of oriented nuclei is given by Morita and Morita [32] (see Eqs. (6.106)-(6.109)). Ambler and his collaborators again performed such an experiment on Mn^{52}, which has a decay scheme

$$6^+ \xrightarrow{\beta^+} 6^+ \xrightarrow{\gamma} 4^+ \xrightarrow{\gamma} 2^+ \xrightarrow{\gamma} 0^+ \qquad [35].$$

They measured (a) the beta asymmetry, (b) the beta–gamma directional correlation with the gamma detectors in the plane of **j** and **p**; (c) the beta-gamma directional correlation with the gamma detectors deployed as in the case of Co^{58} (however, the angle is 45° instead of 37°). The relative phase θ was defined by

$$\frac{C_V \int 1}{C_A \int \sigma} = |R| \exp(i\theta). \tag{3.115}$$

Combining the three experimental data, they obtained θ as a function of the square root of the ratio $|R|$. This is shown in Fig. 3.7, where the cross-hatched region encompasses the values lying between the limits of experimental error. This experiment did not allow any conclusions other than that time-reversal invariance was consistent with the results of the experiment, at that time. It is, however, interesting to note that in a recent experiment on beta-circular gamma angular correlation, Behrens [36] obtained a value of $R = -0.051 \pm 0.014$. From Fig. 3.7 we see that time-reversal invariance prefers a smaller value of $|R|$ and that this value coincides with the minimum value of $|R|$ in the new data.

There is a method of determine the phase θ, free from the ratio R. This can be done in the beta decay of the neutron, where $(\int\sigma/\int 1) = \sqrt{3}$. Jackson *et al.* proposed to measure the electron–neutrino angular correlation from the beta decay of polarized neutrons [37]. The correlation function Eq. (3.116) contains the pseudoscalar $\mathbf{j} \cdot \mathbf{p} \times \mathbf{q}$ under time reversal.

$$W(\mathbf{p}, \mathbf{q}, \mathbf{j})\, dE\, d\Omega_e\, d\Omega_\nu = \frac{F(\pm Z, E)}{2(2\pi)^5}\, pE(E_0 - E)^2\, dE\, d\Omega_e\, d\Omega_\nu$$

$$\times\, \xi \left\{ 1 + a\frac{\mathbf{p}\cdot\mathbf{q}}{EE_\nu} + b\frac{1}{E} + \frac{\langle \mathbf{j} \rangle}{j} \cdot \left(A\frac{\mathbf{p}}{E} + B\frac{\mathbf{q}}{E_\nu} + D\frac{\mathbf{p}\times\mathbf{q}}{EE_\nu} \right) \right\}. \qquad (3.116)$$

Here, $\mathbf{p}$ and $\mathbf{q}$ are the linear momenta of the electron and neutrino; j is $\frac{1}{2}$, which is the spin of the neutron, while $\langle \mathbf{j} \rangle / j$ is the unit vector along the spin

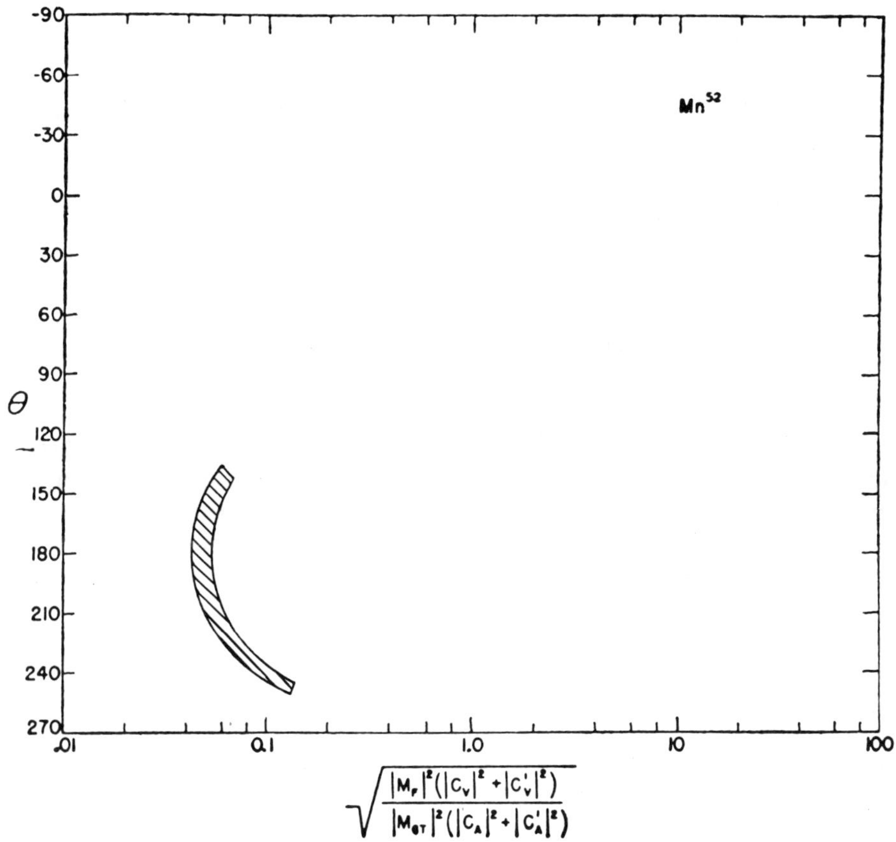

Fig. 3.7 Relative phase θ versus the ratio of the Fermi and Gamow–Teller amplitudes [35].

axis times the degree of polarization. For notatition: ξ, see Eq. (2.181); $a\xi$, see Eq. (2.182); $b\xi$, see Eq. (2.183); $A\xi$, see Eq. (3.78);

$$B\xi = 2\,\mathrm{Re}\left\{\lambda_{jj_1}\left[\frac{\gamma}{E}\,(C_T C_A'^* + C_T' C_A^*) \pm (C_T C_T'^* + C_A C_A'^*)\right](\textstyle\int \sigma)^2 - \delta_{jj_1}\left(\frac{j}{j+1}\right)^{1/2}[(C_S C_T'^* + C_S' C_T^* + C_V C_A'^* + C_A' C_A^*) \pm \frac{\gamma}{E}\,(C_S C_A'^* + C_S' C_A^* + C_V C_T'^* + C_V' C_T^*)](\textstyle\int \sigma)(\textstyle\int 1)\right\} \tag{3.117}$$

with the same λ_{jj_1} as in Eq. (3.78);

$$D\xi = \delta_{jj_1}\left(\frac{j}{j+1}\right)^{1/2}\left[2\,\mathrm{Im}(C_S C_T^* - C_V C_A^* + C_S' C_T'^* - C_V' C_A'^*) \mp \frac{\alpha Z}{p}\,2\,\mathrm{Re}(C_S C_A^* - C_V C_T^* + C_S' C_A'^* - C_V' C_T'^*)\right](\textstyle\int \sigma)(\textstyle\int 1). \tag{3.118}$$

The Chicago group measured the coefficients A, B, and D in the decay of polarized neutrons [38]. Using experimental data on A and B, we can conclude that (1) the beta interaction is a combination of vector and axial vector, and scalar and tensor are excluded; (2) the neutrino is left-handed, i.e., $C_V = C_V'$ and $C_A = C_A'$. The ratio of these two coupling constants is, in general, expressed by

$$C_A/C_V = \alpha \exp{(i\phi)}. \tag{3.119}$$

Using data on D together with those on A and B, we have

$$C_A/C_V = (1.25 \pm 0.07) \exp{[i(175 \pm 10)^\circ]}. \tag{3.120}$$

A precision experiment of the same kind was done by a Berkeley group [39] on the positron decay of $Ne^{19} \to F^{19} + e^+ + \nu$, where Ne^{19} and F^{19} have the same spin and parity, $\frac{1}{2}^+$. From their measured value of D, these authors obtained

$$\phi = (180.2 \pm 1.6)^\circ. \tag{3.121}$$

From Eqs. (3.120) and (3.121), it is most likely that the beta interaction Eq. (3.1) is invariant under time reversal. It is also interesting to note that the ratio α obtained from angular distribution experiments on polarized neutrons agrees with the new value obtained from the ft values of the neutron and Al^{26m}. These are summarized in Table 3.6.

A pseudoscalar

$$\mathrm{Im}(C_V^* C_A + C_V'^* C_A')(\textstyle\int \mathbf{r})(\textstyle\int \boldsymbol{\sigma} \times \mathbf{r})(\boldsymbol{\sigma}\cdot\mathbf{p}\times\mathbf{k})(\mathbf{p}\cdot\mathbf{k}) \tag{3.122}$$

can be measured in the first forbidden transitions. Here σ is the direction of

Table 3.6 Coupling Constants C_V, $C_A = C_V\alpha \exp(i\phi)$ for Beta Decay Based on the Experimental Data a, A, B, D

	n	Ne^{19}	
Decay	Chicago[a]	Berkeley[b]	From ft Values for n and Al^{26m}
a	-0.12 ± 0.04[c]	0.00 ± 0.08[d]	
A	-0.114 ± 0.019	-0.033 ± 0.002	
B	0.88 ± 0.15	-0.90 ± 0.13	
D	0.04 ± 0.05	0.002 ± 0.014	
C_V			(1.4150 ± 0.0011)[e] $\times 10^{-49}$ erg cm^3
α	1.25 ± 0.07		1.239 ± 0.011[f]
$180^\circ - \phi$	$(5 \pm 10)^\circ$	$-(0.2 \pm 1.6)^\circ$	

[a] M. T. Burgy, V. E. Krohn, T. B. Novey, G. R. Ringo, and V. L. Telegdi, *Phys. Rev.* **120**, 1829 (1960).
[b] E. P. Calaprice, E. D. Commins, H. M. Gibbs, G. L. Wick, and D. A. Dobson, *Phys. Rev. Letters* **18**, 918 (1967).
[c] V. V. Vladimirski, V. K. Gregorev, V. A. Ergakov, D. P. Sjarkov, and Y. V. Trebukhovski, *Izv. Akad. Nauk USSR Fys.* **25**, 1121 (1961).
[d] J. S. Allen, R. L. Burman, W. B. Herrmannsfeldt, and P. Stähelin, *Phys. Rev.* **116**, 134 (1959).
[e] R. J. Blin-Stoyle and J. M. Freeman, *Nucl. Phys.* **A150**, 369 (1970).
[f] C. J. Christensen, A. Nielsen, A. Bahnsen, W. K. Brown, and B. M. Rustad, *Phys. Rev.* **D5**, 1628 (1972).

the electron spin. This is a measurement of the transverse polarization of beta particles in coincidence with the following gamma rays in a cascade transition. Simms and Steffen performed the experiment on Au^{198} and the data were again consistent with time-reversal invariance [40].

The beta-ray spectrum of RaE can be used to test time-reversal invariance. If time-reversal invariance does not hold, the $\mathrm{Re}(C_V C_A^*)$ interference term decreases. This is unfavorable for obtaining the strongly energy-dependent shape factor for RaE. Lewis [41] and Yamada *et al.* [42] did a theoretical analysis of the spectrum and obtained

$$\phi = 180^\circ + \Delta\phi \qquad \text{and} \qquad |\Delta\phi| \leqslant 8^\circ.$$

We conclude from these experiments that the time-reversal invariance in beta decay holds. From the *CTP* theorem, it is also concluded that the beta-decay interaction is invariant under the combined operation *CP* or *PC*.

REFERENCES

1 T. D. Lee and C. N. Yang, *Phys. Rev.* **104**, 254 (1956).
2 S. R. de Groot and H. A. Tolhoek, *Physica* **16**, 456 (1950).
3 G. Lüders, *Danske Mat.-fys. Medd.* **28**, No. 5 (1954).
4 W. Pauli, *Niels Bohr and the Development of Physics*, p. 30. Pergamon Press, London, 1955.

5 M. Morita and R. Morita, *Phys. Rev.* **109,** 2048 (1958).
6 C. S. Wu, E. Ambler, R. W. Hayward, D. D. Hoppes, and R. P. Hudson, *Phys. Rev.* **105,** 1413 (1957).
6a S. R. de Groot, H. A. Tolhoek, and W. J. Huiskamp, α, β, γ *Ray Spectroscopy* (K. Siegbahn, ed.), p. 1199. North-Holland, Amsterdam, 1966.
7 E. Ambler, R. W. Hayward, D. D. Hoppes, R. P. Hudson, and C. S. Wu, *Phys. Rev.* **106,** 1361 (1957).
8 M. Morita and R. Morita, *Phys. Rev.* **107,** 1316 (1957).
9 L. Landau, *Nucl. Phys.* **3,** 127 (1957).
10 A. Salam, *Nuovo Cimento* **5,** 299 (1957).
11 T. D. Lee and C. N. Yang, *Phys. Rev.* **105,** 1671 (1957).
12 H. Weyl, *Z. Phys.* **56,** 330 (1929).
13 H. Frauenfelder, R. Bobone, E. von Goeler, N. Levine, H. R. Lewis, R. N. Peacock, A. Rossi, and G. de Pasquali, *Phys. Rev.* **106,** 386 (1957).
14 H. A. Tolhoek, *Rev. Mod. Phys.* **28,** 277 (1956).
15 H. Frauenfelder and R. M. Steffan, α, β, γ *Ray Spectroscopy* (K. Siegbahn, ed.), p. 1431. North-Holland, Amsterdam, 1966.
16 H. Wegener, *Z. Phys.* **154,** 553 (1959).
17 H. Wegener, H. Bienlein, and H. von Issendorff, *Phys. Rev. Letters* **1,** 460 (1958).
18 A. I. Alikhanov, G. P. Eliseyev, and V. A. Liubimov, *Nucl. Phys.* **13,** 541 (1959).
19 V. G. Kulkarni, C. V. Panat, and B. V. Thosar, *Phys. Letters* **7,** 52 (1963).
20 M. Goldhaber, L. Grodzins, and A. W. Sunyar, *Phys. Rev.* **109,** 1015 (1958).
21 S. B. Treiman, *Phys. Rev.* **110,** 448 (1958).
22 M. Morita, R. Morita, and M. Yamada, *Phys. Rev.* **111,** 237 (1958).
23 M. Morita and R. Morita, *Phys. Rev.* **111,** 1130 (1958).
24 R. R. Lewis and R. B. Curtis, *Phys. Rev.* **110,** 910 (1958).
25 A. M. Bincer, *Phys. Rev.* **112,** 244 (1958).
26 N. E. Booth, G. W. Hutchinson, A. M. Seger, G. G. Shute, and D. H. White, *Nucl. Phys.* **11,** 341 (1959).
27 N. A. Burgov and Yu. V. Terekhov, *Nucl. Phys.* **10,** 541 (1959).
28 H. Frauenfelder and R. M. Steffen, α, β, γ *Ray Spectroscopy* (K. Siegbahn, ed.), p. 1453. North-Holland, Amsterdam, 1966.
29 H. Schopper, *Phil. Mag.* **2,** 710 (1957).
30 M. Morita, *Progr. Theoret. Phys.* **10,** 364 (1953).
31 M. Morita and R. Morita, *Phys. Rev.* **110,** 461 (1958).
32 M. Morita and R. Morita, *Phys. Rev.* **107,** 1316 (1957).
33 R. B. Curtis and R. R. Lewis, *Phys. Rev.* **107,** 1381 (1957).
34 E. Ambler, R. W. Hayward, D. D. Hoppes, and R. P. Hudson, *Phys. Rev.* **108,** 503 (1957).
35 E. Ambler, R. W. Hayward, D. D. Hoppes, and R. P. Hudson, *Phys. Rev.* **110,** 787 (1958).
36 H. Behrens; *Kernforschungszentrum Karlsruhe Report KFK* No. 434 (1966).
37 J. D. Jackson, S. B. Treiman, and H. W. Wyld, *Nucl. Phys.* **4,** 206 (1957).
38 M. T. Burgy, V. E. Krohn, T. B. Novey, G. R. Ringo, and V. L. Telegdi, *Phys. Rev. Letters* **1,** 324 (1958); *Phys. Rev.* **120,** 1829 (1960).
39 E. P. Calaprice, E. D. Commins, H. M. Gibbs, G. L. Wick, and D. A. Dobson, *Phys. Rev. Letters* **18,** 918 (1967).
40 P. C. Simms and R. M. Steffen, *Phys. Rev.* **118,** 768 (1960).
41 R. R. Lewis, *Phys. Rev.* **108,** 904 (1957).
42 J. Fujita, M. Yamada, Z. Matumoto, and S. Nakamura, *Progr. Theoret. Phys.* **20,** 287 (1958).

Chapter Four

V—A INTERACTION

4.1 THEORETICAL GROUNDS FOR THE $V-A$ INTERACTION

In section 2.11, we studied experimental grounds supporting the $V-A$ interaction for beta decay. Here we shall introduce several theories for the $V-A$ interaction, all of which were published in 1958. These are the chirality invariance requirement (Sudarshan and Marshak [1]), the two-component formulation of the Dirac spinors (Feynman and Gell-Mann [2]), and mass-reversal invariance (Sakurai [3]).

Let us first consider the chiral transformation

$$\psi(x) \to \psi'(x) = \gamma_5 \psi(x). \tag{4.1}$$

From the Dirac equation (3.7)

$$\gamma_\mu \frac{\partial}{\partial x_\mu} \psi = -m\psi, \qquad \hbar = c = 1, \tag{4.2}$$

we obtain

$$\gamma_\mu \frac{\partial}{\partial x_\mu} (\gamma_5 \psi) = -\gamma_5 \left(\gamma_\mu \frac{\partial}{\partial x_\mu} \psi \right) = m\gamma_5 \psi. \tag{4.3}$$

This means that the Dirac equation is noninvariant under the chiral transformation (unless the sign of the mass term changes simultaneously). Adding and subtracting Eqs. (4.2) and (4.3), we have

$$\gamma_\mu \frac{\partial}{\partial x_\mu} \psi_\pm = -m\psi_\mp \tag{4.4}$$

with the chiral projected wave functions

$$\psi_\pm = \tfrac{1}{2}(1 \pm \gamma_5)\psi \tag{4.5}$$

which are both eigenstates of γ_5,

$$\psi_5 \psi_\pm = \pm \psi_\pm. \tag{4.6}$$

Here ψ_+ and ψ_- are called the positive and negative chirality state, respectively.

The Dirac equation mixes these two states. If we write the four-component wave function ψ with a pair of two-component wave functions u and v, we have

$$\psi = \psi_+ + \psi_- = \tfrac{1}{2}\begin{pmatrix} u+v \\ u+v \end{pmatrix} + \tfrac{1}{2}\begin{pmatrix} -(u-v) \\ u-v \end{pmatrix}, \tag{4.7}$$

or

$$\psi_+ = \begin{pmatrix} \phi_+ \\ \phi_+ \end{pmatrix} \quad \text{and} \quad \psi_- = \begin{pmatrix} -\phi_- \\ \phi_- \end{pmatrix} \tag{4.8}$$

with

$$\phi_\pm = \tfrac{1}{2}(u \pm v), \tag{4.9}$$

since

$$\gamma_5 = \begin{pmatrix} 0 & 1 \\ 1 & 0 \end{pmatrix} \tag{4.10}$$

in our representation of the Dirac matrices (see Eq. (2.56)). If we project with the positive chirality operator on ψ, we can discuss only the two-component wave function $\phi_+ = u + v$, and with the negative chirality operator $\phi_- = u - v$. (In the special case of $m = 0$, the Dirac equation is evidently invariant under the chiral transformation, and Eq. (4.4) reduces to

$$\gamma_\mu \frac{\partial}{\partial x_\mu} \psi_\pm = 0, \tag{4.11}$$

which was studied in section 3.8.)

From foregoing considerations, Sudarshan and Marshak [1] conjectured that the total four-fermion interaction should be invariant under the γ_5 transformation on any of four fermions, $\psi_i \to \gamma_5\psi_i$. Let us write the four-fermion coupling as

$$\sim(\bar{\psi}_1 O \psi_2)(\bar{\psi}_3 O \psi_4), \tag{4.12}$$

which should be invariant under the transformation

$$\psi_1 \to \gamma_5\psi_1 \quad \text{and} \quad \bar{\psi}_1 \to -\bar{\psi}_1\gamma_5, \tag{4.13}$$

since

$$\bar{\psi}_1 = \psi_1^\dagger\gamma_4 \to \psi_1^\dagger\gamma_5\gamma_4 = -\psi_1^\dagger\gamma_4\gamma_5 = -\bar{\psi}_1\gamma_5,$$

or under the transformation

$$\psi_2 \to \gamma_5\psi_2. \tag{4.14}$$

This invariance requirement is given by

$$\bar{\psi}_1 O \psi_2 = \bar{\psi}_1 O \gamma_5 \psi_2 = -\bar{\psi}_1 \gamma_5 O \psi_2. \tag{4.15}$$

From this we have

$$O = O\gamma_5 = -\gamma_5 O. \tag{4.16}$$

This means that the operator O should be odd power terms of the Dirac matrices γ_μ. It has a general form of

$$O = a\gamma_\mu + b\gamma_\mu\gamma_5. \tag{4.17}$$

After substituting Eq. (4.17) into Eq. (4.16), we have

$$a\gamma_\mu + b\gamma_\mu\gamma_5 = (a\gamma_\mu + b\gamma_\mu\gamma_5)\gamma_5 \tag{4.18}$$

and

$$a = b. \tag{4.19}$$

As a consequence, the four-fermion coupling (4.12) is given by

$$\begin{aligned}(\tfrac{1}{2})^{1/2}C_V[\bar{\psi}_1\gamma_\mu(1+\gamma_5)\psi_2][\bar{\psi}_3\gamma_\mu(1+\gamma_5)\psi_4]\\ = (\tfrac{1}{2})^{1/2}C_V[\bar{\psi}_1\gamma_\mu\psi_2][\bar{\psi}_3\gamma_\mu(1+\gamma_5)\psi_4]\\ - (\tfrac{1}{2})^{1/2}C_V[\bar{\psi}_1 i\gamma_\mu\gamma_5\psi_2][\bar{\psi}_3 i\gamma_\mu(1+\gamma_5)\psi_4].\end{aligned} \tag{4.20}$$

This is the parity-nonconserving $V - A$ interaction ($C_V = -C_A$). Here $(\frac{1}{2})^{1/2}$ is required to keep the magnitude of C_V equal to the coupling constant in the old parity-conserving theory.

The Dirac equation (4.2) is evidently invariant under mass reversal

$$\psi_i \to \gamma_5\psi_i \qquad \text{and} \qquad m_i \to -m_i. \tag{4.21}$$

Sakurai [3] required the invariance of four-fermion coupling (4.12) under the mass-reversal transformation on any of the four fermions. The consequence is, therefore, the same as that of the chirality invariance (4.20).

A less restrictive requirement for four-fermion coupling (4.12) is invariance under the mass-reversal transformation on any pair of the four fermions [4]. This brings an approximate form which is nearly equal to $V - A$. It is interesting to note that by this requirement we can prove a similarity of formulas for the nth forbiddenness and the $(n \pm 1)$th. This will be discussed in section 6.2.

Feynman and Gell-Mann introduced the same result, the $V - A$ interaction, on the basis of rather different considerations. Let us consider the Dirac equation (A.62)

$$E\psi = -(\boldsymbol{\alpha}\cdot\mathbf{p} + \beta m)\psi \tag{4.22}$$

for the four-component wave function ψ. This can be expressed by a pair of equations

$$Eu = -\boldsymbol{\sigma}\cdot\mathbf{p}v + mu, \qquad Ev = -\boldsymbol{\sigma}\cdot\mathbf{p}u - mv, \tag{4.23}$$

for two of the two-component wave functions. Adding and subtracting these equations, we have

$$E\phi_+ = -\boldsymbol{\sigma}\cdot\mathbf{p}\phi_+ + m\phi_-, \tag{4.24a}$$

$$E\phi_- = +\boldsymbol{\sigma}\cdot\mathbf{p}\phi_- + m\phi_+. \tag{4.24b}$$

Here ϕ_+ and ϕ_- are as given in Eq. (4.9). For $m = 0$, these two equations are decoupled, and the ϕ_+, ϕ_- are eigenfunctions of the Dirac equation.

For $m \neq 0$, the two equations are coupled. However, the ϕ_+ and ϕ_- satisfy the Klein–Gordon equation. This is seen as follows. From Eq. (4.24a),

$$\phi_- = \frac{1}{m}(E + \boldsymbol{\sigma}\cdot\mathbf{p})\phi_+ \tag{4.25}$$

and we have

$$\begin{aligned} m^2\phi_+ &= m(E - \boldsymbol{\sigma}\cdot\mathbf{p})\phi_- = (E - \boldsymbol{\sigma}\cdot\mathbf{p})(E + \boldsymbol{\sigma}\cdot\mathbf{p})\phi_+ \\ &= [E^2 - (\boldsymbol{\sigma}\cdot\mathbf{p})^2]\phi_+ = (E^2 - p^2)\phi_+. \end{aligned} \tag{4.26}$$

Consequently,

$$\left(\frac{\partial^2}{\partial t^2} - \nabla^2 + m^2\right)\phi_+ = 0. \tag{4.27}$$

This is the Klein–Gordon equation, and it does not include ϕ_-, which is, however, expressed by Eq. (4.25). In this connection, the entire formalism can be expressed by the two-component wave function ϕ_+ which obeys the Klein–Gordon equation.

It is well known that derivative coupling is neglected in the study of the beta-ray spectrum. (Otherwise, the beta-ray spectrum deviates strongly from the statistical shape.) Therefore, an arbitrary interaction that is linear in ψ involves both ϕ_+ and ϕ_-. If it is expressed in terms of ϕ_+ alone, the interaction must contain terms proportional to ϕ_+ and to $\partial\phi_+/\partial x_i$. If we adopt the further restriction that the interaction should include no $\partial\phi_+/\partial x_i$, the interaction formulated with ψ does not contain ψ_- and is expressed by ψ_+ only. This is the same requirement as that of chirality invariance.

Universality of Coupling Constants

Let us again emphasize an analogy of the electromagnetic interaction with the weak interaction. In the former, the electric charge which represents the strength of the interaction is universal for all elementary particles, except for a multiplication factor (of plus or minus integral or zero). The universality holds for the observed charges as well as bare charges. In the weak interaction, the weak charge that is the analog of e is the coupling constant C. Therefore, we may postulate the universality of C for different weak processes. In fact, it has already been pointed out by several authors [5] since 1948. That is, the coupling constants for beta decay, muon capture in complex nuclei, and natural decay of the muon are nearly equal to each other. Since these three processes include four fermions, the couplings are similar to Eq. (4.12) or Eq. (4.20). The latest values are given (after radiative corrections) by

$$C_V = (1.4150 \pm 0.0011) \times 10^{-49} \text{ erg cm}^3 \qquad (4.28)$$

for beta decay and

$$g_\mu = (1.435 \pm 0.001) \times 10^{-49} \text{ erg cm}^3 \qquad (4.29)$$

for muon decay. Muon captures in hydrogen are also consistent with the coupling constant C_V (see section 8.2).

It is, however, reasonable to discuss the universality of the bare coupling constants in different weak interactions because, e.g., C_V for beta decay may receive some renormalization effect due to the meson cloud around the nucleon. g_μ for muon decay is a bare value, since the four fermions involved in this process have no strong interactions. It will be seen in the next section that there is no renormalization effect for the vector coupling in analogy with quantum electrodynamics. And we can postulate the universality of the coupling constants in different weak processes. A slight discrepancy in Eqs. (4.28) and (4.29) will be discussed in section 4.4.

4.2 THEORY OF CONSERVED VECTOR CURRENT

Fermi originally introduced the interaction for beta decay in a close analogy with the electromagnetic interaction. In this section, this analogy is extensively generalized. In the electromagnetic theory, electric charge is conserved. Besides, the positive charges of the proton, positive pion, positron, etc. are exactly the same. This fact is understood as the universality of the bare coupling strength in electromagnetic interactions, since the physical coupling constant (i.e., the observed charge) is equal to the bare charge due to the conservation of the electric current. This is known as no renormalization effect for charge in vector coupling (see, e.g., a textbook on mesons and fields [6]). At the left-hand side of Fig. 4.1a, the physical proton is seen to emit a gamma ray with the coupling constant e. As is shown on the right-hand side, this is the sum of the emission of the gamma ray from the bare proton, from the positive pion which is virtually dissociated from the proton: $p \rightleftarrows n + \pi^+$, etc. Calculation shows that the virtual mesons do not disturb the magnitude of the charge e, and we can use the same strength e on both sides of Fig. 4.1a.

Using this analogy of electrodynamics, Feynman and Gell-Mann [2] and Gerstein and Zeldovich [7] devised a theory of the vector interaction for beta decay which has no renormalization. Now let us define the weak vector and axial vector currents

$$J_\mu^{\mathrm{V}} = \bar{\psi}_{\mathrm{p}}\gamma_\mu\psi_{\mathrm{n}} = \bar{\psi}\gamma_\mu\tau_+\psi, \tag{4.30}$$

$$J_\mu^{\mathrm{A}} = i\bar{\psi}_{\mathrm{p}}\gamma_\mu\gamma_5\psi_{\mathrm{n}} = i\bar{\psi}\gamma_\mu\gamma_5\tau_+\psi. \tag{4.31}$$

These are the counterparts of the electric current

$$J_\mu^{\mathrm{EM}} = \bar{\psi}\gamma_\mu \frac{1+\tau_z}{2}\psi. \tag{4.32}$$

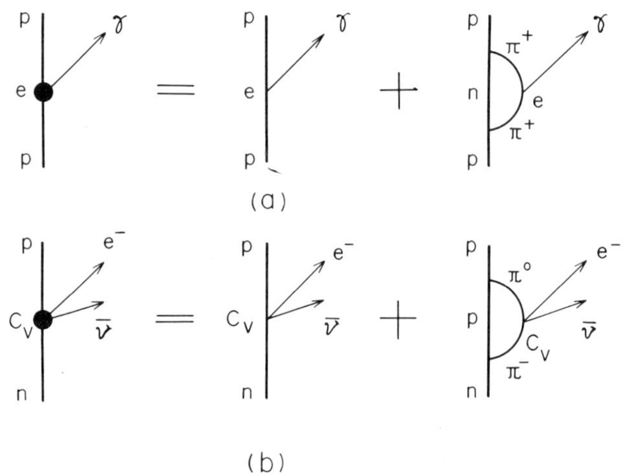

Fig. 4.1 Gamma decay and beta decay in the conserved vector current (CVC) theory. (a) Gamma decay of the physical proton is the sum of contributions from the bare proton and the pion cloud. (b) Beta decay of the physical neutron is the sum of contributions from the bare neutron and the pion cloud.

In the expressions above, ψ is the isospinor

$$\psi = \begin{pmatrix}\psi_{\mathrm{p}}\\ \psi_{\mathrm{n}}\end{pmatrix}.$$

(Readers who are not familiar with the isospin formalism are urged to study Appendix A.8.) J_μ^{EM} consists of the isoscalar current J_μ and isovector current J_μ^z

$$J_\mu^{\mathrm{EM}} = J_\mu + J_\mu^z = \tfrac{1}{2}\bar{\psi}\gamma_\mu\psi + \tfrac{1}{2}\bar{\psi}\gamma_\mu\tau_z\psi. \tag{4.33}$$

The isoscalar part satisfies the continuity condition

$$\frac{\partial J_\mu}{\partial x_\mu} = 0, \tag{4.34}$$

which implies conservation of the nucleon number. The isovector part is not conserved, unless we add the current of pions

$$J_\mu^z = \tfrac{1}{2}\,\bar{\psi}\,\gamma_\mu \tau_z \psi + i[\phi^* \bar{T}_z \nabla_\mu \phi - (\nabla_\mu \phi)^* \bar{T}_z \phi], \tag{4.35}$$

where $\bar{T}_z$ is the 3×3 isospin matrix and ϕ is the pion wave function with three components. J_μ^z is the total isovector current density corresponding to the isospin operator T_z. A generalization of this J_μ^z is the total isovector current $\mathbf{J}_\mu$,

$$\mathbf{J}_\mu = \tfrac{1}{2}\bar{\psi}\gamma_\mu \boldsymbol{\tau}\psi + i[\phi^* \bar{\mathbf{T}} \nabla_\mu \phi - (\nabla_\mu \phi)^* \bar{\mathbf{T}} \phi]. \tag{4.36}$$

Here the boldface characters refer to the vector quantities in the isospace. Apart from small electromagnetic corrections, we have the conservation law

$$\frac{\partial \mathbf{J}_\mu}{\partial x_\mu} = 0.$$

Instead of Eq. (4.30), we adopt the J_μ^+ component of Eq. (4.36) as a new definition of the weak vector current where vector means the Lorentz covariant

$$J_\mu^+ = \bar{\psi}\gamma_\mu \tau_+ \psi + i[\phi^* \bar{T}_+ \nabla_\mu \phi - (\nabla_\mu \phi)^* \bar{T}_+ \phi] \qquad \text{for } \beta^- \tag{4.37}$$

where

$$\tau_+ = \tfrac{1}{2}(\tau_x + i\tau_y) \qquad \text{and} \qquad \bar{T}_+ = \bar{T}_x + i\bar{T}_y.$$

We adopt J_μ^- for positron decay. The assumption that the electromagnetic current and weak vector currents make up the components of the isovector current is called the isotriplet vector current hypothesis.

Since J_μ^+ is conserved, the vector coupling constant is unaffected by the renormalization due to the meson cloud, and the physical coupling constant C_V is identical with its bare value. On the other hand, the coupling constant g_μ for muon decay is the bare value itself, since this process involves the leptons only. As a consequence, C_V should be equal to g_μ, as the experimental data indicate. The theory of beta decay with the conservation of the vector current is referred to hereafter as the theory of vector current or the CVC theory.

The interaction Hamiltonian density for beta decay is obtained from that of the electromagnetic interaction by the following replacement (compare with that described in section 2.5):

(a) electric charge e by the vector coupling constant C_V;
(b) electric current J_μ^z by weak current J_μ^+;
(c) electromagnetic potential A_μ by a bilinear form of leptons $j_\mu = (\tfrac{1}{2})^{1/2}(\bar{\psi}_e \gamma_\mu (1 + \gamma_5)\psi_\nu)$.

As a result, we have

$$C_V J_\mu^+ j_\mu. \tag{4.38}$$

As we see in Eq. (4.35) and Fig. 4.1b, an additional beta decay can take place due to the meson cloud around the nucleon. The effect of this term is most easily understood by considering the electromagnetic form factors. The matrix element of the isovector part of the electromagnetic interaction between two nucleons can be written as

$$M^{\mathrm{EM}} = e\bar{\psi}_1[f^{\mathrm{EM}}\gamma_\mu - ig^{\mathrm{EM}}\sigma_{\mu\nu}p_\nu]\psi_2 \tag{4.39}$$

with the nucleon momentum transfer

$$\mathbf{p} = \mathbf{p}_2 - \mathbf{p}_1 \qquad \text{and} \qquad \sigma_{\mu\nu} = \tfrac{1}{2}(\gamma_\mu\gamma_\nu - \gamma_\nu\gamma_\mu).$$

Here f^{EM} and g^{EM} are the isovector parts of the charge and magnetic form factors for the nucleon, which are the Fourier transforms of the spatial distributions of these quantities. They are, in general, decreasing functions with p^2. In the limit as $p^2 \to 0$, they are normalized as

$$f^{\mathrm{EM}}(0) = 1 \qquad \text{and} \qquad g^{\mathrm{EM}}(0) = (\mu_{\mathrm{p}} - \mu_{\mathrm{n}})/2M. \tag{4.40}$$

The second term comes from the anomalous magnetic moments of the proton and neutron, where

$$\mu_{\mathrm{p}} - \mu_{\mathrm{n}} = 3.7 \tag{4.41}$$

due to the meson cloud. It represents the contribution of the meson current in Eq. (4.35). In the CVC theory, we have the vector matrix element for beta decay

$$M_V = \bar{\psi}_1[f_V\gamma_\mu - ig_V\sigma_{\mu\nu}p_\nu]\psi_2. \tag{4.42}$$

Using the isotriplet vector current hypothesis, we have the following relations in the low energy limit $p^2 \to 0$.

$$g_V/f_V = g^{\mathrm{EM}}/f^{\mathrm{EM}} \tag{4.43}$$

and

$$f_V = C_V \qquad \text{and} \qquad g_V = \frac{\mu_{\mathrm{p}} - \mu_{\mathrm{n}}}{2M} f_V. \tag{4.44}$$

The beta-decay formulas in the CVC theory are, therefore, different from the old Fermi theory, which does not include the contribution of the meson cloud. The correction is straightforward, by studying the interaction which is expressed in the nonrelativistic form for nucleon wave functions. If we perform it, we see that Eq. (4.42) includes the usual Fermi matrix element and the matrix element of the following expression.

$$\frac{C_V}{2M}(1 + \mu_p - \mu_n) i\boldsymbol{\sigma} \cdot \mathbf{p} \times \left(\bar{\psi}_e \boldsymbol{\alpha} \frac{1 + \gamma_5}{2^{1/2}} \psi_\nu \right). \tag{4.45}$$

In the first brackets of expression (4.45), the term 1 represents the contribution of the Dirac moments of nucleons which comes from the retardation expansion of the main vector interaction. On the other hand, the $(\mu_p - \mu_n)$ term represents the effects of the anomalous magnetic moments induced by the meson cloud. The relativistic form of Eq. (4.45) is the second forbidden nuclear matrix element for beta decay $\int \boldsymbol{\alpha} \times \mathbf{r}$, which will be seen in section 6.1. Since all formulas are already given there in terms of $\int \boldsymbol{\alpha} \times \mathbf{r}$ and the other matrix elements, we need only perform the following corrections for obtaining formulas for the CVC theory.

All $\int \boldsymbol{\alpha} \times \mathbf{r}$ terms should be multiplied with a factor $(1 + \mu_p - \mu_n) = 4.7$. (4.46)

The validity of the CVC theory is, therefore, examined by measuring the effects due to the $\int \boldsymbol{\alpha} \times \mathbf{r}$ term experimentally. As an example, Gell-Mann proposed to measure the beta-ray spectrum [8]. This will be discussed in detail in the next section.

4.3 EXPERIMENTAL EVIDENCE FOR CVC THEORY SEEN IN BETA DECAY

The best-known examples of high energy nuclear beta decays are those of B^{12} and N^{12} shown in Fig. 4.2. The ground state of B^{12}, the 15-MeV state of C^{12}, and the ground state of N^{12} all have spin 1 and positive parity, and they make up the members of the isotriplet ($T = 1$). They emit β^-, γ, β^+, respectively, to reach the ground state of C^{12}, which has isospin 0, spin 0 and positive parity. Since the beta decays are pure Gamow–Teller transitions, $1^+ \to 0^+$, these transitions are induced, of course, by the term proportional to the Gamow–Teller matrix

$$C_A \int \sigma$$

together with the relativistic term

$$C_V \int \alpha \times \mathbf{r},$$

which comes from the retardation expansion of $\mu = 1, 2, 3$ terms in the vector interaction. There are three more matrices contributing to these high energy beta decays; they will be mentioned later. The beta-ray spectrum can be derived from Eqs. (6.47), (6.48), (6.62), and (4.46), and the result is

$$P(E) = \text{const} \cdot pE(E_o - E)^2 (1 \pm \tfrac{8}{3} aE) \qquad \text{for } \beta^{\mp} \tag{4.47}$$

with

$$a = \frac{\mu}{2M}\left|\frac{C_V}{C_A}\right| \qquad \text{and} \qquad \mu = 1 + \mu_p - \mu_n = 4.706. \tag{4.48}$$

Here M is the nucleon mass. In the old Fermi theory, where we have no meson correction in Fig. 4.1b, we have simply

$$\mu = 1. \tag{4.49}$$

Therefore, the spectral shape factor

$$C(E) = 1 \pm \tfrac{8}{3}aE \qquad \text{for } \beta^{\mp} \tag{4.50}$$

has a significant deviation from unity in the high energy beta decays of B^{12} and N^{12}. Furthermore, the linear term in the beta energy changes its sign, so

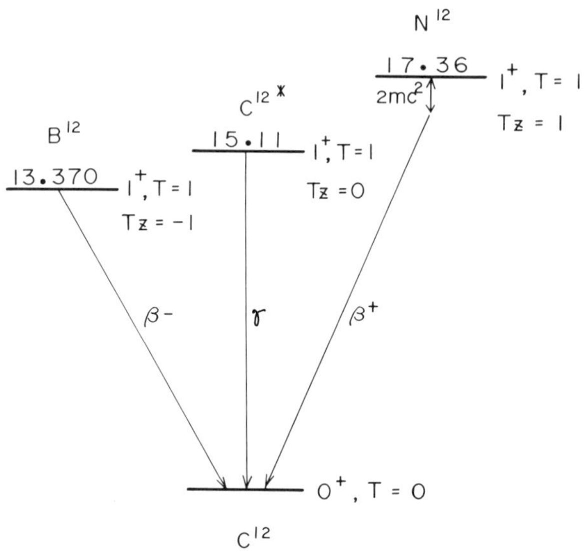

Fig. 4.2 Decay scheme for $A = 12$. Energy is in units of million electron volts (MeV).

that we can compare the beta-ray spectra of B^{12} and N^{12} experimentally. (The C_A term and the C_V term have different charge symmetries so that the cross term changes its sign.) This is the idea of Gell-Mann [8] to test the CVC theory by measuring the beta-ray spectra of B^{12} and N^{12}.

Strictly speaking, μ in Eq. (4.48) is the transition moment, and we can find its magnitude from the corresponding gamma decay $C^{12*} \to C^{12}$. Since the 15-MeV state of C^{12} is the isobaric analog of the B^{12} ground state, the transition moment for gamma decay is $2^{1/2}$ times smaller than that for beta decay.

The gamma width of C^{12*} is the gamma decay rate, which is theoretically given by

$$\Gamma_\gamma = \frac{\mu^2 |\int \sigma|^2}{2 \times 3 \times 137} \frac{E_\gamma^3}{M^2}. \tag{4.51}$$

Here E_γ is the gamma-ray energy. On the other hand, the $ft_{1/2}$ value of the beta decay of B^{12} (or N^{12}) is given, in comparison with that of O^{14}, by

$$\frac{ft_{1/2}(O^{14})}{ft_{1/2}} = \frac{|\int \sigma|^2 C_A^2}{2C_V^2}. \tag{4.52}$$

Strictly speaking, there are several second forbidden matrix elements contributing to the $ft_{1/2}$ value (see Eqs. (6.51) and (6.62)). We omit them. The interference term $(\int \sigma)^*(\int \boldsymbol{\alpha} \times \mathbf{r})$ disappears after integrating over the energy, because it has an energy dependence $[E - \frac{1}{2}E_o - (1/2E)]$ in the limit as $Z \to 0$. Substituting Eqs. (4.51) and (4.52) into Eq. (4.48), we have

$$a = \left[\frac{3 \times 137}{4} \frac{\Gamma_\gamma}{E_\gamma^3} \frac{ft}{ft(O^{14})}\right]^{1/2}. \tag{4.53}$$

Gell-Mann estimated it with the measured value $\Gamma_\gamma = (53 \pm 11)$ eV, and obtained [8]

$$a = \frac{2.34 \pm 0.25}{M}, \tag{4.54}$$

while Eq. (4.48) gives

$$a = \frac{4.70}{2 \times 1.2M} \approx \frac{2}{M}. \tag{4.55}$$

Experiments on high energy beta decays of B^{12} and N^{12} were performed by Mayer-Kuckuk and Michel [9]; Lee, Mo, and Wu [10]; and Glass and Peterson [11]. The experimental data measured by Wu *et al.* [10] are given in Fig. 4.3. Since the beta energy is higher, the correction of the inner bremsstrahlung becomes important. The data are given by the spectral shape factor

$$C(E) = I(E)(1 + A^{\mp} E) \qquad \text{for } \beta^{\mp} \tag{4.56}$$

with $I(E) = 1 + \Delta(E)$. Here $\Delta(E)$ is a correction from the inner bremsstrahlung which was studied by several authors. (For example, it is given by Kinoshita and Sirlin [12].) The ratio of the measured beta intensities for electrons and positrons is proportional to

$$\left[\frac{C}{I(E)}\right]_{B^{12}} \left[\frac{C}{I(E)}\right]_{N^{12}}^{-1} = 1 + (A^- - A^+)E. \tag{4.57}$$

The experimental and calculated values of A are summarized in Table 4.1.

In order to evaluate the coefficient A theoretically, we should take into account the contribution from all the second forbidden matrix elements and the Coulomb corrections. This can be done by using the formulas (6.47), (6.48), and (6.62) [13]. As is shown in Table 4.1, the coefficient A in the shape factor (4.56) or (4.57) is appreciable for the CVC theory, whereas it is

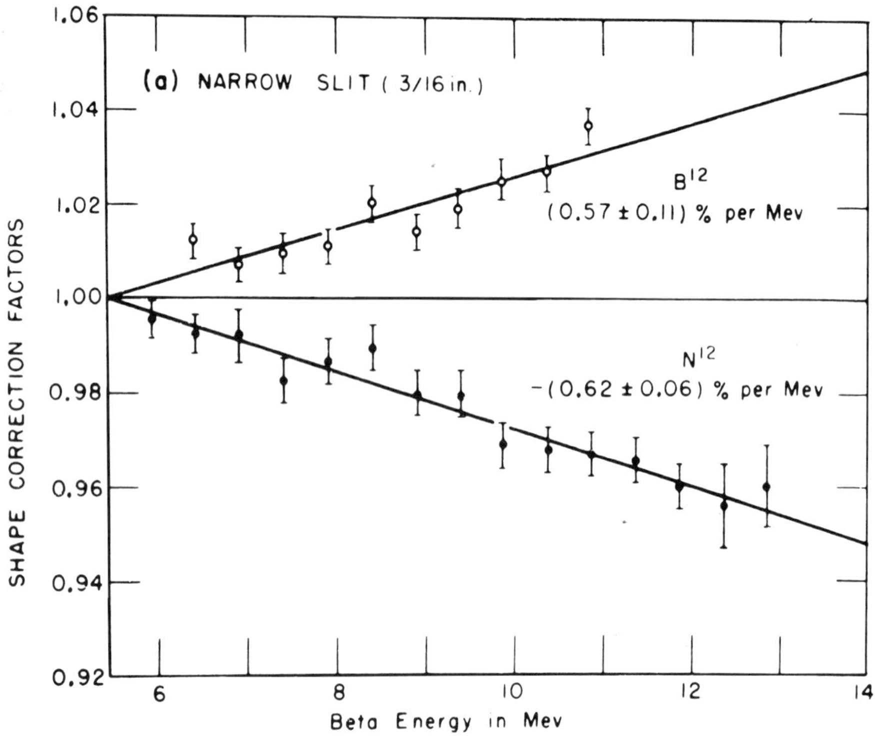

Fig. 4.3 Spectral shape factors $C/I(E)$ for B^{12} and N^{12}. Here $I(E)$ is the correction due to inner bremsstrahlung. Data are given by Wu *et al.* [10].

nearly equal to zero for the Fermi theory. A similar calculation was done by Gell-Mann and Berman [14]. Their result agrees well with that of Morita [13]. All experimental data given by the three groups [9–11], especially on the ratio of beta spectra for B^{12} and N^{12}, are in excellent agreement with the theoretical estimates in favor of the CVC theory.

The validity of the CVC theory can be tested in various beta-decay phenomena. These always involve detection of the cross term $C_A^* C_V \int \sigma \int \alpha \times \mathbf{r}$. If a geometrical coefficient multiplied by this term becomes larger, the test is sensitive. An exhaustive investigation was performed in [4]. Besides beta-ray

Table 4.1 Experimental and Theoretical Values of A in the Spectral Shape Factor $1 + AE$ in the Beta Decays of B^{12} and N^{12}[a]

	A^- for B^{12}	A^+ for N^{12}	$A^- - A^+$	Reference
Experiments				
Mayer-Kuckuk and Michel	1.82 ± 0.09	0.60 ± 0.08	1.30 ± 0.31	*b*
Lee, Mo, and Wu	0.57 ± 0.11	-0.62 ± 0.06	1.19 ± 0.24	*c*
Glass and Peterson			1.62 ± 0.28	*d*
Theories				
Gell-Mann and Berman			1.10 ± 0.17 (0.10*)	*e*
Morita[f]	0.46 ± 0.07	-0.54 ± 0.07	1.00 ± 0.14 (0*)	*g*

[a] The units are percentage per million electron volts.
[b] T. Mayer-Kuckuk and F. C. Michel, *Phys. Rev.* 127, 545 (1962).
[c] Y. K. Lee, L. W. Mo, and C. S. Wu, *Phys. Rev. Letters* **10**, 253 (1963).
[d] N. W. Glass and R. W. Peterson, *Phys. Rev.* 130, 299 (1963).
[e] M. Gell-Mann and S. M. Berman, *Phys. Rev. Letters* 3, 99 (1959).
[f] The E^2 term is included. The numerical values are averaged in an energy region of 6–12 MeV.
[g] M. Morita, *Phys. Rev.* 113, 1584 (1959).
* For the Fermi theory.

spectra, a feasible experiment is the measurement of the beta–alpha angular correlations of

$$Li^8 \xrightarrow{\beta^-} Be^{8*} \xrightarrow{\alpha} He^4 \qquad \text{and} \qquad B^8 \xrightarrow{\beta^+} Be^{8*} \xrightarrow{\alpha} He^4$$

proposed by Bernstein and Lewis [15] and Morita [4, 13]. The anisotropy coefficients in these angular correlations are extremely small. Careful measurements were, however, done by Nordberg, Morinigo, and Barnes [16] and by Gruhle, Lauterjung, and Schimmer [17]. The results fall within the range of the theoretical values of the CVC theory and disagree with the predictions of the old Fermi theory.

For the first forbidden transition, Fujita [18] was able to deduce an important relationship between two matrix elements

$$\int \alpha = [E_o \mp 2.5 \pm 1.2(\alpha Z/R)] \int ir \qquad \text{for } \beta^{\mp}. \tag{4.58}$$

This is a generalization of the Siegert theorem of the electromagnetic theory which is derived from the equation of continuity. It was shown that Eq. (4.58) fits well with a number of the first forbidden beta decays. In connection with this, a discussion of the beta decay of RaE will be given in section 7.9.

Beta Decay of the Positive Pion

According to the CVC theory, the positive pion has a decay process

$$\pi^+ \to \pi^0 + e^+ + \nu_e. \tag{4.59}$$

This is analogous to the $0^+ \to 0^+$ nuclear beta transition discussed in section 2.10. The pions have isospin 1, spin 0, and negative parity. If we consider a pion to be a bound state of a nucleon and an antinucleon, the decay (4.59) looks like a $0^- \to 0^-$ nuclear beta decay, and the probability of the beta-decay rate of the process (4.59) is given by Eq. (2.159)

$$W = \frac{mc^2}{\hbar} \frac{1}{2\pi^3} C_V^2 \left| \int 1 \right|^2 f. \tag{4.60}$$

The magnitude of the Fermi matrix element is given by Eq. (7.9), since Eq. (4.59) is the transition between the members of the isotriplet. We have

$$\left| \int 1 \right|^2 = 2. \tag{4.61}$$

The integrated Fermi function is approximately given by Eq. (2.169),

$$f \approx \tfrac{1}{30} E_o^5, \tag{4.62}$$

since $E_o = 8.8(mc^2) \gg 1$ and there is no Coulomb correction. Finally,

$$W = \frac{mc^2}{\hbar} \frac{E_o^5}{30\pi^3} C_V^2 = 0.43 \text{ sec}^{-1} \tag{4.63}$$

where E_o and C_V are in units of $\hbar = m = c = 1$. The branching ratio of beta decay to natural decay is

$$R = \frac{\pi^+ \to \pi^0 + e^+ + \nu}{\pi^+ \to \mu^+ + \nu} \approx 1.07 \times 10^{-8} \tag{4.64}$$

in comparison with the experimental value [19]

$$R_{\text{exp}} = (1.03 \pm 0.07) \times 10^{-8}. \tag{4.65}$$

This is in favor of the CVC theory.

As a matter of fact, the beta decay of the pion is not forbidden in the old theory. The pion dissociates into $p + \bar{n}$, and after emitting positrons, a pair of nucleons unites again into a neutral pion. The calculation for this process includes, of course, a divergent integration, so that the final result is not conclusive, but it is not impossible to obtain $R \approx 10^{-8}$.

4.4 BETA DECAYS OF HYPERONS AND THE UNIVERSAL FERMI INTERACTION; CABIBBO'S THEORY

Hyperons are heavy fermions with masses that are about equal to nucleon masses. They are listed in Table 5.2. They have different charge Q, isospin T, and strangeness S (or hypercharge Y). There are lambda (Λ^0), sigma (Σ^+, Σ^0, Σ^-), and xi (Ξ^0, Ξ^-) particles. Nucleons (p, n) together with these six hyperons comprise the baryon octet. They can transform with each other by emitting a beta particle and neutrino (or muon and neutrino), if these processes are energetically possible. Therefore, in beta decays of hyperons, there are strangeness changing and nonchanging processes. (There are, however, no processes with $|\Delta S| \geqslant 2$ or with $\Delta S/\Delta Q = -1$, experimentally.) Typical examples are

$$n \to p + e^- + \bar{\nu}_e, \qquad \Sigma^- \to \Lambda^0 + e^- + \bar{\nu}_e \qquad \text{for} \quad \Delta S = 0, \tag{4.66}$$

and

$$\Lambda^0 \to p + e^- + \bar{\nu}_e \qquad \text{for} \quad \Delta S = 1. \tag{4.67}$$

In the process (4.67), the coupling constant C_Λ of the four-fermion interaction is about five times smaller than that for the neutron beta decay [19]. This is also common for all strangeness changing hyperon beta decays. Furthermore, the following relation is nearly satisfied by the three coupling constants g_μ, C_V, and C_Λ.

$$g_\mu^2 = C_V^2 + C_\Lambda^2. \tag{4.68}$$

In this connection, Cabibbo proposed to introduce the following angle θ in the weak current [20].

$$\mathcal{J}_\alpha = j_\alpha^e + j_\alpha^\mu + (\cos\theta) J_\alpha + (\sin\theta) S_\alpha \tag{4.69}$$

with

$$j_\alpha^e = \bar{\psi}_e \gamma_\alpha (1 + \gamma_5) \psi_{\nu_e}, \tag{4.70}$$

$$j_\alpha^\mu = \bar{\psi}_\mu \gamma_\alpha (1 + \gamma_5) \psi_{\nu_\mu}, \tag{4.71}$$

$$J_\alpha = \bar{\psi}_n \gamma_\alpha (1 + \gamma_5) \psi_p, \tag{4.72}$$

$$S_\alpha = \bar{\psi}_\Lambda \gamma_\alpha (1 + \gamma_5) \psi_p. \tag{4.73}$$

The interaction Hamiltonian density is assumed to be the current–current type

$$H = (\tfrac{1}{2})^{1/2} G \mathcal{J}_\alpha^\dagger \mathcal{J}_\alpha. \tag{4.74}$$

(In this section we adopt a convention that the fourth component of the current with a dagger is $-\mathcal{J}_4^*$, instead of $\mathcal{J}_4^*$.) The coupling constants are therefore given by

$$g_\mu = G, \qquad C_V = G\cos\theta, \qquad C_\Lambda = G\sin\theta. \tag{4.75}$$

The reason for the appearance of the angle θ in this theory is not known, and

$$\cos\theta = 0.976 \tag{4.76}$$

from Eqs. (4.28) and (4.29). (The angle θ was also introduced by the Kyoto–Nagoya group based on the particle mixture models for two neutrinos [21].)

Cabibbo [20] also found a systematic way of understanding various processes of beta decays of hyperons by using su_3 symmetry. As we have emphasized, the electromagnetic current and weak currents for nuclear beta decay make up three components of the isotriplet current in su_2. Cabibbo generalized this consideration to the su_3 group. In this theory, J_α and S_α should be members of octet currents. That is, the bilinear quantity of the wave functions of the eight baryons is the direct product of the two octets, so that it is decomposed into irreducible parts. Among those, the currents J_α, S_α, etc. are members of the octets. Since we do not have space to introduce the su_3 computation, we give only the final results in Table 4.2 [20, 22]. The matrix

Table 4.2 Matrix Elements of $\mathscr{J}_\alpha$ in the Limit of su_3 Symmetry (l = e or μ)

Decay	Matrix element
$n \to p + e^- + \bar{\nu}_e$	$G \cos\theta[\gamma_\alpha + (D+F)\gamma_\alpha\gamma_5]$
$\Sigma^- \to \Lambda^0 + e^- + \bar{\nu}_e$	$G \cos\theta[-(\frac{2}{3})^{1/2} D\gamma_\alpha\gamma_5]$
$\Sigma^+ \to \Lambda^0 + e^+ + \nu_e$	$G \cos\theta[-(\frac{2}{3})^{1/2} D\gamma_\alpha\gamma_5]$
$\Sigma^- \to \Sigma^0 + e^- + \bar{\nu}_e$	$G \cos\theta[2^{1/2}\gamma_\alpha + 2^{1/2} F\gamma_\alpha\gamma_5]$
$\Xi^- \to \Xi^0 + e^- + \bar{\nu}_e$	$G \cos\theta[-\gamma_\alpha + (D-F)\gamma_\alpha\gamma_5]$
$\Lambda^0 \to p + l^- + \bar{\nu}_l$	$G \sin\theta[(\frac{3}{2})^{1/2}\gamma_\alpha + (\frac{1}{6})^{1/2}(D+3F)\gamma_\alpha\gamma_5]$
$\Sigma^- \to n + l^- + \bar{\nu}_l$	$G \sin\theta[-\gamma_\alpha + (D-F)\gamma_\alpha\gamma_5]$
$\Xi^- \to \Lambda^0 + l^- + \bar{\nu}_l$	$G \sin\theta[-(\frac{3}{2})^{1/2}\gamma_\alpha + (\frac{1}{6})^{1/2}(D-3F)\gamma_\alpha\gamma_5]$
$\Xi^- \to \Sigma^0 + l^- + \bar{\nu}_l$	$G \sin\theta[(\frac{1}{2})^{1/2}\gamma_\alpha + (\frac{1}{2})^{1/2}(D+F)\gamma_\alpha\gamma_5]$
$\Xi^0 \to \Sigma^+ + l^- + \bar{\nu}_l$	$G \sin\theta[\gamma_\alpha + (D+F)\gamma_\alpha\gamma_5]$

elements refer to the operators which should be folded by the initial wave function ψ_i and the adjoint of the final wave function $\bar{\psi}_f$. For all strangeness changing processes, a factor $\sin\theta$ is multiplied, and for strangeness nonchanging, $\cos\theta$ is used. A parameter F/D appears since there are two octets in the

Table 4.3 Leptonic Decays of Baryons [23]

Decay	Branching Ratio	
	Predicted[a]	Experimental[b]
$n \to p + e^- + \bar{\nu}_e$	1	1
$\Lambda^0 \to p + e^- + \bar{\nu}_e$	$(0.87 \pm 0.03) \times 10^{-3}$	$(0.80 \pm 0.06) \times 10^{-3}$
$\Lambda^0 \to p + \mu^- + \bar{\nu}_\mu$	$(1.48 \pm 0.05) \times 10^{-4}$	$(1.35 \pm 0.6) \times 10^{-4}$
$\Sigma^- \to n + e^- + \bar{\nu}_e$	$(1.29 \pm 0.13) \times 10^{-3}$	$(1.06 \pm 0.05) \times 10^{-3}$
$\Sigma^- \to n + \mu^- + \bar{\nu}_\mu$	$(0.62 \pm 0.06) \times 10^{-3}$	$(0.45 \pm 0.04) \times 10^{-3}$
$\Sigma^- \to \Lambda^0 + e^- + \bar{\nu}_e$	$(0.70 \pm 0.04) \times 10^{-4}$	$(0.60 \pm 0.06) \times 10^{-4}$
$\Sigma^+ \to \Lambda^0 + e^+ + \nu_e$	$(0.21 \pm 0.01) \times 10^{-4}$	$(2.02 \pm 0.47) \times 10^{-5}$
$\Xi^- \to \Lambda^0 + e^- + \bar{\nu}_e$	$(0.43 \pm 0.03) \times 10^{-3}$	$(0.67 \pm 0.23) \times 10^{-3}$
$\Xi^- \to \Lambda^0 + \mu^- + \bar{\nu}_\mu$	$(0.12 \pm 0.01) \times 10^{-3}$	
$\Xi^0 \to \Sigma^+ + e^- + \bar{\nu}_e$	$(0.31 \pm 0.04) \times 10^{-3}$	
$\Xi^0 \to \Sigma^+ + \mu^- + \bar{\nu}_\mu$	$(0.24 \pm 0.04) \times 10^{-5}$	

[a] Sin $\theta = 0.21$, $\sin\theta_A = 0.28$ (Cabibbo angle for axial vector), and $D/F = 2$ are adopted.
[b] A compilation of the new experimental data was kindly supplied by Dr. M. Roos.

direct product of the two octets. Predictions and experimental values for the hyperon beta-decay rates are given in Table 4.3 [23].

Each term of the current–current coupling in Eq. (4.74) can be identified with each process. For example, $J_\alpha^\dagger j_\alpha^e$ is responsible for the beta decay of the neutron, $j_\alpha^{\mu\dagger} J_\alpha$ for the muon capture in hydrogen, $S_\alpha^\dagger j_\alpha^e$ for the beta decay of the lambda particle, $j_\alpha^{\mu\dagger} j_\alpha^e$ for the muon decay, etc. These are the cross terms. For the square terms, we know only the following two processes.

Electron–Positron Annihilation in Stars

The square term $j_\alpha^{e\dagger} j_\alpha^e$ represents a reaction

$$e^- + e^+ \to \nu_e + \bar{\nu}_e. \tag{4.77}$$

If this process takes place, the evolution of stars should be modified. The evolution of stars was studied extensively by Hayashi's group [24], which assumed that the star loses its energy by emitting photons and neutrino pairs (in the process Eq. (4.77)). They obtained the result that observations concerning the population of stars with respect to their ages favor the existence of the process (4.77) with the universal coupling constant, while Chiu [25] obtained a controversial result. According to a recent opinion of Hayashi, these calculations are dependent on the model of the internal structure of stars.

On the other hand, two definite evidences of the existence of the $J_\alpha^\dagger J_\alpha$ term were reported.

Parity-Nonconserving Nuclear Forces

The square term $J_\alpha^\dagger J_\alpha$ gives the four-fermion coupling

$$(\tfrac{1}{2})^{1/2} G(\bar{\psi}_p \gamma_\alpha (1 + \gamma_5) \psi_n)(\bar{\psi}_n \gamma_\alpha (1 + \gamma_5) \psi_p). \tag{4.78}$$

This represents a weak nuclear force which violates parity conservation, while conventional forces due to the meson exchange maintain parity conservation. If the weak nuclear force exists, the nuclear state ψ would no longer be of well-defined parity, but would consist of a mixture

$$\psi = \psi_{\text{normal}} + F\psi_{\text{abnormal}}. \tag{4.79}$$

Here ψ_{normal} has regular parity and ψ_{abnormal} has the opposite parity with a fractional amplitude F. (The magnitude of F is, roughly speaking, the ratio of the universal weak coupling constant to the square of the strong coupling constant. By adopting the pion mass as the standard for the dimension of the mass, we have $F \sim 10^{-7}$.) In the case of the gamma transition with dominant $M1$ (magnetic dipole) and $E2$ (electric quadrupole) radiations, the wrong-parity state induces an $E1$ (electric dipole) transition. As a consequence, the gamma ray has a circular polarization P_γ

$$P_\gamma = -\frac{2RF}{1 + |E2/M1|^2} \tag{4.80}$$

with

$$R = E1/M1. \tag{4.81}$$

Here the transition amplitudes are abbreviated by $E1$, etc. Lobashov and collaborators [26] measured the circular polarization of the gamma rays in Ta^{181} (482 keV $\frac{5}{2}^+ \rightarrow$ ground $\frac{7}{2}^+$) and in Lu^{175} (395 keV $\frac{7}{2}^- \rightarrow$ ground $\frac{7}{2}^+$). The results are summarized in Table 4.4 together with previous data on Ta^{181}. Experimental data are in favor of weak nuclear forces.

Table 4.4 Circular Polarization of Gamma Rays as a Consequence of Parity Nonconservation in Nuclear Forces†

Element	P_γ	R	F	Reference
Ta^{181}				
Exp.	$-(2.0 \pm 0.4) \times 10^{-4}$		$(0.5\text{–}3) \times 10^{-6}$	*a*
Exp.	$+(0.3 \pm 2.1) \times 10^{-4}$		~ 0	*b*
Exp.	$-(6 \pm 1) \times 10^{-6}$		$(0.2\text{–}1) \times 10^{-7}$	*c*
Exp.	$-(1 \pm ?) \times 10^{-5}$			*d*
Theory	$-\left(1.6 \begin{matrix} +1.8 \\ -0.9 \end{matrix}\right) \times 10^2 F$			*e*
Theory	$\|P\| \sim 1 \times 10^{-4}$	$\|R\| \sim 2 \times 10^3$	$\sim 8 \times 10^{-7}$	*f*
Lu^{175}				
Exp.[g]	$+(4 \pm 1) \times 10^{-5}$		$\|F\| \sim (2\text{–}3) \times 10^{-7}$	*c*
Theory	$\|P\| \sim 1 \times 10^{-4}$	$\|R\| \sim 70$	$F \sim 8 \times 10^{-7}$	*f*

† $P_\gamma = -2RF/(1 + |E2/M1|^2)$.
[a] F. Boehm and E. Kankeleit, *Phys. Rev. Letters* **14**, 312 (1965).
[b] P. Bock and H. Schopper, *Phys. Letters* **16**, 284 (1965).
[c] V. M. Lobashov *et al., J. Phys. Soc. Japan Suppl.* **24**, 728 (1968).
[d] F. Boehm, a comment at the International Conference on Nuclear Structure, Tokyo, 1966.
[e] S. Wahlborn, *Phys. Rev.* **138**, B533 (1965).
[f] F. C. Michel, *Phys. Rev.* **133**, B329 (1964).
[g] For Lu^{175}, E and M should be interchanged.

If the gamma transitions following neutron capture are parity violating, then the gamma rays have an asymmetric angular distribution with respect to the polarization axis of the neutron. Abov's group measured this in $Cd^{113}(n,\gamma)Cd^{114}$. Their result is in favor of the existence of the weak nuclear forces [27]. (For a summary, see [28].)

Parity nonconservation in nuclear forces produces an asymmetry of the absorption lines of the Mössbauer-type experiment [29]. This was found in Fe^{57} by Grodzins and others [30].

Universality of the Weak Interactions

Equation (4.74) is now accepted as a universal Fermi interaction of $V-A$ type. The coupling constant G is the universal coupling constant which is given by Eq. (4.29). Since this is the bare interaction, certain renormalizations will be introduced due to the effect of the strong interactions. For example, the axial vector coupling constant is about -1.2 times larger than the vector coupling constant. (The reason for this will be studied in the next section.) Due to this modification, the effective current for nucleons will be

$$J_\alpha = \bar{\psi}_n \gamma_\alpha (1 + 1.2\gamma_5)\psi_p, \tag{4.82}$$

in place of Eq. (4.72). Furthermore, this current will be generalized as a sum of the vector and axial vector currents in Eq. (8.26). Equation (4.74) is also generalized to include the processes in Table 4.2. The universality of the coupling constant and the $V-A$ type hold in processes like $\pi \to \mu + \nu$ (see section 5.4).

4.5 PARTIALLY CONSERVED AXIAL VECTOR CURRENT

In this section, we shall present a theory given by Adler [31] and Weisberger [32] to explain the coupling constant ratio $|C_A/C_V| = 1.24$ in Eq. (2.177). The following assumptions are used in the derivation. (1) The vector and axial vector currents in weak interactions obey the equal-time commutator rules of $su_3 \times su_3$ as proposed by Gell-Mann [33]. (2) The axial vector current is partially conserved (PCAC, i.e., partially conserved axial vector current). Under these assumptions, Alder and Weisberger were able to relate C_A/C_V to integrals over pion-proton total cross sections:

$$\left(\frac{C_V}{C_A}\right)^2 = 1 + \frac{2M^2}{\pi g_{\pi NN}^2} \int_{m_\pi}^{\infty} [\sigma^{\pi^- p}(E) - \sigma^{\pi^+ p}(E)] \frac{p}{E}\, dE. \tag{4.83}$$

Here $\sigma^{\pi^- p}$ and $\sigma^{\pi^+ p}$ are the total cross sections for the $\pi^- p$ and $\pi^+ p$ scatterings. E and p are the pion energy and momentum, respectively, in the laboratory system. M and m_π are the nucleon and pion mass, respectively; $g_{\pi NN}$ is the pion-nucleon coupling constant. All these quantities have been measured. From Eq. (4.83),

$$\begin{aligned} |C_A/C_V| &= 1.24 \quad [31], \\ &= 1.16 \quad [32]. \end{aligned} \tag{4.84}$$

These values are very close to the experimental result, 1.24. The difference between these two numbers comes from details of the theoretical formulation of the PCAC hypothesis, and is a reasonable indication of possible theoretical uncertainties in the result.

Now we follow the article by Weisberger [34]. Contrary to the rest of this book, we do not try to give a self-contained derivation of Eq. (4.83) here, since we do not introduce any systematic method of calculation with the second quantization. The following outline of the derivation should, however, be helpful. We restrict our discussion to the result for the algebraic structure of the $\Delta S = 0$ currents. Assume that the $\Delta S = 0$ axial vector current is a member of an isotriplet, a vector in isospin space. We define charges by the space integrals of the time components of the vector ($V_{\mu,i}$) and axial vector ($A_{\mu,i}$) currents. (Here, the vector and axial vector currents are general forms of Eqs. (4.30) and (4.31). In this subsection, the isospin components are expressed by 1, 2, 3, instead of x, y, z.) Therefore, we have

$$\begin{aligned} I_i(t) &= -i \int V_{4,i}(\mathbf{x}, t)\, d\mathbf{x}, \\ I_i^5(t) &= -i \int A_{4,i}(\mathbf{x}, t)\, d\mathbf{x}, \qquad i = 1, 2, 3. \end{aligned} \tag{4.85}$$

I is the total isospin and $\mathbf{I}^5$ is called the isochirality. (In this subsection, ψ's are field operators in the sense of the second quantitzation, so that the quantities in Eq. (4.85) are also operators.) They obey the equal-time commutation relations

$$[I_i(t), I_j(t)] = i\epsilon_{ijk} I_k(t), \tag{4.86}$$

$$[I_i(t), I_j^5(t)] = i\epsilon_{ijk} I_k^5(t). \tag{4.87}$$

Gell-Mann's proposal [33] is that the commutators of components of the chirality close the algebra in a simple manner

$$[I_i^5(t), I_j^5(t)] = i\epsilon_{ijk} I_k(t). \tag{4.88}$$

If we define the right-handed and left-handed charges by

$$I_i^{\pm} = \tfrac{1}{2}(I_i \pm I_i^5), \tag{4.89}$$

the three members of each of these triplets of chiral charges when commuted among themselves generate on su_2 algebra. Two members of different sets commute with each other. This defines a chiral $su_2 \times su_2$ algebra. (We can generalize it to $su_3 \times su_3$, with the strangeness changing weak current taken into account.)

In order to derive the relation (4.83), we use the commutator

$$2I_3 = [I_+^5, I_-^5] \qquad \text{with} \quad I_\pm^5 = I_1^5 \pm i I_2^5. \tag{4.90}$$

Let us take matrix elements of this commutator between physical one-proton states. The term on the left is the third component of the proton isospin. In the commutator on the right, we insert a complete set of intermediate states and isolate the contribution from the one-neutron state.

$$\langle \mathrm{p}|\mathrm{p}\rangle = \sum_{\mathrm{n\ spins}} \langle \mathrm{p}|I_+^5|\mathrm{n}\rangle\langle \mathrm{n}|I_-^5|\mathrm{p}\rangle + \sum_{\alpha\neq\mathrm{n}} \{\langle \mathrm{p}|I_+^5|\alpha\rangle\langle\alpha|I_-^5|\mathrm{p}\rangle - (I_+^5 \leftrightarrow I_-^5)\}. \tag{4.91}$$

The one-neutron matrix elements are given by

$$\langle \mathrm{p}|I_+^5|\mathrm{n}\rangle = -i\int \langle \mathrm{p}|A_{4,+}|\mathrm{n}\rangle\, d\mathbf{x} = \delta^{(3)}(\mathbf{p}-\mathbf{n})(C_\mathrm{A}/C_\mathrm{V})\bar{u}_\mathrm{p}\gamma_4\gamma_5 u_\mathrm{n}. \tag{4.92}$$

Here u's are the nucleon wave functions in the usual sense.

The second term on the right-hand side of Eq. (4.91) can be expressed in terms of π-nucleon scattering cross sections. To do this it is necessary to introduce the concept of a partially conserved axial vector current. This means that the divergence of the axial vector current is proportional to the pion field, because the axial vector current is the isovector and its divergence is the pseudoscalar under the proper Lorentz transformations and space inversion,

$$\frac{\partial}{\partial x_\mu} A_{\mu,i}(x) = a\phi_{\pi,i}(x). \tag{4.93}$$

We postulate that the matrix element of the corresponding pion current

$$\langle\beta|(m_\pi^2 - \Box)\phi_{\pi^\pm}|\alpha\rangle,$$

is a slowly varying function of the momentum transfer p^2 in the interval $0 < p^2 < m_\pi^2$. The constant a in Eq. (4.93) should not be zero, since the pion decay constant is proportional to a^2. It is given by

$$a = -i2^{1/2}\frac{Mm_\pi^2}{g_{\pi\mathrm{NN}}}\frac{C_\mathrm{A}}{C_\mathrm{V}}, \tag{4.94}$$

which is derived from the matrix element of Eq. (4.93) with respect to the one-neutron state and one-proton state. In the limit of zero momentum transfer, we have $2MC_\mathrm{A}/C_\mathrm{V}$ for the left-hand side and $i2^{1/2}ag_{\pi\mathrm{NN}}m_\pi^{-2}$ for the right-hand side by neglecting the off-shell contributions. The result can be also derived from the assumption that matrix elements of $(\partial/\partial x_\mu)A_{\mu,i}$ obey unsubtracted dispersion relations which are dominated at low momentum transfers by meson poles [31].

To use the PCAC hypothesis we must relate the inelastic matrix elements of the charges in Eq. (4.91) to matrix elements of the divergence as shown in the following equation.

$$\frac{\partial}{\partial t}\langle \mathrm{p}|I_i^5|\alpha\rangle = i(E_p - E_\alpha)\langle \mathrm{p}|I_i^5|\alpha\rangle$$

$$= \int\left\langle \mathrm{p}\left|\frac{\partial}{\partial x_4}A_{4,i}(x)\right|\alpha\right\rangle d\mathbf{x} = \int\left\langle \mathrm{p}\left|\frac{\partial}{\partial x_\mu}A_{\mu,i}(x)\right|\alpha\right\rangle d\mathbf{x}. \tag{4.95}$$

Dividing through by the energy difference and substituting for $(\partial/\partial x_\mu)A_\mu(x)$ in Eq. (4.93), we express $\langle p|I_\pm^5|\alpha\rangle$ in terms of π-proton scattering amplitudes. The result of these substitutions and further mathematical manipulations is the sum rule of Eq. (4.83). For more detail see [35].

REFERENCES

1 E. C. G. Sudershan and R. E. Marshak, *Phys. Rev.* **114,** 1329 (1958).
2 R. P. Feynman and M. Gell-Mann, *Phys. Rev.* **109,** 193 (1958).
3 J. J. Sakurai, *Nuovo Cimento* **7,** 649 (1958).
4 M. Morita, *Nucl. Phys.* **14,** 106 (1959).
5 G. Puppi, *Nuovo Cimento* **5,** 505 (1948); O. Klein, *Nature* **161,** 897 (1948); T. D. Lee, M. Rosenbluth, and C. N. Yang, *Phys. Rev.* **75,** 905 (1949); J. Tiomno and J. A. Wheeler, *Rev. Mod. Phys.* **21,** 21 (1949); S. Nakamura, H. Fukuda, K. Ono, M. Sasaki, and M. Taketani, *Progr. Theoret. Phys.* **5,** 740 (1950).
6 S. S. Schweber, *An Introduction to Relativistic Quantum Field Theory.* Harper and Row, New York, 1964.
7 S. S. Gershtein and J. B. Zeldovich, *Zhur. Eksptl. Teoret. Fiz. USSR* 29, 698 (1955) [translation: *Soviet Phys. JETP* 2, 576 (1957)].
8 M. Gell-Mann, *Phys. Rev.* **111,** 362 (1958).
9 T. Mayer-Kuckuk and F. C. Michel, *Phys. Rev.* **127,** 545 (1962).
10 Y. K. Lee, L. W. Mo, and C. S. Wu, *Phys. Rev. Letters* **10,** 253 (1963).
11 N. W. Glass and R. W. Peterson, *Phys. Rev.* **130,** 299 (1963).
12 T. Kinoshita and A. Sirlin, *Phys. Rev.* **113,** 1652 (1959).
13 M. Morita, *Phys. Rev.* **113,** 1584 (1959).
14 M. Gell-Mann and S. M. Berman, *Phys. Rev. Letters* **3,** 99 (159).
15 J. Bernstein and R. R. Lewis, *Phys. Rev.* **112,** 232 (1958).
16 M. E. Nordberg, Jr., F. B. Morinigo, and C. A. Barnes, *Phys. Rev.* **125,** 321 (1962).
17 W. Grule, K. H. Lauterjung, and B. Schimmer, *Nucl. Phys.* **42,** 321 (1963).
18 J. Fujita, *Phys. Rev.* **126,** 202 (1962); *Progr. Theoret. Phys.* **28,** 338 (1962); J. Eichler, *Z. Phys.* **171,** 463 (1963).
19 A. H. Rosenfeld *et al., Rev. Mod. Phys.* **40,** 77 (1968).
20 N. Cabibbo, *Phys. Rev. Letters* **10,** 531 (1963).
21 Z. Maki, M. Nakagawa, and S. Sakata, *Progr. Theoret. Phys.* **28,** 870 (1962); M. Nakagawa, H. Okonogi, S. Sakata, and A. Toyoda, *Progr. Theoret. Phys.* **30,** 727 (1963); Y. Katayama, K. Matumoto, S. Tanaka, and E. Yamada, *Progr. Theoret. Phys.* **28,** 675 (1962).
22 T. D. Lee and C. S. Wu, *Ann. Rev. Nucl. Sci.* **15,** 381 (1965), especially p. 416.
23 N. Brene, L. Veje, M. Roos, and C. Cronström, *Phys. Rev.* **149,** 1288 (1966).
24 C. Hayashi, R. Hoshi, and D. Sugimoto, *Progr. Theoret. Phys. Suppl.* **22,** 1 (1962).
25 H. Chiu and P. Morrison, *Phys. Rev. Letters,* **5,** 573 (1960); H. Chiu, *Phys. Rev.* **123,** 1040 (1961).
26 V. M. Lobashov, V. A. Nazarenko, L. F. Saenko, L. M. Smotrizky, and G. I. Kharkevitch, *J. Phys. Soc. Japan Suppl.* **24,** 728 (1968).
27 Yu. G. Abov, P. A. Kruptchitsky, and Y. A. Oratovski, *Phys. Letters* **12,** 25 (1965); Yu. G. Abov, P. A. Kruptchitsky, M. I. Bulgakov, O. N. Ermakov, and I. L. Karpikhin, *J. Phys. Soc. Japan Suppl.* **24,** 728 (1968).
28 M. Morita, *J. Phys. Soc. Japan Suppl.* **24,** 419 (1968).
29 M. Morita, *Phys. Rev.* **122,** 1525 (1961).

30 L. Grodzins and F. Genovese, *Phys. Rev.* **121**, 228 (1961); E. Kankeleit, *Proc. Intern. Conf. Nucl. Phys.* (Paris, 1964), p. 1206.

31 S. L. Adler, *Phys. Rev. Letters* **14**, 1051 (1965); *Phys. Rev.* **140**, B736 (1965).

32 W. I. Weisberger, *Phys. Rev. Letters* **14**, 1047 (1965).

33 M. Gell-Mann, *Phys. Rev.* **125**, 1067 (1962).

34 W. I. Weisberger, *Proc. Intern. Conf. Weak Interactions* (Argonne Natl. Lab. Rept. ANL 7130, 1965), p. 409.

35 S. L. Adler and R. F. Dashen, *Current Algebras and Applications to Particle Physics*, Benjamin, New York, 1968.

Chapter Five

RELATED PROBLEMS

5.1 ORBITAL ELECTRON CAPTURE

In this chapter we shall study several weak processes for which the weak interactions are responsible. We will not, however, try to discuss systematically the whole structure of weak interactions or the high energy decay processes. We will just select several decay modes that are closely related to beta decay and the violation of symmetries.

Until now we have studied the most popular beta decay process, that is, the spontaneous emission of either (negative) electrons or positrons. Formulas for the spectrum and angular distribution of positrons have already been given. These formulas are derived from those for electrons by changing signs for some coupling constants and taking their complex conjugates. Orbital electron capture is also a mode of beta decay. It always competes with the emission of positrons. Since only a neutrino is emitted in this process, orbital electron capture takes place even if positron emission is energetically prohibited. The available energy required for the electron capture process was given in section 1.5, and it is

$$\Delta E = ({}_A M^Z - {}_A M^{Z-1})c^2 > E_\mathrm{K} \text{ or } E_\mathrm{L} \tag{5.1}$$

where ${}_A M^Z$ and ${}_A M^{Z-1}$ are the atomic masses of the initial and final states, and E_K and E_L are the binding energies of the orbital electrons. The possibility of electron capture in which the nucleus absorbs one of its own orbital electrons was first suggested by Yukawa and Sakata [1, 2]. Experimental proof was performed first by Alvarez [3] for the case of Ga^{67}.

Calculation of the transition rate for electron capture is similar to that for beta decay. The main difference is the final state, which is now a two-particle state. Therefore, the emitted neutrino has a definite energy, and the final state density is given by

$$\frac{dn}{dE_\nu} = \frac{q^2\, dq\, 4\pi}{(2\pi\hbar)^3\, dE_\nu}, \tag{5.2}$$

where E_ν and q are the energy and momentum, respectively, of the emitted neutrino. The integration over the direction of the neutrino momentum has already been performed, since we do not observe it. Measuring E_ν in mc^2 and q in mc, we have

$$\frac{dn}{dE_\nu} = \frac{E_\nu^2 m^2 c}{2\pi^2 \hbar^3} \qquad \text{with} \quad E_\nu = \Delta E - E_{\mathrm{K}} - E_{\mathrm{R}}. \tag{5.3}$$

Here we have used the relation

$$E_\nu = cq. \tag{5.4}$$

E_{R} stands for the recoil energy of the nucleus, which is usually omitted compared with the other quantities.

The orbital electron capture is induced by the hermitian conjugate term in Eq. (2.55) or (3.1). A simple method for evaluating the absolute square of the Hamiltonian is as follows. The binding energy of the electron is neglected compared with its rest mass, except for very heavy nuclei. We adopt the nonrelativistic approximation and set

$$E \to mc^2 \ (=1), \qquad \mathrm{p} \to 0, \tag{5.5}$$

in the projection operator for the electron, Eq. (2.70). It should be multiplied by

$$|\psi_{\mathrm{K}}(0)|^2 = (\alpha Z)^3/\pi \tag{5.6}$$

where $\psi_{\mathrm{K}}(0)$ is the $r = 0$ value of the K-orbit wave function

$$\psi_{\mathrm{K}}(r) = \left(\frac{1}{\pi}\right)^{1/2} \left(\frac{Z}{a_o}\right)^{3/2} \exp\left(-\frac{Zr}{a_o}\right) \tag{5.7}$$

with

$$a_o = \frac{\hbar^2}{me^2} = \text{Bohr radius,}$$

and Z the atomic number of the parent nuclei. This is because the bound states are essentially due to the Coulomb interaction between the electrons and the nucleus.

As a result, the transition probability P_{K} for the K capture is given by

$$P_{\mathrm{K}} = \frac{E_\nu^2}{2\pi^2} \frac{1}{\pi} (\alpha Z)^3 2\pi |H|^2 \frac{N_{\mathrm{K}}}{2} \qquad \text{in } \hbar = m = c = 1. \tag{5.8}$$

The last factor $N_{\mathrm{K}}/2$ in expression (5.8) comes from averaging over the initial electron spin states, N_{K} being the number of the K electrons. We shall drop $N_{\mathrm{K}}/2$ hereafter. The expression for $|H|^2$ is easily obtained from Eqs. (2.181) and (2.183) with Eq. (5.5), and is given by

$$\begin{aligned} 2|H|^2 = {} & [|C_{\mathrm{S}}|^2 + |C_{\mathrm{V}}|^2 + 2\,\mathrm{Re}(C_{\mathrm{S}}^* C_{\mathrm{V}})] |\textstyle\int 1|^2 \\ & + [|C_{\mathrm{T}}|^2 + |C_{\mathrm{A}}|^2 + 2\,\mathrm{Re}(C_{\mathrm{T}}^* C_{\mathrm{A}})] |\textstyle\int \sigma|^2 + [C_i \to C_i']. \end{aligned} \tag{5.9}$$

Here, $[C_i \to C_i']$ means the same as the expression in the first two sets of brackets in Eq. (5.9) except for replacing C_i by C_i'. As can be seen in Eq. (5.9), there is no $1/E$ factor for the Fierz interferences. This is the consequence of the replacement Eq. (5.5). The K-capture rates are, therefore, more sensitive to the ratio C_S/C_V or C_T/C_A than the beta-ray spectra. In fact, we can obtain the following limits from the experimental data on K capture [4-6].

$$\left(\frac{C_S}{C_V}\right)^2 < 0.001 \qquad \text{and} \qquad \left(\frac{C_T}{C_A}\right)^2 < 0.001. \tag{5.10}$$

These limits are derived under the assumption of the real coupling constant ratios C_S/C_V and C_T/C_A.

Next, let us examine the ratio of the K-capture and positron decay rates. Writing Eq. (5.8) as

$$P_K = \frac{1}{2\pi^3}|H|^2 f_K \qquad \text{with} \qquad f_K = 2\pi(\alpha Z)^3 E_\nu^2, \tag{5.11}$$

we have

$$\frac{P_K}{P_+} = \frac{f_K |H|_K^2}{f_+ |H|_{\beta^+}^2} \approx \frac{2\pi(\alpha Z)^3 E_\nu^2}{f_+}. \tag{5.12}$$

Here f_+ is the integrated Fermi function for positron decay. $|H|_K^2$ is given by Eq. (5.9), and $2|H|_{\beta^+}^2$ is given by Eq. (2.181) plus Eq. (2.183). The right-hand side of Eq. (5.12) is obtained by assuming nonexistence of the Fierz interferences. In the limiting case of high q value, the positron decay is dominant, since

$$f_+ \approx E_o^5/30$$

and

$$\frac{P_K}{P_+} \approx 60\pi\left(\frac{\alpha Z}{E_o}\right)^3 \qquad \text{for} \quad E_o \gg 1. \tag{5.13}$$

The transition probability from the other orbits can be estimated as follows. The orbital electron wave functions at $r = 0$ are given by

$$\begin{aligned} |\psi_{nl}(0)|^2 &= |\psi_K(0)|^2 n^{-3} && \text{for } l = 0, \\ &= 0 && \text{for } l \neq 0, \end{aligned} \tag{5.14}$$

ψ_{nl} being the bound state wave functions with the principal and orbital quantum numbers n and l. Therefore, the ratio of capture rates from the K and L orbits is approximately given by

$$\frac{P_L}{P_K} = \frac{(\Delta E - E_L)^2}{8(\Delta E - E_K)^2} \approx \frac{1}{8}. \tag{5.15}$$

For more detailed calculations, we have to consider the Coulomb field of the finite size of the nucleus and the screening effect due to the other orbital electrons. These have been done by Feenberg and Trigg [7].

5.2 INVERSE BETA DECAY

Bethe and Peierls [8] pointed out the possibility of inverse beta decay. In this process, the nucleon captures either a neutrino or an antineutrino, and emits an electron or positron. The process is expressed by

$$\bar{\nu} + (A, Z) \rightarrow (A, Z-1) + e^+ \qquad \text{or} \qquad \nu + (A, Z) \rightarrow (A, Z+1) + e^-, \tag{5.16}$$

which may be considered the inverse of the direct decay

$$(A, Z-1) \rightarrow (A, Z) + e^- + \bar{\nu}. \quad \text{or} \quad (A, Z+1) \rightarrow (A, Z) + e^+ + \nu \tag{5.17}$$

The cross section for the inverse beta decay is extremely small, of the order of 10^{-44} cm^2, as is shown below.

The transition probability $W(E)$ for the elementary process

$$\bar{\nu} + p \rightarrow n + e^+ \tag{5.18}$$

is given by essentially the same expression as in Eq. (2.14),

$$W(E) = \frac{2\pi}{\hbar} |H_{fi}|^2 \frac{dn}{dE}. \tag{5.19}$$

The cross section is the total probability divided by the velocity of the incoming neutrino, which is equal to the light velocity c,

$$\sigma = \frac{W(E)}{c} = \frac{2\pi}{c\hbar} |H_{fi}|^2 \frac{dn}{dE}. \tag{5.20}$$

The main difference between the direct and inverse beta-decay processes is again in the final-state density, which is similar to Eq. (5.2),

$$\frac{dn}{dE} = \frac{p^2\, dp}{(2\pi\hbar)^3} \frac{4\pi}{dE} = \frac{4\pi}{(2\pi\hbar)^3} \frac{pE}{c^2}. \tag{5.21}$$

Here, integration over the direction of the emitted position momentum has already been performed. The trace of the lepton covariants is given by

$$\begin{aligned} |H_{fi}|^2 &= |C_V|^2 |\textstyle\int 1|^2 + |C_A|^2 |\textstyle\int \sigma|^2 \\ &= \left(G \frac{\hbar^3}{m^2 c}\right)^2 \left[|\textstyle\int 1|^2 + \left(\frac{C_A}{C_V}\right)^2 |\textstyle\int \sigma|^2 \right] \end{aligned} \tag{5.22}$$

for the standard $V - A$ theory. From Eqs. (5.20)–(5.22) we have

$$\sigma(E) = \frac{G^2}{\pi}\left(\frac{\hbar}{mc}\right)^2\left[|\textstyle\int 1|^2 + \left(\frac{C_A}{C_V}\right)^2 |\textstyle\int \sigma|^2\right]\frac{p}{mc}\frac{E}{mc^2}. \tag{5.23}$$

Here we adopt the numerical values

$$G = 3.0 \times 10^{-12},$$

from Eq. (2.170),

$$[\ldots] = [1 + (1.24)^2 \times 3],$$

from Eqs. (2.174) and (2.177), and

$$\frac{\hbar}{mc} = 3.86 \times 10^{-11} \text{ cm}. \tag{5.24}$$

Therefore, the cross section (5.23) is ready to be enumerated in practical units.

The threshold of the reaction (5.18) occurs at an antineutrino energy of approximately 1.8 MeV. The cross section is, therefore, given by

$$\begin{aligned} \sigma &= 0.8 \times 10^{-43} \text{ cm}^2 && \text{for } E = 3.0 \text{ mc}^2 \text{ (or } E_\nu = 4.5 \text{ mc}^2), \\ &= 2 \times 10^{-43} \text{ cm}^2 && \text{for } E = 3.3 \text{ mc}^2 \text{ (or } E_\nu = 5.5 \text{ mc}^2), \\ &= 1.6 \times 10^{-42} \text{ cm}^2 && \text{for } E = 8.3 \text{ mc}^2 \text{ (or } E_\nu = 10.8 \text{ mc}^2). \end{aligned} \tag{5.25}$$

In the derivation of Eq. (5.22), we adopted the two-component neutrino theory and the conservation of leptons (which will be discussed in section 5.3 and 5.8). The trace of the lepton covariants should be multiplied by a factor $\frac{1}{2}$ due to the initial spin average, if the classical four-component theory of the neutrino is adopted; that is

$$\sigma(\text{two-component}) = 2\sigma(\text{classical}). \tag{5.26}$$

This relation was obtained by Lee and Yang [9], using the method of detailed balance.

Reines, Cowan, and co-workers have carried out experiments on the reaction $\bar{\nu} + p \rightarrow n + e^+$. Measurements were made with the liquid scintillator of a toluene–terphenyl mixture loaded with cadmium propionate. This scintillator was located adjacent to the nuclear reactor within a shield designed to absorb other radiation from the reactor. The detection of this reaction is as follows. A prompt pulse results from the annihilation radiations of the positron. A delayed pulse, which appears several microseconds later, is produced by the gamma rays due to the capture of the neutron. In later experiments by Reines and Cowan [11], the signal rate was 36 ± 4 events per hour and the signal-to-noise ratio was 1/5. The cross section per fission $\bar{\nu}$ (assuming 6.1 $\bar{\nu}$ per fission) for the inverse beta decay of the proton was measured and found to be

$$\sigma_{\text{exp}} = (11 \pm 2.6) \times 10^{-44}\ \text{cm}^2, \tag{5.27}$$

in a good agreement with the two-component theory

$$\begin{aligned} \sigma_{\text{theor}} &= 9.5 \times 10^{-44}\ \text{cm}^2 && \text{Carter } et\ al.\ [12], \\ &= 12 \times 10^{-44}\ \text{cm}^2 && \text{Muelhause and Oleksa } [13], \\ &= 15 \times 10^{-44}\ \text{cm}^2 && \text{King and Perkins } [14], \end{aligned} \tag{5.28}$$

where the theoretically expected cross sections are dependent on the antineutrino spectrum from the nuclear reactor.

A well-known example of the reactions (5.16) is the neutrino capture by Cl^{37},

$$\nu + Cl^{37} \rightarrow A^{37} + e^-, \tag{5.29}$$

which is the inverse process of the electron capture in A^{37}

$$A^{37} + e^- \rightarrow Cl^{37} + \nu. \tag{5.30}$$

Davis [15] is working on detecting the reaction (5.29), for which the threshold neutrino energy is approximately 0.8 MeV. His experiment involves irradiation of a large volume of carbon tetrachloride near a nuclear reactor for a long period of time. Afterward, A^{37} is removed and purified by physical methods. The amount of A^{37} is measured by counting the x-rays from its electron-capture decays in a low-background Geiger counter. A very small cross section ($(0.3 \pm 0.4) \times 10^{-45}$ cm^2) has been obtained. This is small in comparison with the theoretical estimate ($\sigma = 2 \times 10^{-45}$ cm^2) by fission neutrinos if $\nu = \bar{\nu}$. Besides, the background due to the muon components in cosmic rays is not known too clearly.

In his lecture at Tokyo in 1967, Goldhaber described the status of Davis's experiment. Using a deep underground shield against cosmic rays, Davis set a 100,000-gallon C_2Cl_4 counting vessel at a depth of 4700 ft in a mine called the Homestake Mine in South Dakota. Reaction (5.29) was found to take place for the solar neutrinos, and reaction (5.30) to follow. Each time, a hole is made in the K shell and the Auger electron is detected with high efficiency. Figure 5.1 shows the counts obtained by Davis from a sample of argon from a 100,000-gallon tank that was exposed for one month in the mine and then counted for 39.7 days. The background was counted for only 11.5 days.

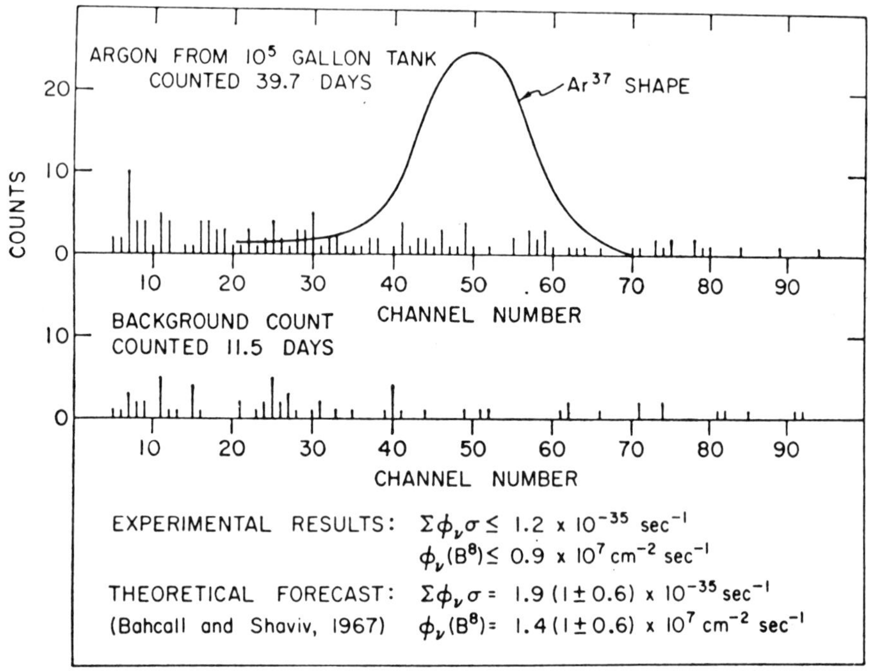

Fig. 5.1 Counts of A^{37} in the Davis experiment referred in [16].

The solar neutrino spectrum was studied by Bahcall [17] theoretically. In particular, he studied the chain of reactions that leads to B^8,

$$
\begin{aligned}
&H + H \to D + e^+ + \nu, \qquad && E_\nu(\max) = 0.42\ \text{MeV},\\
&H + D \to He^3 + \gamma, &&\\
&He^3 + He^3 \to He^4 + 2H, &&\\
&He^3 + He^4 \to Be^7 + \gamma, &&\\
&Be^7 + e^- \to Li^7 + \nu, && E_\nu = 0.861\ \text{MeV}\\
&Li^7 + H \to 2He^4, &&\\
&Be^7 + H \to B^8 + \gamma, &&\\
&B^8 \to Be^{8*} + e^+ + \nu, && E_\nu(\max) = 14\ \text{MeV},\\
&Be^{8*} \to 2He^4. &&
\end{aligned}
\tag{5.31}
$$

The energy spectrum of these three neutrinos is shown in Fig. 5.2 with that from the famous CNO cycle. The dominant source of the solar neutrinos is the primary reaction $H + H \to D + e^+ + \nu$ for which the neutrino maximum energy is 0.42 MeV, and it is below the threshold of reaction (5.29). Figure 5.1 shows the theoretical estimate for the cross section times the neutrino flux ϕ ($cm^{-2}\ sec^{-1}$). The experimental data are preliminary, and any con-

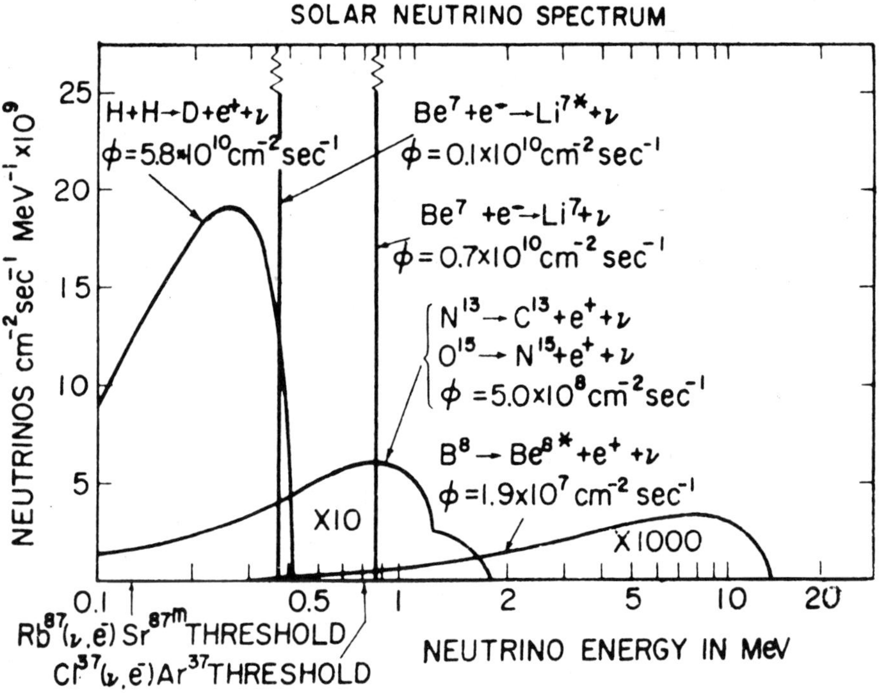

Fig. 5.2 Solar neutrino spectrum [16].

clusion would be premature. It is, however, possible to say that the Davis experiment improves a definite upper limit on the CNO part of the energy production of the sun. This is because $\Sigma\phi\sigma$ would be 3.5×10^{-35} sec^{-1} if we assume the total energy produced in the sun comes from the CNO cycle.

Goldhaber made a table of possible nuclear reactions which may detect the solar neutrinos [16]. This is shown in Table 5.1.

Table 5.1 Reactions for Studying the Solar Neutrino Spectrum [16][a].

Reaction (ν, e$^-$)	E_t (keV)	$t_{1/2}$	$\Sigma\phi\sigma \times$ (10^{35} sec^{-1})	Detecting material (tons)	Captures from various solar neutrino processes (%)			
					p–p	Be^7	CN	B^8
$Rb^{87} \to Sr^{87m}$	115	2.8 hours	64	99	65	27	5	3
$Mn^{55} \to Fe^{55}$	231	2.6 years	0.43	2600	56	33	6	5
$Ga^{71} \to Ge^{71}$	233	11 days	31	115	57	33	5	5
$V^{51} \to Cr^{51}$	752	28 days	1.2	810	0	12	3	85
$Cl^{37} \to A^{37}$	814	35 days	3.0	930	0	7	2	91
$Li^7 \to Be^7$	862	53 days	11	14	0	0	14	86

[a] The second column gives the threshold neutrino energy, and the fifth column gives the quantity of the detecting material in tons, to get 10 counts a day. The neutrino fluxes used are 5.9×10^{10} for p–p, 0.69×10^{10} for Be^7, 0.60×10^9 for CN, and 2.0×10^7 for B^8, in cm^{-2} sec^{-1}.

5.3 DOUBLE BETA DECAY AND LEPTON NUMBER CONSERVATION

Double beta decay is the second-order process of the weak interaction

$$(A, Z) \to (A, Z+2) + e^- + e^- + \bar{\nu}_e + \bar{\nu}_e. \tag{5.32}$$

This decay mode is only possible under the restricted conditions: (A, Z), $(A, Z+1)$, and $(A, Z+2)$ make an isobaric triplet; and the state for $(A, Z+1)$

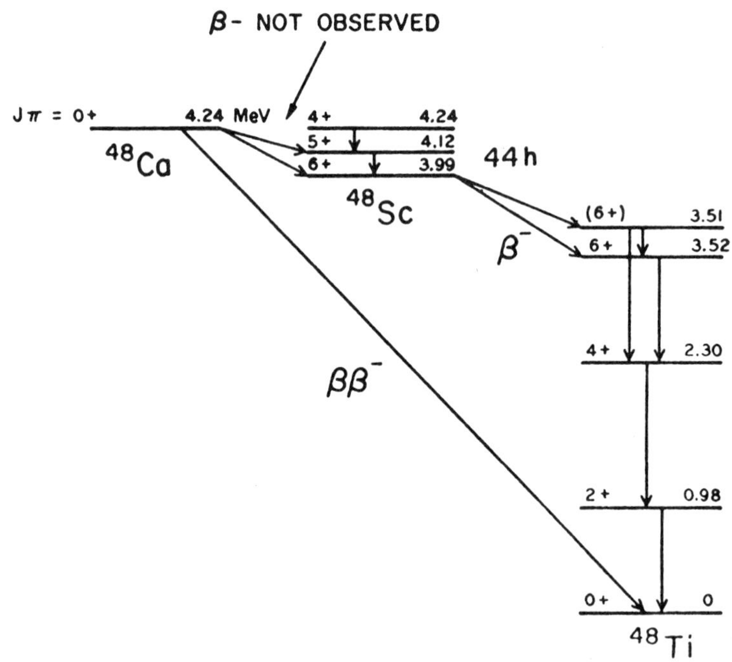

Fig. 5.3 Decay scheme of Ca^{48} [25].

has a higher mass than the others. Therefore, the initial (A, Z) and final $(A, Z+2)$ states are the stable even-even (i.e., N = even and Z = even) nuclei. An example is the pair Ca^{48} and Ti^{48}; the decay scheme of Ca^{48} is given in Fig. 5.3. The double beta decays of Zr^{96}, Mo^{100}, Cd^{116}, Sn^{124}, Te^{130}, Nd^{150}, and U^{238} are also known.

Investigations of double beta decay have always been related to properties of the neutrino. Formerly, this phenomenon was examined in connection with the question whether the neutrino is the Dirac particle or the Majorana neutrino. The Dirac neutrino has the same wave function as that of Eq. (A.85), possibly with a nonvanishing rest mass. A crude approximation for the double beta-decay rate $W_{\beta\beta}$ is obtained simply by squaring the single beta-decay rate W_β

$$W_{\beta\beta} = (W_\beta)^2 = \left\{ \frac{C_V^2}{2\pi^3} \left[|\int 1|^2 + \left(\frac{C_A}{C_V}\right)^2 |\int \boldsymbol{\sigma}|^2 \right] \frac{E_o^5}{30} \right\}^2 \tag{5.33}$$

in units of

$$\frac{mc^2}{\hbar} = 0.7763 \times 10^{21}\ \mathrm{sec}^{-1}. \tag{5.34}$$

Here, we have assumed $E_o \gg 1$. Adopting numerical values for the coupling constants and matrix elements in Eqs. (2.170), (2.174), and (2.177), we have the half life for double beta decay

$$t_{1/2} \sim 10^{20}\ \text{years} \qquad \text{for} \quad E_o \sim 5mc^2. \tag{5.35}$$

On the other hand, the Majorana neutrino is a Fermi particle whose antiparticle is identical with itself

$$\nu = \bar{\nu}. \tag{5.36}$$

This possibility was considered by Majorana [18] in 1937. If this is the case, the neutrino is virtually emitted in the beta decay of the first neutron in the decaying nucleus, and is then reabsorbed by the second neutron in its beta decay. The energy sum of these two electrons is constant in this case, while it is not so in the case of two-neutrino emission. The transition probability for the Majorana theory increases greatly, since the integration for the momentum of a virtually emitted neutrino should be taken into account. In fact, a numerical calculation shows that the estimated lifetime is about 10^5 times shorter than that for the Dirac neutrino (see Eqs. (5.41) and (5.42)).

Since the discovery of parity nonconservation, all kinds of beta-decay phenomena seem to be explained by the two-component neutrino. It also appears certain that only one kind of neutrino, a left-handed one (with helicity -1), exists. Its antiparticle is right-handed (with helicity $+1$). Consequently, the neutrinoless mode of double beta decay

$$(A, Z) \rightarrow (A, Z+2) + e^- + e^- \tag{5.37}$$

should not take place, because the antineutrino emitted in the first step has a definite helicity, which is opposite to the helicity of the neutrino absorbed in the second step. The two-neutrino process (5.32) is only one possible mode for double beta decay. Conversely, for the neutrinoless mode to take place, the neutrino cannot be fully polarized. For this case, the neutrino may have a rest mass, or there may exist two kinds of two-component neutrinos, that is, left-handed and right-handed neutrinos. Therefore, it is extremely important to discover the neutrinoless mode. If the neutrinoless mode is forbidden, we can define conservation of the lepton number L [19]. In particular, we add the subscript e to L, since the conservation holds for the electron and muon, separately (see section 5.8). We assign the lepton number L_e as follows.

$$\begin{aligned} L_e &= +1 && \text{for } e^- \\ &= -1 && \text{for } e^+ \\ &= +1 && \text{for } \nu_e \\ &= -1 && \text{for } \bar{\nu}_e \\ &= 0 && \text{for all other particles.} \end{aligned} \tag{5.38}$$

The conservation of lepton numbers refers to the fact that the sum of L_e for a system is conserved in the reaction. This conservation holds in Eq. (5.32) and is violated in Eq. (5.37).

On the other hand, it might be thought that these results do not necessarily imply lepton conservation [20, 21]. According to the two-component neutrino theory, neutrinos emitted in beta decay are right-handed. In the second step of the double beta-decay process the neutrino might be absorbed, but according to the two-component theory it would have to be left-handed in order to be absorbed. Thus even if the neutrino and antineutrino are the same particle, neutrinoless double beta decay still cannot occur.

For this argument, we recall a theoretical work by Ryan and Okubo [22], who found a unitary transformation which connects the Majorana theory with the two-component massless-neutrino theory, and showed a possibility for assigning the lepton numbers. (There is a two-component theory with a non-vanishing mass [19], so that we emphasize the massless neutrino in the description above.)

These considerations indicate the extreme importance of measuring the neutrinoless beta decay to test lepton number conservation. According to the detailed calculations of Primakoff and Rosen [20], the half lives are

$$t_{1/2} \sim 6 \times 10^{19 \pm 2} (60/Z)^2 [1 - \exp(\mp 2\pi\alpha Z)]^2 (8/E)^{10} \text{ years} \tag{5.39}$$

for two-neutrino double beta decay, and

$$t_{1/2} \sim 1 \times 10^{15 \pm 2} (60/Z)^2 [1 - \exp(\mp 2\pi\alpha Z)]^2 (A/150)^{2/3} (8/E)^6 \text{ years} \tag{5.40}$$

for neutrinoless double beta decay. Here, E is the energy release in units of mc^2. The negative sign in the exponential factors refers to electron emission, and the positive sign refers to positron emission. The numerical factor $10^{\pm 2}$ indicates an ambiguity in the estimation of the nuclear matrix elements. From these formulas we obtain the following numerical results.

For $Ca^{48} \rightarrow Ti^{48}$

$$\begin{aligned} t_{1/2} &= 4 \times 10^{20 \pm 2} \text{ years} && \text{for the two-neutrino mode} \\ &= 3 \times 10^{15 \pm 2} \text{ years} && \text{for the neutrinoless mode;} \end{aligned} \tag{5.41}$$

and for $Te^{130} \rightarrow Xe^{130}$

$$t_{1/2} = 2 \times 10^{21\pm 2} \text{ years} \quad \text{for the two-neutrino mode}$$
$$= 8 \times 10^{15\pm 2} \text{ years} \quad \text{for the neutrinoless mode.} \qquad (5.42)$$

Increasing of the neutrinoless double beta-decay rate without lepton conservation makes it possible to obtain the same information on lepton conservation by simply observing the decay rate. Recently, convincing evidence of double beta decay was obtained by mass-spectrometric analyses of tellrium and selenium ores of known ages [23, 24]. A large excess of Xe^{130} was found, unaccompanied by any other anomalies. This excess could only be from the double beta-decay rate calculated from the age of the ore, and the concentrations of Te^{130} and Xe^{130} in the ores were consistent with lepton conservation. An upper limit could be set on the fraction of the beta interaction amplitude for the violation of the lepton conservation to be $\alpha \lesssim 3 \times 10^{-3}$. Ogata and his co-workers [24] obtained a new value

$$t_{1/2} = \left(8.0^{+0.4}_{-0.9}\right) \times 10^{20} \text{ years} \quad \text{for } Te^{130} \qquad (5.43)$$

and set $\alpha \lesssim 5 \times 10^{-3}$. These experiments detected the presence of the daughter nucleus from double beta decay, and gave no information on whether neutrinos were emitted. For a conclusive proof of the violation of lepton conservation, the following experiment was performed.

These two modes of double beta decay can be identified by observing the spectrum of the sum of two electron energies. In neutrinoless decay, they must add up to the available decay energy. If neutrinos are emitted, the energy will be statistically distributed among all four particles and the electron sum energy spectrum will be a broad distribution peaking at about half the available energy. An attempt to measure the electron spectrum from double beta decay of Ca^{48} was made by Wu and her collaborators in a deep salt mine in the Fairport Harbor, Ohio, using a streamer chamber in a magnetic field [25]. In 1150 hours of observation with a 10.6-gram source, only one event was seen that might be neutrinoless double beta decay. This implies a half life of more than 2×10^{21} years at the 80% confidence level. The theoretically estimated $t_{1/2}$ for the neutrinoless mode is about 10^{15} years, Eq. (5.41), so that the violation of lepton conservation α must be of the order of 10^{-3} or less.

$$\alpha^2 < \frac{t_{1/2}\,(\text{theoretical})}{t_{1/2}\,(\text{experimental})} \sim \frac{10^{15}\,\text{years}}{10^{21}\,\text{years}}$$

or

$$\alpha < 10^{-3}. \qquad (5.44)$$

A half-life limit for the two-neutrino mode of Ca^{48} double beta decay was also obtained.

$$t_{1/2} > 3.6 \times 10^{19} \text{ years.} \tag{5.45}$$

Number of events versus the sum of two electron energies is called the sum energy spectrum of events, and is shown in Fig. 5.4, which is from [25].

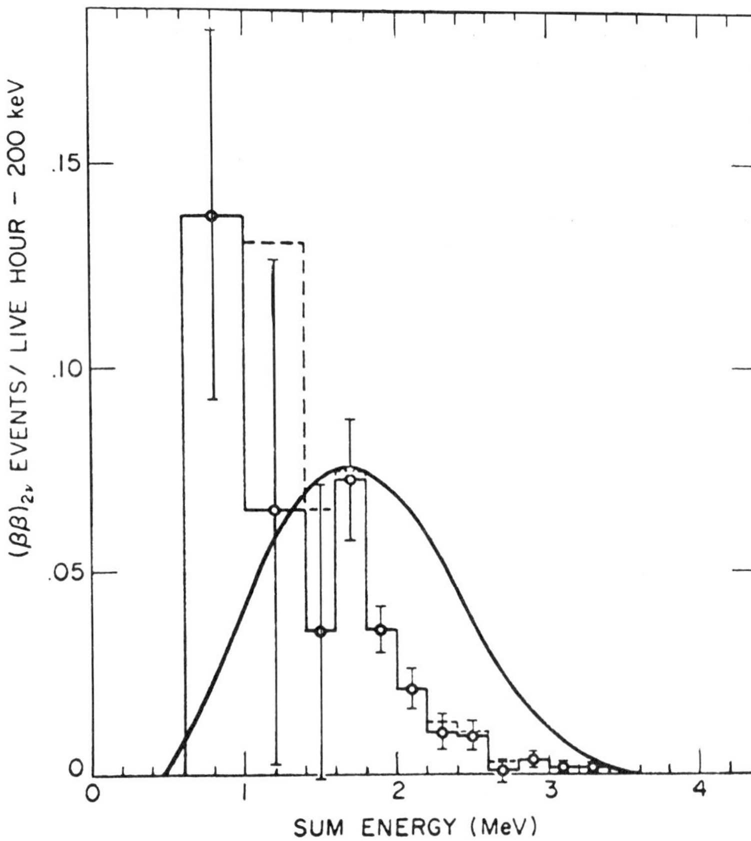

Fig. 5.4 The sum energy spectrum of events satisfying two-neutrino double beta-decay criteria for Ca^{48}. The upper (dashed) histogram represents the spectrum of events from the Ca^{48} source; the lower (solid) histogram represents the data with Ca^{40} background subtracted. The smooth curve, normalized to the same net number of events, is the theoretical sum energy spectrum with the apparatus efficiency included [25].

5.4 PION DECAY

Charged pions normally decay through the processes

$$\pi^+ \to \mu^+ + \nu_\mu, \qquad \pi^- \to \mu^- + \bar{\nu}_\mu. \tag{5.46}$$

The lifetime for this decay is

$$\tau_{\pi^\pm} = (2.55 \pm 0.03) \times 10^{-8} \text{ sec.} \tag{5.47}$$

Neutral pions, however, have a much shorter lifetime, and ordinarily decay through the process

$$\pi^0 \rightarrow 2\gamma. \tag{5.48}$$

The lifetime for this decay is

$$\tau_{\pi^0} = (1.05 \pm 0.18) \times 10^{-16} \text{ sec.} \tag{5.49}$$

Because of the value of the lifetime for neutral pions exhibited in (5.49), its measurement presents great practical difficulties. We do not want to enter into the details here, and will discuss only the decay of charged pions.

We consider first the decay of the positive pion. As is shown in the Feynman diagram (Fig. 5.5), the positive pion dissociates into a proton and antineutron at

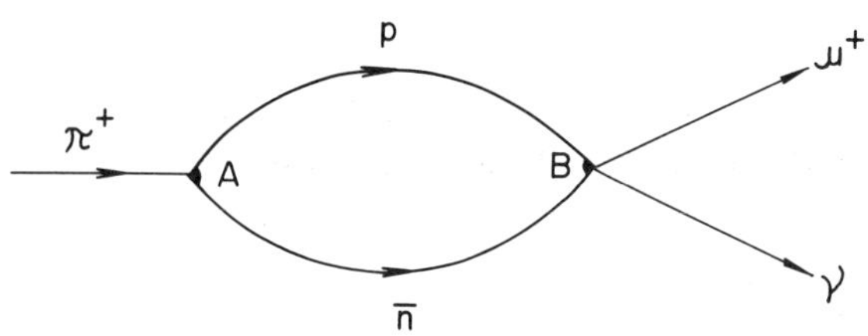

Fig. 5.5 Pion decay.

point A under the influence of the strong interaction, and the nucleon pair decays into a positive muon and a neutrino at point B due to the weak interaction. The latter part is a reaction analogous to muon capture in the proton, as we can see by changing the sense of the appropriate fermion lines. The exact calculation of this diagram gives, of course, a divergent result which seems to be meaningless.

There is, however, another decay mode for pions into an electron and a neutrino,

$$\pi^+ \rightarrow e^+ + \nu_e, \qquad \pi^- \rightarrow e^- + \bar{\nu}_e. \tag{5.50}$$

Until 1957, this process had been not detected experimentally, and it was believed that the process did not take place. However, the branching ratio for $\pi \rightarrow e$ decays to $\pi \rightarrow \mu$ decays was found in 1958 [26]. The latest value [27] is

$$R = \frac{W(\pi^+ \rightarrow e^+ + \nu_e)}{W(\pi^+ \rightarrow \mu^+ + \nu_\mu)} = (1.25 \pm 0.03) \times 10^{-4}. \tag{5.51}$$

Figure 5.5 can be drawn more precisely as Fig. 5.6, where the $\pi \rightarrow$ e and $\pi \rightarrow \mu$ decays are sketched in the same manner, and the (p, $\bar{\text{n}}$) is replaced by a black box, which indicates the complex of strong interactions. The black box is difficult to evaluate, but it represents the function of the pion mass and the coupling constant of the strong interaction. In the branching ratio of processes (5.46) and (5.50), the contribution of the black box cancels in the denominator and numerator. Here we neglect a small difference due to the change in magnitude of the momentum transfer.

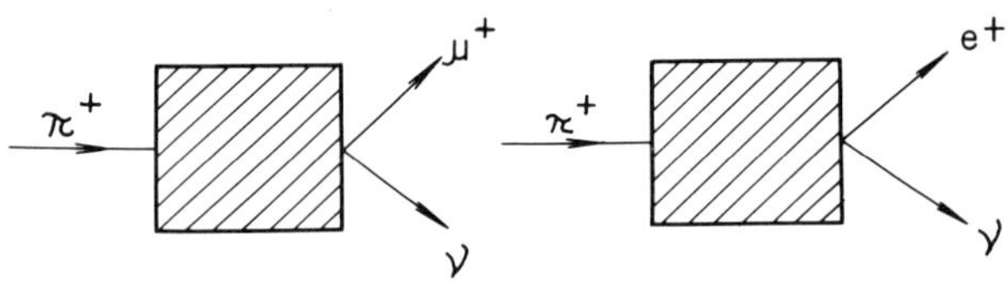

Fig. 5.6 $\pi \rightarrow \mu$ and $\pi \rightarrow$ e modes.

Now, let us assume the weak interactions have the same form for electrons and muons. Then the interaction Hamiltonian for pion decays can be written as

$$H = g\left(\frac{\partial}{\partial x_\mu}\phi_\pi\right)(\bar{\psi}_l\gamma_\mu(1+\gamma_5)\psi_\nu) + \text{h.c.} \tag{5.52}$$

Here l refers to either a muon or an electron; g is the coupling constant, which includes the effect of the black box; and ϕ_π is the pion field. Using the Dirac equations (3.7),

$$\left(\gamma_\mu \frac{\partial}{\partial x_\mu} + \frac{m_l c}{\hbar}\right)\psi_l = 0 \quad \text{and} \quad \gamma_\mu \frac{\partial}{\partial x_\mu}\psi_\nu = 0, \tag{5.53}$$

we have

$$H = -g\phi_\pi \frac{m_l c}{\hbar}(\bar{\psi}_l(1+\gamma_5)\psi_\nu) + \text{h.c.} \tag{5.54}$$

From this, we have a rough estimate for the branching ratio

$$R \sim m_e^2/m_\mu^2 = 2.3 \times 10^{-5}, \tag{5.55}$$

which agrees with the experimental data (5.51) qualitatively.

For a precise evaluation of R, we need to calculate the final state density ρ for the process $\pi^+ \rightarrow l^+ + \nu$. The number of final states between the energy range E and $E + dE$ is

$$\rho\, dE = \frac{p^2}{2\pi^2\hbar^3}\frac{dp}{dE}\,dE. \tag{5.56}$$

Here E is the total energy of the final state and is equal to $m_\pi c^2$. The conservation of energy and momentum gives

$$E = m_\pi c^2 = (m_l^2 c^4 + p^2 c^2)^{1/2} + pc \tag{5.57}$$

and

$$pc = p_l c = p_\nu c = E_\nu, \tag{5.58}$$

so that

$$E_l = \frac{m_\pi^2 + m_l^2}{2m_\pi} c^2, \tag{5.59}$$

$$pc = \frac{m_\pi^2 - m_l^2}{2m_\pi} c^2, \tag{5.60}$$

and

$$\frac{dp}{dE} = \frac{m_\pi^2 + m_l^2}{2m_\pi^2} \frac{1}{c} \qquad \text{for} \quad E = m_\pi c^2. \tag{5.61}$$

The final state density is now given by

$$\rho = \frac{p^2}{2\pi^2 \hbar^3} \frac{dp}{dE} = \frac{(m_\pi^2 - m_l^2)^2 (m_\pi^2 + m_l^2)}{8m_\pi^4} \frac{c}{2\pi^2 \hbar^3}. \tag{5.62}$$

An additional factor $(1 - v_l/c)$ should be multiplied by the transition rate for charged pion decay, as is shown in the following considerations. In the rest system of the pion, the neutrino and the lepton go out antiparallel to satisfy the conservation of momentum. The neutrino is left-handed, so that the lepton should also be left-handed, from the conservation of magnetic quantum numbers, since the pion has a zero spin. We have studied that positive leptons are mostly right-handed. And we might be aware of the forbidden decay of $\pi^+ \to l^+ + \nu$. Positive leptons have, however, a small probability of being left-handed, which is $\frac{1}{2}[1 - (v_l/c)]$, because of their nonvanishing mass. As a consequence, the decay probability is proportional to

$$\frac{1}{2}\left(1 - \frac{v_l}{c}\right) \times \frac{(m_\pi^2 - m_l^2)^2 (m_\pi^2 + m_l^2)}{8m_\pi^4} \tag{5.63}$$

with

$$1 - \frac{v_l}{c} = 1 - \frac{p_l c}{E_l} = \frac{2m_l^2}{m_\pi^2 + m_l^2}. \tag{5.64}$$

This comes from Eqs. (5.59) and (5.60). Finally, the branching ratio becomes

$$R = \frac{W(\pi^+ \to e^+ + \nu_e)}{W(\pi^+ \to \mu^+ + \nu_\mu)} = \frac{m_e^2}{m_\mu^2} \frac{(m_\pi^2 - m_e^2)^2}{(m_\pi^2 - m_\mu^2)^2} = 1.3 \times 10^{-4}, \tag{5.65}$$

which agrees quite well with the experimental data (5.51). Thus, the weak interaction of the $V - A$ type is again strongly supported by pion decay.

Here, the vector part does not introduce the decay of the pion because it is pseudoscalar.

If the lepton combinations were S and P, the interaction Hamiltonian would be

$$H = g\phi_\pi(\bar{\psi}_l(1+\gamma_5)\psi_\nu) + \text{h.c.} \qquad (5.66)$$

In this case the branching ratio is essentially the same as the phase space ratio

$$R = \frac{(m_\pi^2 - m_e^2)^2(m_\pi^2 + m_e^2)}{(m_\pi^2 - m_\mu^2)^2(m_\pi^2 + m_\mu^2)} = 3.5. \qquad (5.67)$$

Such a possibility can be safely ruled out [28].

5.5 MUON DECAY

The muon and the electron have very similar properties except for their mass difference,

$$m_\mu = (206.768 \pm 0.003)m_e. \qquad (5.68)$$

In fact, these two particles are fermions of spin $\frac{1}{2}$. They have a unit charge of either positive or negative sign. The gyromagnetic ratios [29] are

$$g_e = 2 \times (1.001159622 \pm 0.000000027)$$

and (5.69)

$$g_\mu = 2 \times (1.001164 \pm 0.000003).$$

Here, the anomalies of gyromagnetic ratios are well explained by the quantum electrodynamic calculations with the sixth-order correction [30].

Since the electromagnetic interactions of muons and electrons are identical, it would be natural to assume that the weak interactions of these two particles are also the same. A partial support of this hypothesis is seen in the decay of pions into muons and electrons. In this section, we wish to describe the weak interaction between muons and electrons. Muons normally decay under emissions of one electron and two neutral particles. The kinematics of the decay is such that the two neutral particles both have very small masses or zero masses. As we shall learn in the two-neutrino experiments (see section 5.8), there are muon neutrinos ν_μ and electron neutrinos ν_e, so that muon decays can be written as

$$\mu^+ \to e^+ + \nu_e + \bar{\nu}_\mu \qquad \text{and} \qquad \mu^- \to e^- + \bar{\nu}_e + \nu_\mu. \qquad (5.70)$$

The lifetime for this decay is [29]

$$\tau_\mu = (2.199 \pm 0.001) \times 10^{-6} \text{ sec}. \qquad (5.71)$$

In the early days, it was asked whether these two neutral particles are identical or not. The correct answer for this is given by $\pi \to \mu \to e$ decay. In the rest system of the positive pion, the emitted positive muon is left-handed. We know from experiments that the positrons come out dominantly in the direction opposite to the muon momentum, and that they are right-handed. Furthermore, the main part of the momentum of the muon is carried away by two neutrinos, which are, therefore, emitted in the forward direction. Conservation of the angular momentum requires that these two neutrinos have different helicities. Consequently, these two particles are not identical, and are at least in different quantum states (see Fig. 5.7).

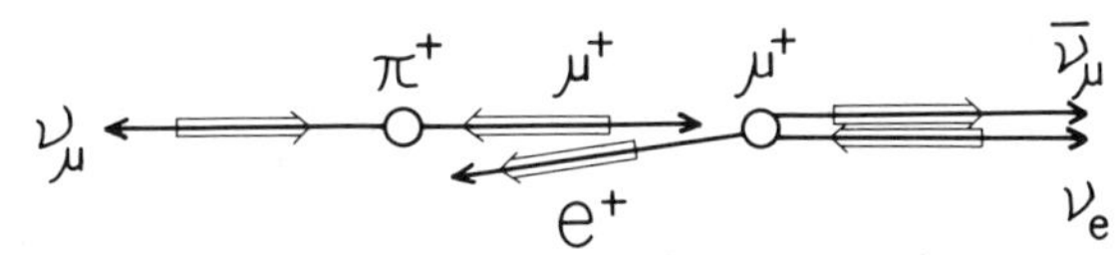

Fig. 5.7 $\pi \to \mu \to e$ **decay in the rest system of the pion. The positive muon is left-handed, since the neutrino is left-handed. In muon decay, the momentum of the muon is mostly transferred to the two neutrinos, so that these two particles are emitted forward. The positron is emitted preferentially in the backward direction. Since the neutrino spins are antiparallel to each other, the muon spin is transferred to the positron. This means that the positron is right-handed, as is confirmed in experiments.**

It thus seems reasonable to postulate, by analogy with beta decay, the four-fermion interactions of (5.70) with the Hamiltonian density

$$H = g_\mu 2^{-1/2}(\bar{\psi}_e \gamma_\alpha (1+\gamma_5)\psi_{\nu_e})(\bar{\psi}_{\nu_\mu}\gamma_\alpha(1+\gamma_5)\psi_\mu) + \text{h.c.} \tag{5.72}$$

Here, the $V - A$ coupling is assumed for the lepton covariants, since this coupling is supported in nuclear beta decay. It is interesting to note here that the most general quadrilinear covariant form

$$H = \sum_i (\bar{\psi}_e O_i (C_i + \gamma_5 C_i')\psi_{\nu_e})(\bar{\psi}_{\nu_\mu} O_i \psi_\mu) + \text{h.c.} \tag{5.73}$$

reduces to the form (5.72) by requiring that the charge exchange-type interaction of H in (5.73) be identical to the charge retention-type interaction of H in (5.74),

$$H = \sum_j (\bar{\psi}_e O_j \psi_\mu)(\bar{\psi}_{\nu_\mu} O_j (D_j + \gamma_5 D_j')\psi_{\nu_e}) + \text{h.c.} \tag{5.74}$$

In Eqs. (5.73) and (5.74), i and j run over S, V, T, A, and P. We will discard here another solution $S - A - P$, since it is excluded by the data on nuclear beta decays. The permutation of fields is called the Fierz rearrangement [31], which is left as an exercise for the reader.

Using the trace method for estimation of the transition rate introduced in Chapter 2 and neglecting the terms of m/E, we have the positron spectrum,

$$P(x) = \frac{m_\mu^5 g_\mu^2 c^4}{16\pi^3 \hbar^7} [(1-x) + \tfrac{2}{9}\rho(4x-3)]x^2 \qquad \text{with} \quad x = \frac{2p_e}{m_\mu c}. \tag{5.75}$$

In particular, we have

$$\rho = \tfrac{3}{4} \tag{5.76}$$

for the $V-A$ interaction, (5.72). For the more general interaction (5.73), ρ varies from 0 to 1 depending on ten coupling constants, and it is called the Michel parameter [32].

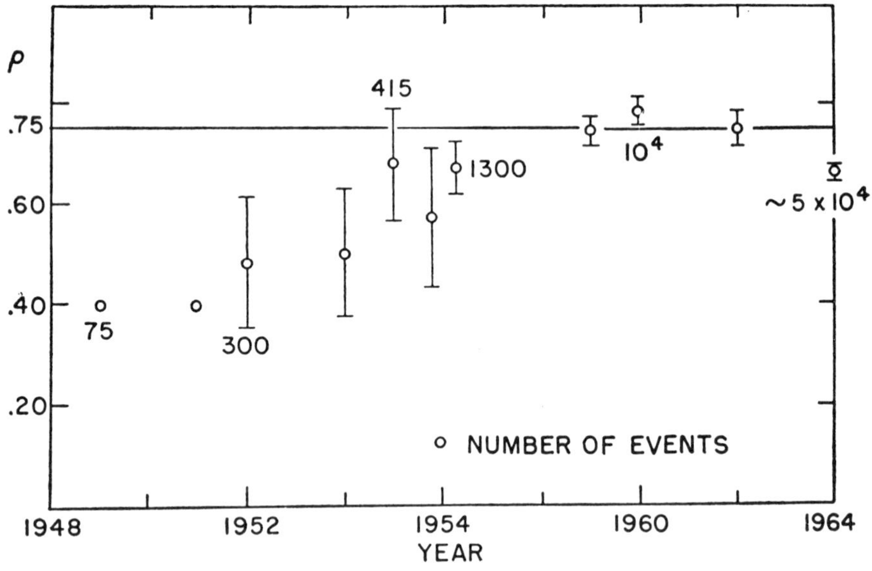

Fig. 5.8 Historical account of the measured ρ value in the positron spectrum, Eq. (5.75) [19].

Old data on the positron spectrum gave ρ values considerably lower than $\frac{3}{4}$. The historical account of the measured ρ value is best illustrated as in Fig. 5.8, by Lee and Wu [19]. Since 1959, the curve has finally approached the asymptotic value $\rho = 0.75$. The theoretical spectrum for different ρ values is given in [32], and the experimental results measured by the Columbia group [33] are shown in Fig. 5.9. These data have a best fit with $\rho = 0.747 \pm 0.005$. The curve is calculated for $\rho = \frac{3}{4}$, taking into account the instrumental resolution. This is again in favor of the $V-A$ interaction.

The mean life of the muon can be calculated from Eq. (5.75) by integrating over the positron energy,

$$\tau_\mu = \frac{192\pi^3\hbar^7}{g_\mu^2 m_\mu^5 c^4}. \tag{5.77}$$

The radiative correction for muon decay can be unambiguously calculated by

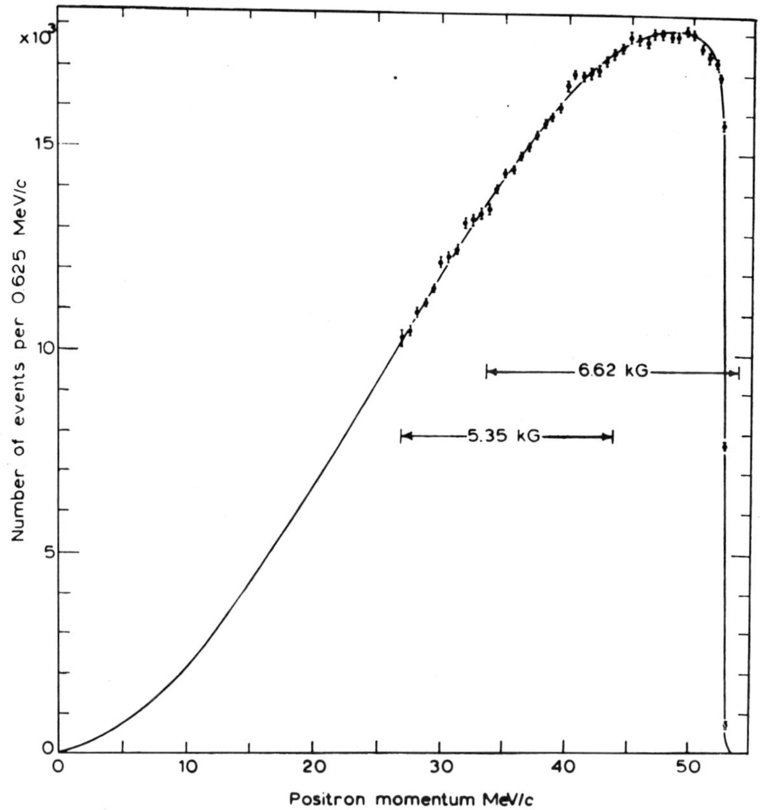

Fig. 5.9 Positron spectrum in muon decay [33]. The curve is calculated for $\rho = \frac{3}{4}$, taking into account the instrumental resolution.

using the $V - A$ theory. And the lifetime of the muon, including lowest-order radiative corrections [34, 35], is given by

$$\tau_\mu = \frac{192\pi^3\hbar^7}{g_\mu^2 m_\mu^5 c^4}\left[1 + \frac{\alpha}{2\pi}\left(\pi^2 - \frac{25}{4}\right)\right]. \tag{5.78}$$

Using recent experimental data (5.68) and (5.71) on muon mass and lifetime, we have

$$g_\mu = (1.435 \pm 0.001) \times 10^{-49} \text{ erg cm}^3. \tag{5.79}$$

As is shown in Fig. 5.7, the positive muons produced in the positive pion decay are longitudinally polarized, and positrons are emitted preferentially in the direction opposite to the muon momentum. This should be expected if parity conservation is violated in muon decay. By observing this asymmetry, Garwin, Lederman, and Weinrich [36] showed, first, that space reflection invariance does indeed break down in pion and muon decays. In both cases the positive and negative muon decays, the positrons and electrons come out preferentially in the backward direction with respect to the muon momentum, as observed by Friedman and Telegdi [37].

The angular distribution of electrons from polarized muons at rest is given by

$$W(x,\theta)=\frac{1}{\tau_\mu}\frac{1}{4\pi}12x^2\{(1-x)+\tfrac{2}{9}\rho(4x-3)$$
$$\mp\tfrac{1}{3}\xi P\cos\theta[(1-x)+\tfrac{2}{3}\delta(4x-3)]\}. \qquad (5.80)$$

Here the upper (lower) sign refers to the electron (positron). P is the degree of muon polarization, θ is the emission angle of electron with respect to the muon polarization, and ξ and δ are again functions of the coupling constants appearing in the Hamiltonian (5.73). For example,

$$\xi=\frac{2\,\mathrm{Re}(C_V^*C_A)}{|C_V|^2+|C_A|^2} \qquad (5.81)$$

for the VA interaction. The two-component $V-A$ theory (5.72) gives us

$$\rho=\tfrac{3}{4},\qquad \xi=-1,\qquad \text{and}\qquad \delta=\tfrac{3}{4}. \qquad (5.82\text{a})$$

The signs of ξ and δ are not determined experimentally, without knowing the helicities of the muons. The helicity of the negative muons was, however, shown to be positive [38].

The latest values of the parameters appearing in Eq. (5.80) are summarized as follows [39].

$$\rho=0.752\pm0.003,\quad \xi=-(0.973\pm0.014),\quad \text{and}$$
$$\delta=0.752\pm0.009. \qquad (5.82\text{b})$$

5.6 NONLEPTONIC DECAYS OF HYPERONS AND THE $|\Delta T| = \frac{1}{2}$ RULE

There is a group of decay processes which are caused by weak interactions, but which include no leptons. These processes are called the nonleptonic decays, and are classified in three categories:

$$\text{I.}\quad \Lambda\to N+\pi,\quad \Sigma\to N+\pi,\quad \Xi\to\Lambda+\pi;$$
$$\text{II.}\quad K\to\pi+\pi,\quad K\to\pi+\pi+\pi;$$
$$\text{III.}\quad \Omega^-\to\Xi+\pi,\quad \Omega^-\to\Lambda+K. \qquad (5.83)$$

A common characteristic of these three processes is parity nonconservation. In fact, category II was once called the θ–τ puzzle, which was the motivation for the discovery of parity nonconservation. These processes are also related to the *CP* violation, which is described in the next section. (Violation of the invariance of the physical law under the product transformation of space inversion and charge conjugation is abbreviated to *CP* violation.) The processes in the category III are the decays of one member of the spin $\frac{3}{2}$ baryon decuplet into a spin $\frac{1}{2}$ baryon octet and a pseudoscalar meson octet. These processes are not often found. We will concentrate on category I in this section.

The processes I are decays of one member of the spin $\frac{1}{2}$ baryon octet into the other member of the same octet plus a meson octet. Before further discussion, we introduce some terminology that may not be familiar.

Classification of Elementary Particles

There are four classes of elementary particles.

Baryons: heavy fermions, nucleons, and hyperons, etc.
Mesons: pions, kaons, etc.
Photons: photons only.
Leptons: light fermions, muons, electrons, neutrinos.

Baryons and mesons interact strongly with each other, so that they are called hadrons. Photons and leptons have no strong interactions, and these classes have very few members. On the other hand, there are about 200 hadrons and their number continues to grow as the energy of particle accelerators increases. To enable us to understand hadrons systematically, they are arranged in the multiplets. The members of a multiplet have almost the same masses, but different quantum numbers, except for spin and parity. For example, there are eight baryons of spin and parity $\frac{1}{2}^{+}$: p, n, Λ, Σ^+, Σ^0, Σ^-, Ξ^0, and Ξ^-. To identify each particle, we need to specify the isospin T, its z component T_z, and strangeness S (or equivalently, hypercharge Y). The charge of hadrons is given by

$$Q = T_z + \tfrac{1}{2}(S + B) = T_z + \tfrac{1}{2}Y \tag{5.84}$$

with

$$Y = S + B \tag{5.85}$$

where B is the baryon number, which is 1, -1, and 0 for baryons, antibaryons, and the other particles, respectively. The isospin is the angular momentum in the hypothetical three-dimensional charge space, and is explained in Appendix A.8. The strangeness S (or hypercharge Y) was introduced by Nakano and

Nishijima [40], and by Gell-Mann [41] in connection with the associated production of baryons and kaons, phenomenologically. This quantum number can distinguish the nucleons N, Λ and Σ particles, and Ξ particles. Similarly it distinguishes pions and kaons. As is shown in Table 5.2, three independent quantum numbers are necessary to identify the member of the multiplet. To do this, we use the group of three-dimensional unitary unimodular matrices.

Table 5.2 Quantum Numbers for the $\frac{1}{2}^+$ Baryon Octet and the 0^- Meson Octet[a]

Baryons		Quantum Numbers				Mesons	
Name	Member	Q	T	T_z	Y	Name	Member
N	p	1	$\frac{1}{2}$	$\frac{1}{2}$	1	K	K^+
	n	0	$\frac{1}{2}$	$-\frac{1}{2}$	1		K^0
Λ	Λ^0	0	0	0	0	η	η^0
Σ	Σ^+	1	1	1	0	π	π^+
	Σ^0	0	1	0	0		π^0
	Σ^-	-1	1	-1	0		π^-
Ξ	Ξ^0	0	$\frac{1}{2}$	$\frac{1}{2}$	-1	$\bar{K}$	$\bar{K}^0$
	Ξ^-	-1	$\frac{1}{2}$	$-\frac{1}{2}$	-1		K^-

[a] The charge Q is given by $Q = T_Z + \frac{1}{2}Y$. The strangeness S is given by $S = Y - B$, B being 1 and 0 for baryons and mesons, respectively.

These matrices transform the basic spinors of the three dimensions. There are three commuting generators whose eigenvalues correspond to T, T_z, and Y, respectively. The irreducible representations of this group correspond to the multiplets of the elementary particles. There are two fundamental representations which are three dimensional. The relevant multiplets are called quarks and antiquarks. The charges of these particles (if we can call them so) are fractional, and the existence of quarks is still open to question, although many attempts to detect them are going on. (The ordinary angular momentum refers to the irreducible representation of the su_2 group, which is the group of two-dimensional unitary unimodular matrices. From this group, only two quantum numbers can be derived, and they are the angular momentum J and its z component J_z. The basic representation is two dimensional, and the eigenfunctions are the two-dimensional spinors. The situation is quite analogous to the case of su_3.) Now the lowest irreducible representation which has integral charges is eight dimensional and is called the octet. Typical examples are the $\frac{1}{2}^+$ baryon octet and the pseudoscalar (0^-) meson octet, whose members are K^+, K^0, η^0, π^+, π^0, π^-, $\bar{K}^0$, K^- (see Table 5.2). The next lowest representation is the decuplet. The famous example of this multiplet is the $\frac{3}{2}^+$ baryon decuplet. Nine of ten members were known previously, and the last member was named Ω^-. It should be an isosinglet with $Y = -2$. The theoretical prediction for Ω^- by Gell-Mann and Okubo [42] based on su_3 and the subsequent finding [43] of this particle in an experiment are the basic support for the su_3 group. We cannot discuss this particular algebra further here, but recommend reading one of the basic books for courses in advanced quantum mechanics [86].

Parity Nonconservation in Process I

There are seven different modes of the nonleptonic decay of the $\frac{1}{2}^+$ baryons. These decay modes and their amplitudes are listed below.

Decay	Amplitude
$\Lambda \to p + \pi^-,$	Λ^0_-
$\Lambda \to n + \pi^0,$	Λ^0_0
$\Sigma^+ \to n + \pi^+,$	Σ^+_+
$\Sigma^+ \to p + \pi^0,$	Σ^+_0
$\Sigma^- \to n + \pi^-,$	Σ^-_-
$\Xi^- \to \Lambda + \pi^-,$	Ξ^-_-
$\Xi^0 \to \Lambda + \pi^0.$	Ξ^0_0

(5.86)

Here, the superscripts and subscripts for the amplitudes refer to the charges of the initial baryon and the final pion, respectively. Though not used in this subsection, these symbols are used later. The following discussion of the decay process

$$\Lambda \to p + \pi^- \tag{5.87}$$

is, however, equally valid for all the modes in (5.86).

Production of the Λ particle is called associated production according to the reaction

$$\pi^- + p \to \Lambda + K^0. \tag{5.88}$$

In the center of mass system of the initial state, π^- has a momentum $\mathbf{k}_\pi$, the produced Λ has momentum $\mathbf{k}_\Lambda$, and π^- from the decay of Λ has momentum $\mathbf{p}_\pi$. The momenta of all four particles in Eq. (5.88) are in a plane, and the normal to this plane is defined by $\mathbf{k}_\pi \times \mathbf{k}_\Lambda$. Since the associated production is due to the strong interaction, the spin $\boldsymbol{\sigma}_\Lambda$ of Λ is polarized in the direction of this normal. If parity is not conserved in the decay process, (5.87), the expression of the transition probability includes a term which is proportional to $\boldsymbol{\sigma}_\Lambda \cdot \mathbf{p}_\pi$ or equivalently $(\mathbf{k}_\pi \times \mathbf{k}_\Lambda) \cdot \mathbf{p}_\pi$. That is, the intensity of π^- which goes up to the production plane of Λ is different from the intensity of π^- which goes down to this plane. The measurement of this asymmetry was proposed by Lee and Yang [44]. The first successful experiment was performed by Crawford *et al.* [45], and in fact, parity nonconservation was found in this process.

The asymmetric distribution of π^- can be understood as the result of interference between the *s*- and *p*-wave amplitudes, as follows. Let us denote the angle θ which refers to the direction of pion momentum with respect to the spin of Λ. Since the spin of Λ is $\frac{1}{2}$, there are two possible final nucleon–pion states. The spin-angular parts of the wave functions for these states are

$$\psi_{s_{1/2}} = Y_{0,0}(\theta,\phi)\chi_{1/2} = \left(\frac{1}{4\pi}\right)^{1/2}\chi_{1/2}, \tag{5.89}$$

$$\begin{aligned}\psi_{p_{1/2}} &= -(\tfrac{1}{3})^{1/2}Y_{1,0}(\theta,\phi)\chi_{1/2} + (\tfrac{2}{3})^{1/2}Y_{1,1}(\theta,\phi)\chi_{-1/2}\\ &= -\left(\frac{1}{4\pi}\right)^{1/2}[\chi_{1/2}\cos\theta + \chi_{-1/2}\sin\theta\exp(i\phi)],\end{aligned} \tag{5.90}$$

where χ_μ is the spin function of the nucleon, and $Y_{l,m}(\theta,\phi)$ is the angular part of the pion wave function. The transition amplitude for the decay (5.87) is given by

$$\begin{aligned}\langle N\pi|H_w|\Lambda\rangle &\propto S\psi_{s_{1/2}} + P\psi_{p_{1/2}}\\ &= \left(\frac{1}{4\pi}\right)^{1/2}[S\chi_{1/2} - P(\chi_{1/2}\cos\theta + \chi_{-1/2}\sin\theta\exp(i\phi))].\end{aligned} \tag{5.91}$$

Here, H_w refers to the weak interaction and S and P are transition amplitudes for the *s* and *p* waves.

The angular distribution of π is obtained from the square of Eq. (5.91),

$$\begin{aligned}|\langle N\pi|H_w|\Lambda\rangle|^2 &\propto |S - P\cos\theta|^2 + |P\sin\theta\exp(i\phi)|^2\\ &= |S|^2 - 2\,\mathrm{Re}(S^*P)\cos\theta + |P|^2\\ &= (|S|^2 + |P|^2)\left[1 - \frac{2\,\mathrm{Re}(S^*P)}{|S|^2+|P|^2}\cos\theta\right].\end{aligned} \tag{5.92}$$

The coefficient of the $\cos\theta$ term is called the asymmetry parameter. If the polarization of Λ is not 100%, the degree of polarization should be multiplied by this term.

The amplitude S (P) connects the initial and final states, both of which have the different (or same) parities. The existence of the nonzero asymmetry indicates the coexistence of the $s_{1/2}$ and $p_{1/2}$ states, and it is the sign of the parity nonconservation. These amplitudes are complex numbers. Since, however, a common phase factor is physically meaningless, we have three independent parameters to specify these amplitudes. (If we do not discuss the decay rate, we need only two parameters to specify the angular distribution.) These are the three commonly used parameters α, β, and γ, and they are called the decay parameters for the nonleptonic decay of hyperons:

$$\alpha = \frac{2\ \mathrm{Re}(S^*P)}{|S|^2 + |P|^2}, \tag{5.93}$$

$$\beta = \frac{2\ \mathrm{Im}(S^*P)}{|S|^2 + |P|^2}, \tag{5.94}$$

$$\gamma = \frac{|S|^2 - |P|^2}{|S|^2 + |P|^2}, \tag{5.95}$$

with a condition

$$\alpha^2 + \beta^2 + \gamma^2 = 1. \tag{5.96}$$

Here α is the asymmetry parameter, as before.

The precision experiment on α can be performed not by measuring the angular distribution of pion directly, but by measuring the longitudinal polarization of the nucleon which comes out to the spin direction of Λ. The principle of such an experment is as follows. Let us consider the case $\theta = 0°$ in Eq. (5.92). In this particular geometry, the protons go out antiparallel to the spin direction of Λ, with relative intensity $1 - \alpha$. From the conservation of momentum and that of the magnetic quantum number, the linear and angular momenta of the proton are antiparallel. That is, the proton is in the helicity -1 state. If $\theta = 180°$, the relative intensity of the proton is $1 + \alpha$, and it is in the helicity $+1$ state (see Fig. 5.10). As a consequence, the longitudinal polarization is given by

$$P_L = \frac{(1+\alpha) - (1-\alpha)}{(1+\alpha) + (1-\alpha)} = \alpha, \tag{5.97}$$

which is nothing but the asymmetry parameter α.

The parameter β is known by measuring the term $\boldsymbol{\sigma}_p \cdot (\boldsymbol{\sigma}_\Lambda \times \mathbf{p}_p)$, $\boldsymbol{\sigma}_p$ and $\mathbf{p}_p$ being the spin and momentum of the decay proton, respectively. This is nonzero only if the time reversal invariance is violated. (We should, of course, correct the phase due to the final state interaction.) This can be done by measuring the polarization of proton, with respect to the plane made of the spin of Λ and the proton momentum.

The third quantity γ is given by measuring the polarization of the proton in the x–y plane if the z axis is chosen to be $\boldsymbol{\sigma}_\Lambda$. In the case of the pure s wave, the proton spin is up in this plane (as is seen in Eq. (5.91) with $\theta = 90°$), and $\gamma > 0$. Similarly, in the case of the pure p wave, the proton spin is down, and $\gamma < 0$. Measurement of γ will be difficult if the amplitudes S and P are almost equal. This is, in fact, the case of $\Sigma^+ \to p + \pi^0$.

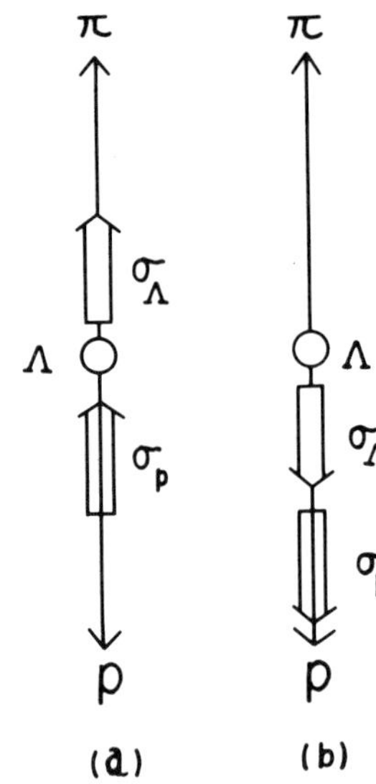

Fig. 5.10 Angular distribution of pion and proton with respect to the spin of the lambda particle σ_Λ and the helicity of the proton. In Fig. 5.10a, the pion is emitted in the direction $\theta = 0°$, and the proton is emitted antiparallel to the pion momentum, so that the proton is in the helicity -1 state with relative intensity $1 - \alpha$. In Fig. 5.10b, the pion is emitted in the direction $\theta = 180°$ and the proton is in the helicity $+1$ state with the relative intensity $1 + \alpha$.

$|\Delta T| = \frac{1}{2}$ Rule

From the Nakano-Nishijima-Gell-Mann rule, the charge of the elementary particle is given by

$$Q = T_z + \tfrac{1}{2}(S + B). \tag{5.98}$$

On the other hand, the charge and baryon number are conserved in the elementary process, so that

$$\Delta Q = \Delta B = 0, \tag{5.99}$$

where Δ means the difference of the relevant quantity in the initial and final states. In the nonleptonic process, the change of the strangeness is

$$\Delta S = \pm 1, \tag{5.100}$$

and taking Δ of Eq. (5.98), we have

$$\Delta Q = \Delta T_z + \tfrac{1}{2}(\Delta B + \Delta S), \tag{5.101}$$

so that we obtain

$$\Delta T_z = \pm\tfrac{1}{2} \qquad \text{or} \qquad |\Delta T_z| = \tfrac{1}{2}, \tag{5.102}$$

from Eqs. (5.99)–(5.101). This relation means that there are the following possible values for the change of the isospin T.

$$|\Delta T| = \tfrac{1}{2}, \tfrac{3}{2}, \tfrac{5}{2}, \cdots. \tag{5.103}$$

The $|\Delta T| = \frac{1}{2}$ rule states that only the lowest value in Eq. (5.103) is allowed in the nonleptonic processes. Some of the theoretical results from this rule and the experimental status will be given as follows.

As given in (5.86), there are seven modes of the nonleptonic decay of the hyperons. For the symbols of decay amplitudes, see also Eq. (5.86). To specify the final state, we use the symbol, e.g., $|p\pi^-\rangle$, for the wave function of the proton and meson of minus charge, etc. In this subsection, we also use $|T, T_z\rangle$ in place of the isospin wave function ψ_{T, T_z}. An example of the results from the $|\Delta T| = \frac{1}{2}$ is related to the decay of Λ particle.

Example 1

For the process $\Lambda \to N + \pi$, there are two final states,

$$|p\pi^-\rangle = |\tfrac{1}{2}, \tfrac{1}{2}\rangle|1, -1\rangle = (\tfrac{1}{3})^{1/2}|\tfrac{3}{2}, -\tfrac{1}{2}\rangle - (\tfrac{2}{3})^{1/2}|\tfrac{1}{2}, -\tfrac{1}{2}\rangle, \tag{5.104}$$

and

$$|n\pi^0\rangle = |\tfrac{1}{2}, -\tfrac{1}{2}\rangle|1, 0\rangle = (\tfrac{2}{3})^{1/2}|\tfrac{3}{2}, -\tfrac{1}{2}\rangle + (\tfrac{1}{3})^{1/2}|\tfrac{1}{2}, -\tfrac{1}{2}\rangle. \tag{5.105}$$

Since Λ has $T = 0$, we have the decay amplitude Λ^0_- from Eq. (5.104)

$$\Lambda^0_- = \langle p\pi^-|H_w|\Lambda\rangle = -(\tfrac{2}{3})^{1/2}A_1 + (\tfrac{1}{3})^{1/2}A_3, \tag{5.106}$$

where A_1 and A_3 refer to the partial amplitudes for $\Delta T = \frac{1}{2}$ and $\frac{3}{2}$, respectively. Similarly, from Eq. (5.105) we have

$$\Lambda^0_0 = \langle n\pi^0|H_w|\Lambda\rangle = (\tfrac{1}{3})^{1/2}A_1 + (\tfrac{2}{3})^{1/2}A_3. \tag{5.107}$$

If we assume

$$|\Delta T| = \tfrac{1}{2} \qquad \text{and therefore} \qquad A_3 = 0,$$

we obtain the following relation from Eqs. (5.106) and (5.107).

$$\Lambda^0_- = -2^{1/2}\Lambda^0_0. \tag{5.108}$$

This holds for the s- as well as p-wave amplitudes, so that the decay modes

$$\Lambda \to p + \pi^- \qquad \text{and} \qquad \Lambda \to n + \pi^0$$

have the same values for α, β, γ, and the decay rate for the former process is twice as large as for the latter.

Example 2

The respective isospin functions of the final states for the decay modes $\Sigma^+ \to n + \pi^+$, $\Sigma^+ \to p + \pi^0$, and $\Sigma^- \to n + \pi^-$ are

$$
\begin{aligned}
|\tfrac{1}{2}, -\tfrac{1}{2}\rangle|1, 1\rangle &= (\tfrac{1}{3})^{1/2}|\tfrac{3}{2}, \tfrac{1}{2}\rangle + (\tfrac{2}{3})^{1/2}|\tfrac{1}{2}, \tfrac{1}{2}\rangle, \\
|\tfrac{1}{2}, \tfrac{1}{2}\rangle|1, 0\rangle &= (\tfrac{2}{3})^{1/2}|\tfrac{3}{2}, \tfrac{1}{2}\rangle - (\tfrac{1}{3})^{1/2}|\tfrac{1}{2}, \tfrac{1}{2}\rangle, \\
|\tfrac{1}{2}, -\tfrac{1}{2}\rangle|1, -1\rangle &= |\tfrac{3}{2}, -\tfrac{3}{2}\rangle.
\end{aligned}
\tag{5.109}
$$

Let us assume $|\Delta T| = \frac{1}{2}$ rule, and describe A_1 and A_3 for the amplitudes which lead the final states to be isospin $\frac{1}{2}$ and $\frac{3}{2}$, respectively. From Eqs. (5.109) and (5.111) we have the following decay amplitudes.

$$
\Sigma_+^+ = \tfrac{2}{3}A_1 + \tfrac{1}{3}A_3, \qquad \Sigma_0^+ = (2^{1/2}/3)(-A_1 + A_3), \qquad \Sigma_-^- = A_3. \tag{5.110}
$$

This can be derived by using a spurion s of isospin $\frac{1}{2}$ for the initial states. The hypothetical particlelike object called the spurion refers to a mathematical tool that carries the isospin only in the following way.

$$
\begin{aligned}
|\Sigma^+ s\rangle &= |1, 1\rangle|\tfrac{1}{2}, -\tfrac{1}{2}\rangle = (\tfrac{1}{3})^{1/2}|\tfrac{3}{2}, \tfrac{1}{2}\rangle + (\tfrac{2}{3})^{1/2}|\tfrac{1}{2}, \tfrac{1}{2}\rangle, \\
|\Sigma^- s\rangle &= |1, -1\rangle|\tfrac{1}{2}, -\tfrac{1}{2}\rangle = |\tfrac{3}{2}, -\tfrac{3}{2}\rangle.
\end{aligned}
\tag{5.111}
$$

From Eq. (5.110), we can find the relation

$$
\Sigma_-^- - \Sigma_+^+ = \tfrac{2}{3}(A_3 - A_1) = 2^{1/2}\Sigma_0^+. \tag{5.112}
$$

If we consider these amplitudes as the two-dimensional vectors whose components are the s- and p-wave amplitudes, Eq. (5.112) indicates that the three vectors Σ_-^-, Σ_+^+, and $2^{1/2}\Sigma_0^+$ make a triangle.

Another relation

$$
\Xi_-^- = 2^{1/2}\Xi_0^0 \tag{5.113}
$$

is obtained for the decay modes $\Xi^- \to \Lambda + \pi^-$ and $\Xi^0 \to \Lambda + \pi^0$, in a similar way.

Some of the experimental data are given in comparison with theoretical predictions as follows [46].

$$
\begin{aligned}
& & & |\Delta T| = \tfrac{1}{2} \text{ rule} \\
&\frac{\alpha(\Lambda_0^0)}{\alpha(\Lambda_-^0)} = 1.10 \pm 0.27, & & 1 \\
&\frac{\gamma(\Lambda_0^0)}{\gamma(\Lambda_-^0)} = 1.04 \begin{array}{l} +0.33 \\ -0.21 \end{array} & & 1 \\
&\frac{W(\Lambda_-^0)}{W(\Lambda_-^0) + W(\Lambda_0^0)} = 0.640 \pm 0.014, & & \tfrac{2}{3} \\
&\frac{\alpha(\Xi_0^0)}{\alpha(\Xi_-^-)} = 0.82 \pm 0.19, & & 1 \\
&\frac{W(\Xi_0^0)}{W(\Xi_-^-)} = 0.548 \pm 0.036. & & \tfrac{1}{2}
\end{aligned}
\tag{5.114}
$$

Here W's are the decay rates. The experimental values are in good agreement with the $|\Delta T| = \frac{1}{2}$ rule. The experimental status of relation (5.112) in a two-dimensional vector diagram is shown in Fig. 5.11. Σ_+^+ is mostly p wave, while Σ_-^- is mostly s wave. Σ_0^+ is an s- and p-wave mixture of almost the same amount. The edge of the vector of Σ_0^+ coincides approximately with that of Σ_-^-. (As a matter of fact, two cases for Σ_-^- are plotted, since the sign of γ is undetermined.) The areas encircled in dotted lines indicate the experimental uncertainty. In these connections, we conclude that the $|\Delta T| = \frac{1}{2}$ rule is in

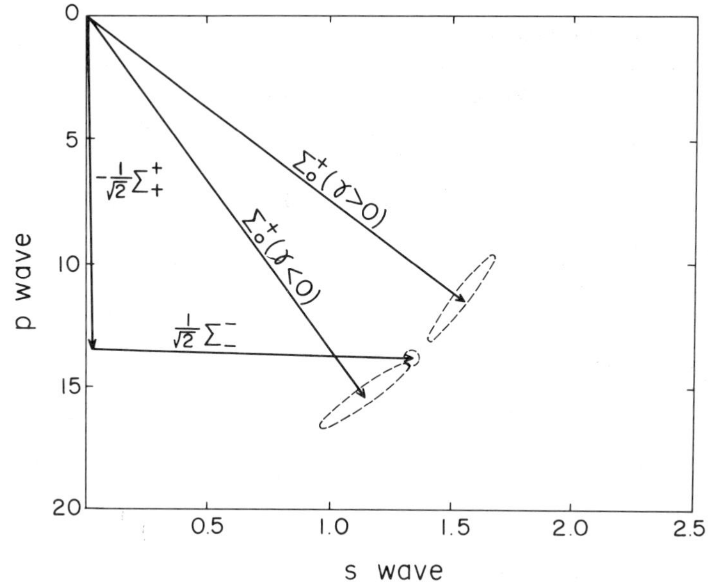

Fig. 5.11 **Triangular relation of the decay amplitudes for Σ [46] (see Eq. (5.112)). Abscissa and ordinate indicate the s- and p-wave amplitudes, respectively, in units of 10^5 sec$^{-1/2}$. Since the sign of γ is undetermined, two cases are drawn. Dotted lines enclose experimental errors.**

good agreement with experimental data, while the experiment, e.g., the last line of Eq. (5.114), does not necessarily confirm the $|\Delta T| = \frac{1}{2}$ rule, since we have the same result from

$$\Xi_-^- = -2^{1/2}\Xi_0^0 \qquad (5.115)$$

instead of Eq. (5.113).

5.7 *CP* VIOLATION IN $K_L^0 \to 2\pi$

In the last section, we mentioned category II of the nonleptonic decays, which includes the decay mode called the θ–τ puzzle. We know this puzzle was the motivation for the discovery of parity nonconservation. Recently, another important finding was reported in this category. Christenson, Cronin, Fitch, and Turlay [47] reported that K_L^0 decays to two pions with a branching ratio

$$R = \frac{K_L^0 \to \pi^+ + \pi^-}{K_L^0 \to \text{all charged modes}} = (2.0 \pm 0.4) \times 10^{-3}. \tag{5.116}$$

(Abashian *et al.* also reported a similar result, simultaneously, as is shown in Table 5.3.) In this measurement, K_L^0 mesons with mean momentum 1.1 BeV/c

Table 5.3 Data on the Branching Ratio of $K_L^0 \to \pi^+ + \pi^-$ [a]

Experiment	Medium	Momentum (BeV/c)	γ^2 [b]	Distance from Target[c]	$\dfrac{K_L^0 \to \pi^+ + \pi^-}{K_L^0 \to \text{all charged modes}} \times 10^3$
Princeton (1964)[d]	He	1.1	5.9	0.55	2.0 ± 0.4
Illinois (1964)[e]	Vacuum	1.5	10	0.38	2–3
CERN (1965)[f]	Vacuum	10.7	463	0.15	3.5 ± 1.4
Rutherford (1965)[g]	Vacuum	3.25	44	0.39	2.08 ± 0.35
CERN (1966) I[h]	Vacuum	4.8	93	0.27	1.78 ± 0.18
CERN (1966) II[i]	Vacuum Carbon	4.5	82	0.29	1.93 ± 0.15 1.993 ± 0.080
CERN (1967)[j]	Vacuum	10.7	463	0.16	2.12 ± 0.18
Princeton (1967)[k]	Vacuum	1.55	10.6	0.49	1.97 ± 0.16

[a] This table was prepared by Y. Yokoo.
[b] The Lorentz factor $\gamma^2 = [1 - (v/c)^2]^{-1}$.
[c] Measured in units of the lifetime of K_L^0 meson (5.3×10^{-8} sec).
[d] J. H. Christenson *et al., Phys. Rev. Letters* **13**, 138 (1964).
[e] A. Abashian *et al., Phys. Rev. Letters* **13**, 243 (1964).
[f] X. de Bouard *et al., Phys. Letters* **15**, 58 (1965).
[g] W. Galbraith *et al., Phys. Rev. Letters* **14**, 383 (1965).
[h] M. Bott-Bodenhausen *et al., Phys. Letters* **20**, 212 (1966).
[i] M. Bott-Bodenhausen *et al., Phys. Letters* **23**, 277 (1966).
[j] X. de Bouard *et al., Nuovo Cimento* **52**, 662 (1967).
[k] V. L. Fitch *et al., Phys. Rev.* **164**, 1711 (1967).

were produced at the Brookhaven alternating gradient synchrotron in an internal Be target bombarded by 30-BeV protons. The detector was located at a distance 57 ft from the internal target to avoid possible contamination of K_S^0. The existence of the mode

$$K_L^0 \to \pi^+ + \pi^- \tag{5.117}$$

would imply violation of the invariance of the physical law under the product transformation of space inversion and charge conjugation. This is usually abbreviated to *CP* violation, and is seen in the following [48].

First, the antiparticle state $|\bar{K}^0\rangle$ of the neutral kaon state $|K^0\rangle$ is usually defined by

$$|\bar{K}^0\rangle = CP|K^0\rangle. \tag{5.118}$$

On the other hand, the decaying kaons K_S^0 and K_L^0 are known as the states of the linear combination of K^0 and $\bar{K}^0$,

$$|K_S^0\rangle = (\tfrac{1}{2})^{1/2}[|K^0\rangle + |\bar{K}^0\rangle] \quad \text{and} \quad |K_L^0\rangle = (\tfrac{1}{2})^{1/2}[|K^0\rangle - |\bar{K}^0\rangle]. \tag{5.119}$$

Hence, we have

$$CP|K_S^0\rangle = |K_S^0\rangle \qquad \text{and} \qquad CP|K_L^0\rangle = -|K_L^0\rangle. \tag{5.120}$$

That is, K_S^0 and K_L^0 are the eigenstates of *CP* and have eigenvalues +1 and −1, respectively. The typical nonleptonic decay modes for these particles are

$$\begin{aligned} K_S^0 &\to \pi^+ + \pi^- \quad \text{or} \quad \pi^0 + \pi^0, \\ K_L^0 &\to \pi^+ + \pi^- + \pi^0 \quad \text{or} \quad \pi^0 + \pi^0 + \pi^0. \end{aligned} \tag{5.121}$$

They have some leptonic decay modes, too. The lifetimes are approximately 1×10^{-10} sec and 5×10^{-8} sec for K_S^0 and K_L^0, respectively. The eigenvalues for *CP* in the final states are obtained by considering the states of the neutral pions. π^0 is pseudoscalar for *P* and even for *C*, so the π^0 has the eigenvalue −1 for *CP*. Therefore, the 2π and 3π states of zero total angular momentum are the eigenstates with the eigenvalues $+1[= (-1)^2]$ and $-1[= (-1)^3]$ for *CP*. This means that the *CP* invariance holds in the decay modes (5.121), while there is *CP* violation in the decay mode (5.117). The experiments for this mode have been repeatedly performed by several groups (see Table 5.3). The branching ratio (5.116) is almost constant for different momenta of the incoming K_L^0 (see Table 5.3), e.g.,

$$R = (2.12 \pm 0.18) \times 10^{-3} \qquad \text{at } 10.7 \text{ BeV}/c. \tag{5.122}$$

In general, the ratio of the decay amplitudes of K_L^0 and K_S^0 is expressed by two real parameters η and θ,

$$\frac{\text{amp}(K_L^0 \to \pi^+ + \pi^-)}{\text{amp}(K_S^0 \to \pi^+ + \pi^-)} = \eta e^{i\theta}. \tag{5.123}$$

The experimental data (5.116) and (5.122) correspond to

$$|\eta| \approx 2 \times 10^{-3}. \tag{5.124}$$

The phase factor θ can be measured by the time dependence of the production rate of 2π mode, which has an interference effect, because K_S and K_L decay into the same mode. The latest datum on θ is [39]

$$\theta \approx 45°. \tag{5.125}$$

For the decay mode $K_L^0 \rightarrow 2\pi^0$, the experimental value of η is

$$|\eta_{00}| = 2 \times 10^{-3}. \tag{5.126}$$

Here η is defined similarly to Eq. (5.123). In this case, θ is still unknown.

There are a number of theoretical attempts to explain the experimental data (5.124)-(5.126). Here we only try to classify these ideas in several groups. They are divided into two classes: (A) theories in the framework of *CP* invariance, and (B) theories based on the assumption of *CP* violation. Some of the typical ideas follow.

A1. Possible Existence of New Vector Fields. The theory assumes a vector field which produces an energy difference for the K^0 and $\bar{K}^0$ states. For this theory, the amplitude ratio η in Eq. (5.123) depends strongly on energy, in contradiction to the experiments.

A2. Two-Step Process. This theory assumes the existence of the hyperphoton, and the decay is two-step transition,

$$K_L^0 \rightarrow K_S^0 + \text{hyper } \gamma.$$
$$\qquad \hookrightarrow 2\pi$$

Here the hyperphoton is a particle, like a photon, that is obtained from a quantization of a vector field.

A3. Violation of Exponential Law. Exponential law $\exp(-\gamma t)$ for decay is assumed to be violated at larger t, so that a small amount of K_S^0 still remains at the counter area in experiments on $K_L^0 \rightarrow 2\pi$.

A4. Particle Mixture. This theory introduces an additional K' meson, which is neutral. Its mass is equal to that of K_L^0. It is assumed to be an even state for *CP*, and its lifetime is about 10^{-8} sec.

Since interference phenomena were revealed, the four kinds of hypotheses just described have been discarded, because A1–A3 do not introduce interference. A4 gives interference, but with a reversed sign from that in experiments. There have been many ambitious attempts in the area of class A, but none were completely successful. In this connection, *CP* violation seems to be established. Theories based on the assumption of *CP* violation are classified as follows.

B1. Superweak Interaction. This theory assumes the possible existence of an interaction with a coupling constant $10^{-3} \times G^2$, where G is given in Eq. (2.266), with $\Delta S = 2$. K^0 and $\bar{K}^0$ can transform mutually through this interaction.

B2. Miliweak Interaction. This theory assumes the possible existence of an interaction with a coupling constant $10^{-3} \times G$.

B3. Semiweak Interaction. The theory assumes the possible existence of an interaction with a coupling constant $g = G^{1/2}$, and it is a *CP*-odd interaction. In this theory, the first-order interaction is forbidden by a proper condition. The second-order interaction is assumed to be the usual *CP*-even weak interaction, and the third-order one is the *CP*-odd interaction, which is responsible for $K_L^0 \to 2\pi$.

Of all these theories, Wolfenstein's theory [49] explains the experimental evidence most successfully. It should be noticed here that the *CP* violation also implied *T* violation under the assumptions of *CPT* theorem (see Chapter 3).

5.8 EXISTENCE OF TWO NEUTRINOS AND LEPTON NUMBER CONSERVATION

In 1942, Sakata and Inoue [50] introduced two kinds of neutrinos in a study of the possible existence of two kinds of mesons, theoretically. They calculated the lifetimes of the boson (now called the π meson) and the fermion (μ meson). They wrote, e.g.,

$$Y^{\pm} \to m^{\pm} + n \qquad \text{in place of} \qquad \pi^{\pm} \to \mu^{\pm} + \nu_{\mu}, \tag{5.127}$$

and

$$m^{\pm} \to e^{\pm} + \nu + n \qquad \text{in place of} \qquad \mu^{\pm} \to e^{\pm} + \nu_e + \nu_{\mu}. \tag{5.128}$$

Here, ν_{μ} (or ν_e) refers to the neutrino associated with the muon (or electron). The antiparticles of the neutrinos are not explicitly indicated. A remarkable distinction between the pions and the muons was already recognized at that time. (Pions were understood to be responsible for nuclear forces, and muons were found in the cosmic rays.) A possible existence of two kinds of neutrino had, however, been forgotten since.

The problem was recently revived due to a major difficulty of the Fermi theory at high energies. In a reaction like

$$\nu + n \to p + \mu^{-} \tag{5.129}$$

the cross section is given by

$$\sigma \sim G^2 \left(\frac{\hbar}{mc}\right)^2 \left(\frac{E_\nu}{mc^2}\right)^2. \tag{5.130}$$

Thus, the cross section increases with the neutrino energy E_ν in the center-of-mass system. On the other hand, s-wave scattering gives it maximum cross section by

$$\sigma \sim \pi \lambda\!\!\!^{-2} = \pi\left(\frac{\hbar}{mc}\right)^2\left(\frac{E_\nu}{mc^2}\right)^{-2}. \tag{5.131}$$

Violation of this unitarity condition occurs at

$$E_\nu \lesssim 300 \text{ BeV}. \tag{5.132}$$

This difficulty may be remedied if an intermediate boson mediates the weak interactions, as Yukawa studied it in his meson theory of beta decay [51].

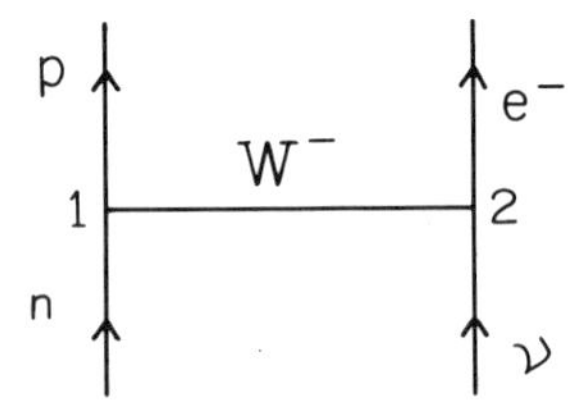

Fig. 5.12 Feynman diagram for the process $\nu + n \to p + e^-$ through an intermediate boson.

For example, the process (5.129) is understood as a two-step transition (see Fig. 5.12)

$$n \to W^- + p, \qquad \nu + W^- \to e^-. \tag{5.133}$$

The intermediate boson W^- transmits the action of a current at a space-time point 1 to another current at a different point 2, and an effective nonlocality is generated. As a consequence, the weak coupling constant C or equivalently the form factor can reduce its magnitude at high energy, and the cross section does not violate the unitarity condition.

It is, however, pointed out by Feinberg [52] that such a boson implies a branching ratio

$$\frac{\mu \to e + \gamma}{\mu \to e + \nu + \bar{\nu}} \tag{5.134}$$

of the order of 10^{-4}, unless the neutrinos associated with muons are different from those associated with electrons. Inasmuch as the branching ratio is measured [53] to be less than 10^{-8}, the hypothesis that the two neutrinos may be different has found some favor. Lee and Yang [54] calculated the cross section of the process (5.129) using the vector form factor deduced from electron scattering results and assuming the axial vector form factor to be the same as the vector form factor. It is expected that if there is only one type of neutrino, then neutrino interactions should produce muons and electrons in equal abundance. In the event that there are two neutrinos, there is no reason to expect any electrons at all.

The first experiment was performed by the Columbia group [55] at the Brookhaven AGS with the high energy neutrinos produced as the result of the decay of the pions,

$$\pi^+ \to \mu^+ + \nu \qquad \text{and} \qquad \pi^- \to \mu^- + \bar{\nu}. \tag{5.135}$$

The pions are produced by 15-BeV protons striking a beryllium target. The neutrinos or antineutrinos are introduced in a huge spark chamber. If we assume only one type of neutrino, the following processes may take place.

$$\begin{aligned} &\nu + \text{n} \to \text{p} + \text{e}^-, \qquad \bar{\nu} + \text{p} \to \text{n} + \text{e}^+, \\ &\nu + \text{n} \to \text{p} + \mu^-, \qquad \bar{\nu} + \text{p} \to \text{n} + \mu^+. \end{aligned} \tag{5.136}$$

The cross sections for these processes should be of the order of 10^{-38} cm^2 at about 1 BeV. In the spark chamber experiment, the electrons and muons can be easily distinguished, since electrons make showers and muons make single lines. The Columbia group found 29 single muons and 6 showers. A few of the showers are suspected to be either the backgrounds due to the neutron events or the electron–neutrinos from the decay of kaons. As a consequence, the existence of at least two kinds of neutrino is concluded. That is, $\nu_e \neq \nu_\mu$.

Later on, the CERN group at Geneva [56] studied the neutrino interactions in the 500-liter freon chamber using the neutrino beam from proton synchrotron, and they found 459 events. Of these, 454 contain a negative muon candidate, and 5 a negative electron of energy exceeding 400 MeV. In this experiment, we expect 3.3 electron events from the ν_e flux resulting from kaon decays like

$$\text{K}^+ \to \text{e}^+ + \pi^0 + \nu_e \qquad \text{or} \qquad \text{K}^0 \to \text{e}^+ + \pi^- + \nu_e.$$

Thus, the data confirm the results of the Brookhaven group regarding the two-neutrino hypothesis. Of the muon events, 236 contain no pions

$$\nu_\mu + \text{n} \to \text{p} + \mu^-. \tag{5.137}$$

This is called the elastic process. Of the remaining events, 209 contain pions and 9 contain pions and strange particles. These are called the inelastic processes.

Conservation of Lepton Numbers

The high energy neutrino experiments provide us with another important datum, i.e., the conservation of lepton numbers. For example, for the incoming neutrino, we may expect, without such a rule, the following two reactions.

$$\nu_\mu + \text{n} \to \text{p} + \mu^-, \qquad \text{or} \quad \mu^- + \text{hadrons}, \tag{5.138a}$$

$$\nu_\mu + \text{p} \to \text{n} + \mu^+, \qquad \text{or} \quad \mu^+ + \text{hadrons}, \tag{5.138b}$$

since the target nuclei consist of neutrons and protons. (Here, μ^- + hadrons means the processes $\mu^- + p$ + pions, $\mu^- + \Lambda + K^0$, etc.) Experimentally, we can find the charge of the produced muons by applying the magnetic field on the detectors. In 1963, the spark chamber experiment with a neutrino beam produced result [56, 57]

$$\frac{\text{number of positive leptons}}{\text{number of negative leptons}} \lesssim 6\%. \tag{5.139}$$

It was suggested that the positive leptons are partially due to a contamination of $\bar{\nu}_\mu$ in the ν_μ beam. Recent data [58] give a stricter limit for the processes (5.138),

$$\frac{(5.138\text{b})}{(5.138\text{a})} \lesssim 4.6 \times 10^{-3} \tag{5.140}$$

with 95% confidence level.

With these experimental data, we define the muon numbers as

$$\begin{aligned} L_\mu &= +1 \quad \text{for} \quad \mu^- \text{ and } \nu_\mu, \\ &= -1 \quad \text{for} \quad \mu^+ \text{ and } \nu_\mu, \\ &= 0 \quad\ \ \text{for } e^\pm, \nu_e, \bar{\nu}_e, \text{ and all nonleptons.} \end{aligned} \tag{5.141}$$

In elementary processes like (5.138a), the muon number of the system is conserved. It is not, however, in Eq. (5.138b). The conservation of the muon number prohibits the following processes automatically,

$$\begin{aligned} &\mu^\pm \not\rightarrow e^\pm + \gamma, && \mu^\pm \not\rightarrow e^\pm + e^+ + e^-, \\ &\nu_\mu + n \not\rightarrow p + e^-, && \pi^- \not\rightarrow \mu^- + \nu_\mu, \\ &\nu_\mu + p \not\rightarrow n + \mu^+, && K^+ \not\rightarrow \mu^+ + \nu_e. \end{aligned} \tag{5.142}$$

Experimentally, this is very probable.

Similarly, the electron number L_e (defined in Eq. (5.38)) is conserved with greater accuracy, as is shown in the double beta-decay experiments. We thus have lepton number conservation for both L_e and L_μ.

5.9 $\Delta S/\Delta Q = 1$ RULE

The $\Delta S/\Delta Q = 1$ rule has been found in the strangeness changing leptonic decays of hyperons or kaons. The rule states that in the elementary processes the change of strangeness for the hadronic part of the system should be equal to that of the charge. This is easily seen in the following decay modes.

	ΔS	ΔQ	$\Delta S/\Delta Q$
$\Sigma^+ \to n + e^+ + \nu_e$	$+1$	-1	-1
$\Sigma^- \to n + e^- + \bar{\nu}_e$	$+1$	$+1$	$+1$
$K^0 \to \pi^+ + e^- + \bar{\nu}_e$	-1	$+1$	-1
$K^0 \to \pi^- + e^+ + \nu_e$	-1	-1	$+1$
$K^+ \to \pi^+ + \pi^+ + e^- + \bar{\nu}_e$	-1	$+1$	-1
$K^+ \to \pi^+ + \pi^- + e^+ + \nu_e$	-1	-1	$+1$

(5.143)

In the first four processes, there are also energetically possible modes where the e-ν_e pair is replaced by the μ-ν_μ pair. The experimental evidence for the processes above shows that there are only a few uncertain events for $\Delta S/\Delta Q = -1$.

If the world data are combined to examine the limit on $\Delta Q = -\Delta S$ transitions by observation of Σ^+ decays, we find the limits [39]

$$\frac{W(\Sigma^+ \to n + e^+ + \nu_e)}{W(\Sigma^- \to n + e^- + \bar{\nu}_e)} \leqslant 0.01 \tag{5.144}$$

and

$$\frac{W(\Sigma^+ \to n + \mu^+ + \nu_\mu)}{W(\Sigma^- \to n + \mu^- + \bar{\nu}_\mu)} \leqslant 0.05 \tag{5.145}$$

with 90% confidence. Here W means the decay rate. As a matter of fact, only one e^+ event and two μ^+ events have been seen, but these are consistent with the expected background.

In the K_{e4} experiment, 264 $K^+ \to \pi^+ + \pi^- + e^+ + \nu_e$ events have been found with no $K^+ \to \pi^+ + \pi^+ + e^- + \bar{\nu}_e$, giving an upper limit on the branching ratio

$$\frac{W(K^+ \to \pi^+ + \pi^+ + e^- + \bar{\nu}_e)}{W(K^+ \to \pi^+ + \pi^- + e^+ + \nu_e)} \leqslant 0.02 \tag{5.146}$$

with 90% confidence. We therefore find with increasing accuracy that there are no $\Delta Q = -\Delta S$ transitions in these days.

5.10 POSSIBLE *G*-PARITY NONCONSERVATION

The G transformation is defined by the product of the charge conjugation and the charge symmetry, which is the rotation about the y axis through 180°, in the charge space [59]. (For the charge space or isospace, see Appendix A.8.) We shall see the physical significance of the G transformation from an example of pions.

The pion wave functions can be represented by linear combinations of three real wave functions ϕ_x, ϕ_y, ϕ_z;

$$\phi^+ = (\tfrac{1}{2})^{1/2}(\phi_x + i\phi_y) \qquad \text{for } \pi^+,$$
$$\phi^- = (\tfrac{1}{2})^{1/2}(\phi_x - i\phi_y) \qquad \text{for } \pi^-,$$
$$\phi^0 = \phi_z \qquad \text{for } \pi^0. \tag{5.147}$$

The charge conjugation C transforms these wave functions as follows.

$$C\phi^+ = \phi^-, \qquad C\phi^- = \phi^+, \qquad C\phi^0 = \phi^0, \tag{5.148}$$

because the charged pions are mutually transformed, and the neutral pion is self conjugate. Phase factors, factors with the absolute value of unity, may appear in Eqs. (5.148). Those for the charged pions can be absorbed in the definition of wave functions, and that for the neutral pion should be unity if one consults the decay of π^0 into two photons. From Eqs. (5.147) and (5.148) we have the transformation rule for real wave functions under C

$$\phi_x \to \phi_x, \qquad \phi_y \to -\phi_y, \qquad \phi_z \to \phi_z. \tag{5.149}$$

That is, the operator C has a function to reflect in the y axis, and to break the symmetry of the charge space. Furthermore, the charged pions can not be the eigenstates of C, although they are those for T_z. A symmetrical operation in the charge space is obtained by reflecting in the x and z axes after the C transformation. The former is performed by the rotation about the y axis through 180°. We call this symmetrical operation the G transformation,

$$G = \exp(i\pi T_y)C. \tag{5.150}$$

From the considerations above, we see that the pion wave functions change their signs under the G transformation. That is, the pions are eigenstates with G parity odd [$G = -1$]. The G parity is assigned only for the meson states or the states with baryon number $B = 0$. The states with nonzero B are not the eigenstates of G because B is transformed to $-B$ under the G transformation. The G parity of the system is conserved if the Hamiltonian is invariant under C and the isospin is conserved. If this is the case, the transition from the state of even pions to the state of odd pions is forbidden. Violation of the G-parity invariance is of the order of α (fine structure constant) by the electromagnetic interactions. They consist of the isoscalar and isovector parts, and the latter violates the charge independence.

The G transformation for single nucleons can be studied by remembering the isodoublet property

$$\begin{pmatrix} \psi_\text{p} \\ \psi_\text{n} \end{pmatrix} \tag{5.151}$$

where ψ_p and ψ_n are the proton and neutron wave functions, respectively.

(For details, see Appendix A.8.) The operator for the rotation about the y axis through 180° is given by

$$\exp\left(i\frac{\pi}{2}\tau_y\right) = \cos\frac{\pi}{2} + i\tau_y \sin\frac{\pi}{2} = i\tau_y = \begin{pmatrix} 0 & 1 \\ -1 & 0 \end{pmatrix} \tag{5.152}$$

for the isodoublet. The Pauli matrix τ_y is given in Eq. (A.49), which has exactly the same form as σ_y. The G transformation for the nucleon wave functions is, therefore, given by

$$G\begin{pmatrix} \psi_{\rm p} \\ \psi_{\rm n} \end{pmatrix} = \begin{pmatrix} 0 & 1 \\ -1 & 0 \end{pmatrix} \begin{pmatrix} \psi_{\bar{\rm p}} \\ \psi_{\bar{\rm n}} \end{pmatrix} = \begin{pmatrix} \psi_{\bar{\rm n}} \\ -\psi_{\bar{\rm p}} \end{pmatrix}. \tag{5.153}$$

From Eqs. (3.35) and (3.39), we have the wave functions for the antiparticles $\bar{\rm p}$ and $\bar{\rm n}$ as follows.

$$\psi_{\bar{\rm p}} = \gamma_2 \psi_{\rm p}^* \qquad \text{and} \qquad \psi_{\bar{\rm n}} = \gamma_2 \psi_{\rm n}^*. \tag{5.154}$$

We are now ready to discuss G-transformation properties for the bilinear forms of the nucleon wave functions in weak interactions. We can assign G parities for these relativistic covariants. (They have a baryon number zero, because they represent, e.g., annihilation of a neutron and creation of a proton in the second quantized field theory. This is also seen in Fig. 5.5, where a pion dissociates into a proton–antineutron pair at point A.) The G transformations for the relativistic covariants are expressed by the simultaneous replacement

$$\psi_{\rm p} \to \psi_{\bar{\rm n}} \qquad \text{and} \qquad \psi_{\rm n} \to -\psi_{\bar{\rm p}}. \tag{5.155}$$

As a result, we have

$$\begin{aligned} (\bar{\psi}_{\rm p} O_i \psi_{\rm n}) &\to -(\bar{\psi}_{\bar{\rm n}} O_i \psi_{\bar{\rm p}}) \\ &= (\bar{\psi}_{\rm p} (\gamma_2 \gamma_4 O_i \gamma_4 \gamma_2)^T \psi_{\rm n}) \\ &= \eta_i^G (\bar{\psi}_{\rm p} O_i \psi_{\rm n}). \end{aligned} \tag{5.156}$$

Comparing Eq. (5.156) with Eq. (3.45), we have a relation

$$\eta_i^G = -\eta_i^C, \tag{5.157}$$

where η_i^C are the sign functions for charge conjugation. (They are given in Table 3.1. As a matter of fact, we adopt the values which are the product of η_i^C in Table 3.1 with -1, from a field-theoretic consideration. See the discussion following Eq. (3.45).) Finally, we have

$$\begin{aligned} \eta_i^G &= -1 \qquad \text{for} \quad i = S, A, P, \\ &= +1 \qquad \text{for} \quad i = V, T. \end{aligned} \tag{5.158}$$

From now on, we shall restrict our discussion to the $V-A$ theory for four-fermion interactions. That is, the beta-decay interaction is expressed

$$H = (\tfrac{1}{2})^{1/2}[V_\alpha(\bar{\psi}_e\gamma_\alpha(1+\gamma_5)\psi_\nu) + A_\alpha(\bar{\psi}_e i\gamma_\alpha\gamma_5(1+\gamma_5)\psi_\nu)] + \text{h.c.} \quad (5.1$$

with the following forms for nucleon covariants. (See the remark below Eq. (2.56).)

$$V_\alpha = C_V\bar{\psi}_p\gamma_\alpha\psi_n \qquad \text{and} \qquad A_\alpha = C_A(\bar{\psi}_p i\gamma_\alpha\gamma_5\psi_n). \quad (5.1$$

We call these the vector and axial vector currents or, considered together t weak currents. V_α and A_α in Eq. (5.160) are given in the limit of vanishing four-momentum transfer $p_\alpha = 0$. In the case of higher-momentum transfer the weak currents are more complicated. As we have studied regarding the CVC theory, meson clouds around nucleons make several contributions. Consequently, the vector and axial vector currents are made up of the Dira matrices as well as the four-momentum transfer. The most general forms a therefore, given by

$$\begin{aligned} V_\alpha &= (\bar{\psi}_p(f_V\gamma_\alpha - ig_V\sigma_{\alpha\beta}p_\beta + ih_V p_\alpha)\psi_n),\\ A_\alpha &= (\bar{\psi}_p(if_A\gamma_\alpha\gamma_5 - g_A p_\alpha\gamma_5 - h_A\sigma_{\alpha\beta}\gamma_5 p_\beta)\psi_n), \end{aligned} \quad (5.1$$

with

$$\sigma_{\alpha\beta} = \tfrac{1}{2}(\gamma_\alpha\gamma_\beta - \gamma_\beta\gamma_\alpha) \qquad \text{and} \qquad p_\alpha = (p^{(n)} - p^{(p)})_\alpha.$$

Here, the momentum transfer p is the difference between the proton and neutron momentum, $p^{(p)}$ and $p^{(n)}$. The form factors f, g, and h are functio of p^2, and they are real if time-reversal invariance holds. In the limit as $p \rightarrow$ we set

$$f_V(0) = C_V \qquad \text{and} \qquad f_A(0) = C_A. \quad (5.16$$

The detailed analysis of each term in Eqs. (5.161) will be performed in Chapter 8 (see especially Eq. (8.59)). From this analysis, we call the g_V, h_V g_A, and h_A, the weak magnetism, induced scalar, induced pseudoscalar, an induced tensor interaction, respectively. (The last includes a differential operator.) Using Eqs. (5.156) and (5.157) and the discussion following Eq. (5.157)), we derive G parities for six terms in Eq. (5.161),

$$\begin{aligned} G &= +1 \qquad \text{for } f_V, g_V, h_A,\\ G &= -1 \qquad \text{for } f_A, g_A, h_V. \end{aligned} \quad (5.16$$

In each current, the G parity for f and g terms is different from that for th h term. If we assume each current of a definite G parity, h terms should vanish [60], since the existence of f terms is experimentally verified. Adopt the conserved vector current theory in Chapter 4, $\partial_\alpha V_\alpha = 0$, we also have a vanishing h_V term, $h_V = 0$. We cannot, however, set a similar condition for axial vector current A_α, since the pion decay to muon is strictly forbidden under the condition $\partial_\alpha A_\alpha = 0$. (See Eqs. (4.109) and (4.110).)

Weinberg first introduced the current of the type (5.160) [61]. He called f and g terms first-class interactions, and h terms second-class interactions, and suggested several experiments for proving the possible existence of the second-class interactions. These experiments involve measurements of the $ft_{1/2}$ values of the mirror nuclei, decay rates of charged sigma particles in $\Sigma^+ \to \Lambda + e^+ + \nu_e$ and $\Sigma^- \to \Lambda + e^- + \bar{\nu}_e$, β–α angular correlations, etc. Since the leptonic decays of charged sigma particles are analogous to the mirror transitions, we concentrate our discussion on the $ft_{1/2}$ values. In order to simplify the problem, we discuss the beta decay of B^{12} and N^{12}. These two are pure Gamow–Teller transitions ($1^+ \to 0^+$). For these, the effective interaction is derived from Eq. (8.59) in Chapter 8, which is given in the nonrelativistic approximation for nucleons,

$$(C_A + C_T)(u_p^\dagger \sigma u_n)(\psi_e^\dagger \sigma \psi_\nu) + \text{h.c.} \tag{5.164}$$

with $C_T = -E_o h_A$. That is, a maximum energy-dependent Gamow–Teller coupling constant is introduced by the effect of the tensor interaction. The vector current may contribute to the Gamow–Teller transitions through the nuclear matrix element

$$(1 + \mu_p - \mu_n) \int \boldsymbol{\alpha} \times \mathbf{r}.$$

In fact, this term was the proof for the CVC theory in beta-ray spectra. We can omit this term here, since the energy integral for this term vanishes in the limit of $Z = 0$. The conventional form for $ft_{1/2}$ values in Eq. (2.165)

$$ft_{1/2} = \frac{2\pi^3 \log 2}{|C_A \int \sigma|^2}$$

should now be replaced by

$$ft_{1/2} = \frac{2\pi^3 \log 2}{|(C_A \pm C_T) \int \sigma|^2} \qquad \text{for } \beta^{\mp}. \tag{5.165}$$

The ratio of the $ft_{1/2}$ values for N^{12} and B^{12} is unity if Eq. (2.165) is valid and the nuclear matrix elements are the same. It is, however, not the case, as is seen in Eq. (5.165). Instead,

$$\frac{(ft_{1/2})_{\beta^+}}{(ft_{1/2})_{\beta^-}} = \left| \frac{(C_A + C_T)(\int \sigma)_{\beta^-}}{(C_A - C_T)(\int \sigma)_{\beta^+}} \right|^2. \tag{5.166}$$

Under the assumptions $C_T \ll C_A$ and equal nuclear matrix elements, we obtain a difference

$$\delta = \left[\frac{(ft_{1/2})_{\beta^+}}{(ft_{1/2})_{\beta^-}} - 1 \right] = -2 \frac{h_A}{C_A} (E_o^+ + E_o^-), \tag{5.167}$$

where E_o^+ and E_o^- are maximum total energies (in units of mc^2) of positive and negative beta particles. In Eq. (5.167), the energy sum $(E_o^+ + E_o^-)$ is very large, i.e., about 60. (A more detailed consideration gives a factor of $\frac{4}{3}$ instead of 2 in Eq. (5.167).)

The same consideration holds for the beta decay of charged sigma particles. The ratio of the decay rates is now found experimentally [39]

$$\frac{W(\Sigma^+ \to \Lambda + e^+ + \nu_e)}{W(\Sigma^- \to \Lambda + e^- + \bar{\nu}_e)} = 0.67 \pm 0.15, \qquad (5.168)$$

compared to an expected phase-space ratio of 0.61 if G invariance holds. A deviation would indicate the presence of the second-class interactions. A detailed investigation was also performed in [62].

For the B^{12} and N^{12} beta decays, the experimental data on δ are given by

$$\begin{aligned} \delta &= (16 \pm 3)\% \qquad && \text{Peterson and Glass [63],} \\ &= (14 \pm 2.5)\% && \text{Fisher [64],} \\ &= (10 \pm 2)\% && \text{Fisher and Whaling [65],} \\ &= (11 \pm 1)\% && \text{Kavanagh [66].} \end{aligned} \qquad (5.169)$$

Huffaker and Greuling [67] investigated this ratio of $ft_{1/2}$ values and the muon capture rate in C^{12} theoretically. They found an interaction consistent with the experimental data; this interaction includes $m_\mu g_A \sim 8$ and $h_A/f_A \sim -0.002$. Blin-Stoyle and Rosina [68] performed a theoretical investigation for possible sources of the nonzero δ in Eq. (5.167) other than the h terms. These involve

1. the effect of the second forbidden nuclear matrices, $\sim -1.0\%$;
2. electromagnetic corrections, $\sim -2.0\%$;
3. the difference in overlapping radial integrals due to the difference in the distribution of nuclear matter, +10%;
4. the effect of configuration mixing due to the charge-dependent nuclear forces, plus a few tenths of a percent;

as the most favorable values (i.e., those leading to the largest positive δ). The sum of these four kinds of corrections is 7%, which is still too small to explain the experimental data on δ. On the basis of this analysis Blin-Stoyle and Rosina concluded possible G-parity nonconservation. Wilkinson and Alburger are now working systematically on the half-life measurements of the mirror nuclei, having obtained the data [69]

$$\frac{ft_{1/2}(O^{13} \to N^{13})}{ft_{1/2}(B^{13} \to C^{13})} = 1.166 \pm 0.026, \qquad \frac{ft_{1/2}(Ne^{18} \to F^{18})}{ft_{1/2}(O^{18} \to F^{18})} = 0.992 \pm 0.015,$$

$$\frac{ft_{1/2}(Na^{20} \to Ne^{20})}{ft_{1/2}(F^{20} \to Ne^{20})} = 1.054 \pm 0.023,$$

$$\frac{ft_{1/2}(Si^{25} \to Al^{25})}{ft_{1/2}(Na^{25} \to Mg^{25})} = 1.187 \pm 0.076. \qquad (5.170)$$

The induced tensor interaction (h_A term) has an increasing contribution with the maximum energy E_o and is, in fact, significant in the muon capture in hydrogen or complex nuclei, since $E_o \sim m_\mu = 206m$. In this connection, a formalism with possible *G* nonconservation has been developed on the theory of muon capture [70]. A slight indication in favor of a possible admixture of the induced tensor interaction was found in a theoretical analysis of the muon capture in gaseous and liquid hydrogen by Ohtsubo and Fujii [71].

In principle, it is rather difficult to assume a large second-class interaction, since the nuclear forces are *G* invariant. Furthermore, the algebra of the currents, the algebra for the bilinear quantities of field operators, loses its validity if *G* parity is not conserved. Based on the usefulness of this algebra in high energy hadronic reactions as well as in weak processes, many physicists are reluctant to accept possible *G* parity nonconservation. Nevertheless, interesting data are being accumulated for the open problem of *G* invariance. (See Section 10.2 for a recent status of experiments.)

5.11 INTERMEDIATE BOSONS

As we have repeatedly emphasized, the four-fermion coupling of the beta-decay interaction was originally introduced by Fermi. He formulated it (in 1934) in analogy with the electromagnetic interaction in such a way that the electromagnetic vector potential is replaced by the vector covariant consisting of the electron and neutrino spinors, and the proton current is simultaneously replaced by the vector covariant made up of the neutron and proton spinors. We call this sort of four-fermion coupling the direct interaction or the Fermi-type interaction.

There is, however, another sort of coupling, called the indirect interaction (or, frequently, the Yukawa-type interaction). And it is possible to construct the theory of beta decay with the indirect interaction. In 1935, Yukawa published his theory of nuclear forces, and he predicted the existence of pions [51]. He thought that nuclear forces are transmitted through pion fields (Fig. 5.13B) in the way that electromagnetic forces are transmitted though electromagnetic fields. For example, in Møller scattering, electrons interact by exchanging photons, see Fig. 5.13A.) Yukawa noted in this original paper the possibility that beta decay is induced by the exchange of a pion between nucleon and lepton currents. Consequently, the coupling constant for the nucleon vertex and that for the lepton vertex should be different. Later on, the same coupling constant was adopted for both vertices and a heavy intermediate boson W was introduced in place of the pion (see Fig. 5.13C). A variant form of the Yukawa theory was proposed by Tanikawa [72], who considered beta decay in such a way that the nucleon absorbs a lepton and emits a heavy boson which has a baryon number as well as a lepton number, and subsequently the boson decays into a nucleon and a lepton (see Fig. 5.13D). (The Tanikawa bosons carry both lepton and nucleon numbers, whereas the weak bosons of the Yukawa type do not. In the Tanikawa theory, the muon (electron) couples with the electron (muon) neutrino in muon decay.) Since then, many theories have been published on various kinds of intermediate bosons: Ogawa [73], Schwinger [74], Feynman and Gell-Mann [75], and Lee and Yang [76] have discussed the vector bosons,

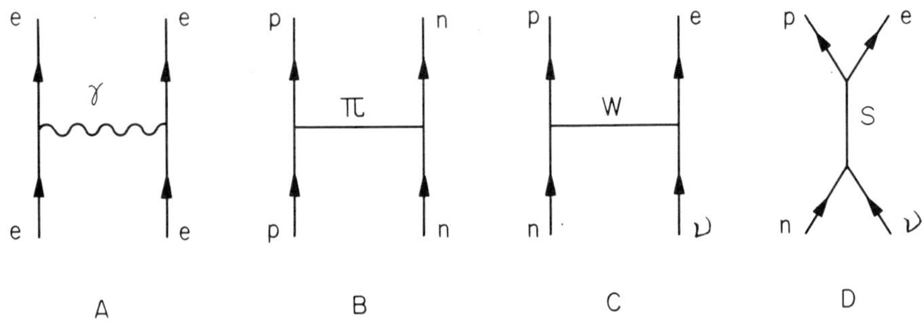

Fig. 5.13 Indirect couplings. (A) Møller scattering. (B) Nucleon–nucleon scattering. (C) Beta decay. (D) Tanikawa-type beta decay.

while Tanikawa and Watanabe have dealt with the scalar bosons [77]. Several conclusive tests for intermediate bosons, promising in high energy neutrino experiments, were proposed by Pontecorvo [78], Schwarz [79], Lee and Yang [54], Yamaguchi [80], and Cabibbo and Gatto [81] for the vector bosons, and by Kinoshita [82] and Glashow [83] for the scalar bosons.

The status of high energy neutrino experiments is as follows. The limit for the mass of the W meson is obtained as

$$M_W c^2 \gtrsim 2 \text{ Bev}, \tag{5.171}$$

if it exists. Resonances in the cross sections of neutrino reactions were not conclusive for the Tanikawa bosons. It should, however, be recalled that the Fermi theory will violate the unitarity condition around the neutrino energy 300 BeV in the center-of-mass system, as was discussed in section 5.8. And this was the main reason for introducing the intermediate bosons. In this connection, the neutrino experiments in higher energies are important.

The beta-ray spectra in the Fermi, Yukawa, and Tanikawa theories are different. This will be discussed in the following subsections.

Yukawa-Type Beta Decay

In this theory, beta decay is a two-step transition

$$\text{n} \rightarrow \text{p} + \text{W}^-, \qquad \text{W}^- \rightarrow \text{e}^- + \bar{\nu}_\text{e}. \tag{5.172}$$

The charged vector boson interacts with leptons and baryons with the same coupling constant g, which is dimensionless,

$$H_{\text{int}} = igW_\alpha[\bar{\psi}_e\gamma_\alpha(1+\gamma_5)\psi_{\nu_e} + \bar{\psi}_\mu\gamma_\alpha(1+\gamma_5)\psi_{\nu_\mu}],$$

$$H'_{\text{int}} = igW_\alpha[\bar{\psi}_n\gamma_\alpha(1+\gamma_5)\psi_p + \cdots]. \tag{5.173}$$

Here, the vector fields for W are expressed by W_α ($\alpha = 1, 2, 3, 4$). The propagator for this vector boson is given by

$$\frac{\delta_{\alpha\beta} + (K_\alpha K_\beta/M_W^2)}{K^2 + M_W^2}. \tag{5.174}$$

Here K_α and M_W are the four-momentum transfer and mass of the vector boson, respectively. This theory is equivalent to the direct-coupling Fermi theory, which is a combination of S, V, T, A, P with the following magnitudes of the coupling constants [84].

$$C_V = -C_A = \frac{2^{-1/2}g^2}{M_W^2 - m^2},$$

$$\frac{C_S}{C_V} \approx -\frac{2.54m^2}{M_W^2}, \qquad \frac{C_T}{C_V} \approx \frac{2mM}{M_W^2}, \qquad C_P = 0. \tag{5.175}$$

Here m and M stand for the electron and nucleon mass, respectively. The beta-decay spectrum [85] is obtained from the evaluation of the Feynman diagram in Fig. 5.12,

$$P(E)\,dE = C_V^2[|\textstyle\int 1|^2 + |\int \sigma|^2]\Big\{1 + \frac{4}{M_W^2}\,[E(E_o - E) - a(E_o^2 + \tfrac{1}{3}E(E_o - E))]$$
$$+ \frac{2}{M_W^2}\left[1 + \tfrac{1}{3}a\left(5\frac{E_o}{E} - 2\right) + \tfrac{1}{3}b\left(1 - 2\frac{E_o}{E}\right)\right]\Big\}\frac{1}{2\pi^3}$$
$$\times (E_o - E)^2\,Ep\,dE, \tag{5.176}$$

with

$$a = \frac{|\int 1|^2 - \frac{1}{3}|\int \sigma|^2}{|\int 1|^2 + |\int \sigma|^2} \qquad \text{and} \qquad b = \frac{|\int \sigma|^2}{|\int 1|^2 + |\int \sigma|^2},$$

in units of $\hbar = m = c = 1$. The quantity in the curly brackets is the spectral shape factor, which is unity for the Fermi theory. The energy-dependent term in this factor is of the order of $E_o^2 M_W^{-2}$, which is

$$\frac{E_o^2}{M_W^2} \lesssim \left(\frac{30}{4000}\right)^2 \sim 1 \times 10^{-4} \tag{5.177}$$

for the most energetic beta decay of $E_o \sim 30mc^2$ and the boson mass $M_W c^2 \sim$ 2 BeV. Therefore, the nonlocal effect due to the weak boson will be negligibly small, particularly in the beta-ray spectrum. However, the effect could be observable in the beta-ray spectrum in the leptonic process, such as $\Sigma^- \to \Lambda^0$ (or n) + e^- + $\bar{\nu}_e$, since $E_o^2 M_W^{-2}$ becomes about 1%.

Tanikawa-Type Beta Decay

In this theory, beta decay is a two-step transition

$$\mathrm{n} \to \mathrm{S} + \bar{\nu}_{\mathrm{e}}, \qquad S \to \mathrm{p} + \mathrm{e}^-. \tag{5.178}$$

Here S is the scalar boson of charge zero. It has both lepton and baryon numbers and the coupling

$$H_{\mathrm{int}} = g_{\mathrm{S}} \mathrm{S} [\bar{\psi}_{\bar{\nu}_{\mathrm{e}}}(1 + \gamma_5)\psi_{\mathrm{n}} + \bar{\psi}_{\mathrm{e}^+}(1 + \gamma_5)\psi_{\mathrm{p}} + \cdots]. \tag{5.179}$$

This theory is almost equivalent to the $V - A$ theory of the direct Fermi coupling. In Eq. (5.179), g_{S} and S are the dimensionless coupling constant and the field of the S meson, respectively. The beta-ray spectrum can be calculated with the Feynman diagram in Fig. 5.13D. The spectral shape factor is given by

$$\left[1 \mp \frac{4M}{M_{\mathrm{S}}^2 - M^2}(E_o - E)\right] \qquad \text{for } \beta^{\mp}. \tag{5.180}$$

The energy dependence of this factor is again negligibly small, unless the scalar meson mass M_{S} is close to the nucleon mass.

REFERENCES

1 H. Yukawa and S. Sakata, *Proc. Phys. Math. Soc. Japan* **17,** 467 (1935).
2 H. Yukawa and S. Sakata, *Proc. Phys. Math. Soc. Japan* **18,** 128 (1936).
3 L. W. Alvarez, *Phys. Rev.* **54,** 486 (1938).
4 R. Sherr and R. H. Miller, *Phys. Rev.* **93,** 1076 (1954).
5 R. W. P. Drever, A. Moljk, and J. Scobie, *Phil. Mag.* **1,** 942 (1956).
6 J. Scobie and G. M. Lewis, *Phil. Mag.* **2,** 1089 (1957).
7 E. Feenberg and G. Trigg, α, β, γ *Ray Spectroscopy* (K. Siegbahn, ed.) p. 1361. North-Holland, Amsterdam, 1968.
8 H. A. Bethe and R. Peierls, *Nature* **133,** 532 (1934).
9 T. D. Lee and C. N. Yang, *Phys. Rev.* **105,** 1671 (1957).
10 F. Reines and C. L. Cowan, *Phys. Rev.* **92,** 830 (1953); C. L. Cowan, F. Reines, F. B. Harrison, H. W. Kruse, and A. D. Mcguire, *Science* **3213,** 103 (1956).
11 F. Reines and C. L. Cowan, *Phys. Rev.* **113,** 273 (1959).
12 R. E. Carter, F. Reines, J. J. Wagner, and M. E. Wyman, *Phys. Rev.* **113,** 280 (1959).
13 C. O. Muelhause and S. Oleksa, *Phys. Rev.* **105,** 1332 (1957).
14 R. W. King and J. F. Perkins, *Phys. Rev.* **112,** 963 (1958).
15 R. Davis, *Phys. Rev.* **97,** 766 (1955); *Bull. Amer. Phys. Soc.*, Ser. II, **1,** 219 (1956).
16 M. Goldhaber, *J. Phys. Soc. Japan Suppl.* **24,** 1 (1968).
17 J. N. Bahcall, W. A. Fowler, I. Iben, and R. L. Sears, *Astrophys.* **137,** 244 (1963).
18 M. Majorana, *Nuovo Cimento* **14,** 171 (1937).
19 T. D. Lee and C. S. Wu, *Ann. Rev. Nucl. Sci.* **15,** 381 (1965).
20 H. Primakoff and S. P. Rosen, *Rept. Progr. Phys.* **22,** 121 (1959).
21 E. J. Konopinski, *Ann. Rev. Nucl. Sci.* **9,** 99 (1959).
22 C. Ryan and S. Okubo, *Nuovo Cimento Suppl.* I, **2,** 234 (1964).

23 T. Kristen, E. Norton, O. A. Schaeffer, and R. W. Stoenner, *Phys. Rev. Letters* **20**, 1300 (1968).
24 N. Takaoka, Ph.D. thesis, Osaka University, 1967; N. Takaoka and K. Ogata, *Mass Spectroscopy* **16**, 113 (1968); N. Takaoka, J. Okano, and K. Ogata, *Advances in Mass Spectroscopy*, Vol. IV, p. 943. Elsevier, Amsterdam, 1968.
25 B. K. Bardin, P. J. Gollon, J. D. Ullman, and C. S. Wu, *J. Phys. Soc. Japan Suppl.* **24**, 548 (1967); *Nucl. Phys.* **A158**, 337 (1970).
26 G. Impeduglia, R. Plano, A. Prodell, N. Samios, M. Schwarz, and J. Steinberger, *Phys. Rev. Letters* **1**, 249 (1958); T. Fazzini, G. Fidecare, A. W. Merrison, H. Paul, and A. V. Tollestrup, *Phys. Rev. Letters* **1**, 247 (1958).
27 E. di Capua, R. Garland, L. Pondrom, and A. Sterelzoff, *Phys. Rev.* **133B**, 1333 (1964); H. L. Anderson, T. Fujii, R. H. Miller, and L. Tau, *Phys. Rev.* **119**, 2050 (1960).
28 R. Finkelstein and M. Ruderman, *Phys. Rev.* **76**, 1458 (1949).
29 A. H. Rosenfeld, A. Barbaro-Galtieri, W. J. Podolsky, L. R. Price, Matts Roos, P. Soding, W. J. Willes, and C. G. Wohl, *Rev. Mod. Phys.* **39**, 1 (1967).
30 M. J. Levine and J. Wright, *Phys. Rev. Letters* **26**, 1351 (1971); S. J. Brodsky and T. Kinoshita, *Phys. Rev.* D3, 356 (1971); T. Kinoshita and P. Cvitanovic, *Phys. Rev. Letters* **29**, 1534 (1972).
31 M. Fierz, *Z. Phys.* **104**, 533 (1937).
32 L. Michel, *Proc. Roy. Soc. (London)* **A63**, 514 and 1371 (1950).
33 M. Bardon, P. Norton, J. Peoples, A. Sachs, and J. Lee-Franzini, *Phys. Rev. Letters* **14**, 449 (1965).
34 T. Kinoshita and A. Sirlin, *Phys. Rev.* **113**, 1652 (1959).
35 S. M. Berman, *Phys. Rev.* **112**, 267 (1958).
36 R. L. Garwin, L. M. Lederman, and M. Weinrich, *Phys. Rev.* **105**, 1415 (1957).
37 J. I. Friedman and V. L. Telegdi, *Phys. Rev.* **105**, 1681 (1957).
38 M. Bardon, P. Franzini, and J. Lee, *Phys. Rev. Letters* **7**, 23 (1961).
39 A review article by J. W. Cronin, *Proceedings of the 14th International Conference on High-Energy Physics* (CERN), Geneva 1968, p. 281.
40 T. Nakano and K. Nishijima, *Progr. Theoret. Phys.* **10**, 581 (1953). K. Nishijima, *Progr. Theoret. Phys.* **12**, 107 (1954); **13**, 285 (1955).
41 M. Gell-Mann, *Phys. Rev.* **92**, 833 (1953).
42 M. Gell-Mann, *Phys. Rev.* **125**, 1067 (1962), and *Proceedings of the International Conference on High-Energy Nuclear Physics* (CERN), Geneva 1962; S. Okubo, *Progr. Theoret. Phys.* **27**, 949 (1962).
43 V. E. Barnes *et al.*, *Phys. Rev. Letters* **12**, 204 (1964).
44 T. D. Lee and C. N. Yang, *Phys. Rev.* **104**, 254 (1956).
45 F. S. Crawford, Jr., M. Cresti, M. L. Good, K. Gottstein, E. M. Lyman, E. T. Solmitz, M. L. Stevenson, and H. K. Ticho, *Phys. Rev.* **108**, 1102 (1957).
46 A review paper by H. Filthuth, *Proc. Topical Conf. Weak Interactions*, CERN 69-7, p. 131, Geneva, 1969.
47 J. H. Christenson, J. W. Cronin, V. L. Fitch, and R. Turlay, *Phys. Rev. Letters* **13**, 138 (1964).
48 For details of the discussions on $K_L^0 \to 2\pi$, see e.g., T. D. Lee and C. S. Wu, *Ann. Rev. Nucl. Sci.* **16**, 511 (1966). For experimental data see e.g., [39].
49 L. Wolfenstein, *Phys. Rev. Letters* **13**, 562 (1964).
50 S. Sakata and T. Inoue, *Proc. Phys. Math. Soc. Japan* **16**, 232 (1942) and *Progr. Theoret. Phys.* **1**, 143 (1946); J. A. Wheeler, *Rev. Mod. Phys.* **21**, 133 (1949).
51 H. Yukawa, *Proc. Phys. Math. Soc. Japan*, **17**, 48 (1935).
52 G. Feinberg, *Phys. Rev.* **110**, 1482 (1958).

53 D. Bartell, S. Devons, and A. Sachs, *Phys. Rev. Letters* **8**, 120 (1962); S. Frankel, J. Halpern, L. Holloway, W. Wales, M. Yearian, O. Chamberlain, A. Lemonick, and F. M. Pipkin, *Phys. Rev. Letters* 8, **123**, (1962); M. Conversi, L. di Lella, G. Penso, M. Toller, and C. Rubbia, *Phys. Rev. Letters* **8**, 125 (1962).
54 T. D. Lee and C. N. Yang, *Phys. Rev. Letters* **4**, 307 (1960).
55 G. Danby, J.-M. Gaillard, K. Goulianos, L. M. Lederman, N. Mistry, M. Schwartz, and J. Steinberger, *Phys. Rev. Letters* **9**, 36 (1962).
56 M. M. Block *et al.*, *Phys. Letters* **12**, 281 (1964).
57 H. H. Binham *et al.*, *Intern. Conf. Elementary Particles* (Sienna), p. 555, 1963.
58 A review paper by D. H. Perkins, *Proc. Topical Conf. Weak Interactions*, CERN 69-7, p. 1, Geneva, 1969.
59 T. D. Lee and C. N. Yang, *Nuovo Cimento* **3**, 749 (1956); A. Pais and R. Jost, *Phys. Rev.* **87**, 871 (1952); L. Michel, *Nuovo Cimento* **10**, 319 (1953).
60 M. L. Goldberger and S. B. Treiman, *Phys. Rev.* **111**, 354 (1958).
61 S. Weinberg, *Phys. Rev.* **112**, 1375 (1958).
62 M. Morita, *Progr. Theoret. Phys. Suppl.* **26**, 1 (1963).
63 R. W. Peterson and N. W. Glass, *Phys. Rev.* **130**, 292 (1963).
64 T. R. Fisher, *Phys. Rev.* **130**, 2388 (1963).
65 T. R. Fisher and W. Whaling, *Phys. Rev.* **133**, B1502 (1964).
66 R. W. Kavanagh, *Phys. Rev.* **133**, B1504 (1964).
67 J. N. Huffaker and E. Greuling, *Phys. Rev.* **132**, 738 (1963).
68 R. J. Blin-Stoyle and M. Rosina, *Nucl. Phys.* **70**, 321 (1965).
69 D. H. Wilkinson, *Phys. Letters* **31B**, 447 (1970); D. H. Wilkinson and D. E. Alburger, *Phys. Rev. Letters* **24**, 1134 (1970); D. E. Alburger and D. H. Wilkinson, *Phys. Letters* **32B**, 190 (1970); D. H. Wilkinson, D. E. Alburger, D. R. Goosman, K. W. Jones, E. K. Warburton, G. T. Garvey, and R. L. Williams, *Nucl. Phys.* **A166**, 661 (1971).
70 M. Morita and R. Morita, *J. Phys. Soc. Japan* **19**, 1759 (1964); M. Morita, R. Morita, and T. Shirafuji, *Progr. Theoret. Phys. Suppl.* Commemoration Issue 96 (1965).
71 H. Ohtsubo and A. Fujii, *Nuovo Cimento* **42**, 109 (1966).
72 Y. Tanikawa, *Proc. Phys. Math. Soc. Japan* **34**, 610 (1942); *Progr. Theoret. Phys.* **3**, 314 (1948); *Phys. Rev.* **108**, 1615 (1957).
73 S. Ogawa, *Progr. Theoret. Phys.* **15**, 487 (1956).
74 J. Schwinger, *Ann. Phys.* **2**, 407 (1957).
75 R. P. Feynman and M. Gell-Mann, *Phys. Rev.* **109**, 193 (1958).
76 T. D. Lee and C. N. Yang, *Phys. Rev.* **119**, 1410 (1960).
77 Y. Tanikawa, *Phys. Rev.* **108**, 1615 (1957); Y. Tanikawa and S. Watanabe, *Phys. Rev.* **113**, 1344 (1957).
78 B. Pontecorvo, *J. Exp. Theoret. Phys. USSR* **37**, 1751 (1959); *Soviet Phys. JETP* **10**, 1236 (1960).
79 M. Schwarz, *Phys. Rev. Letters* **4**, 306 (1960).
80 Y. Yamaguchi, *Progr. Theoret. Phys.* **23**, 1117 (1960).
81 N. Cabibbo and R. Gatto, *Nuovo Cimento* **15**, 304 (1960).
82 T. Kinoshita, *Phys. Rev. Letters* **4**, 378 (1960).
83 S. L. Glashow, *Phys. Rev.* **118**, 316 (1960).
84 J. D. Childress, *Phys. Rev.* **123**, 1729 (1961).
85 R. Morita referred to on p. 40 of M. Morita, *Progr. Theoret. Phys. Suppl.* **26**, 1 (1963).
86 For example see P. Caruthers, *Introduction to Unitary Symmetry.* Interscience, New York 1966. M. Gell-Mann and Y. Ne'eman, *The Eightfold Way*, Benjamin, New York 1964.

Chapter Six

THEORY OF FORBIDDEN TRANSITIONS

6.1 SELECTION RULES FOR SPIN AND PARITY, AND CLASSIFICATION OF NUCLEAR MATRIX ELEMENTS

As is shown in Chapter 1, beta decays are classified in the allowed and forbidden transitions, by their spin and parity changes. For later use, we write it again as follows.

Name of Transition	$\Delta J = J_i - J_f$	Parity Change
Allowed transition	$0, \pm 1$	No
First forbidden transition	$0, \pm 1, \pm 2$	Yes
Second forbidden transition	$\pm 2, \pm 3$	No
nth forbidden transition	$\pm n, \pm(n+1)$	$(-)^n$

(6.1)

Here, J_i and J_f are the spins of the initial and final states. Parity change No (or Yes) means that the initial and final states have the same (or different) parities. In the nth forbidden transition, the parity change is No (or Yes) for n = even (or odd).

In the following, we will sketch the way of deriving various nuclear matrix elements from the four terms of the interaction Hamiltonian

$$C_V[(\psi_p^\dagger \psi_n)(\psi_e^\dagger \psi_\nu) - (\psi_p^\dagger \alpha \psi_n)(\psi_e^\dagger \alpha \psi_\nu)] + C_A[(\psi_p^\dagger \sigma \psi_n)(\psi_e^\dagger \sigma \psi_\nu) - (\psi_p^\dagger \gamma_5 \psi_n)(\psi_e^\dagger \gamma_5 \psi_\nu)] \quad (6.2)$$

for negatron decay. This Hamiltonian is parity conserving. But a generalization to the parity-nonconserving two-component neutrino theory is straightforward. Furthermore, we neglect the Coulomb correction to the electron. These simplifications do not preclude introducing nuclear matrix elements.

As is seen in section 2.7, we have assumed the s-wave leptons in the theory of allowed transitions. Now we have to consider the higher-wave leptons in the forbidden transitions. The simplest case is the first term in the vector coupling,

$$\psi_p^\dagger \psi_n \psi_e^\dagger \psi_\nu \propto \psi_p^\dagger \psi_n \exp^{(-i(\mathbf{p}+\mathbf{q})\cdot\mathbf{r})} \tag{6.3}$$

where the spatial parts of leptons are written with the plane waves. We can expand the exponential function in a power series of $(\mathbf{p}+\mathbf{q})\cdot\mathbf{r}$, since the lepton momenta are of the order of 1 or less and the nuclear radius is about 0.01 (in units of $\hbar = m = c = 1$), in usual beta decays. After expanding Eq. (6.3) and integrating over space coordinates, we have the following nuclear matrix elements.

$$\begin{aligned}&\int \psi_p^\dagger \psi_n \, d\mathbf{r} - i(\mathbf{p}+\mathbf{q})\cdot\int \psi_p^\dagger \mathbf{r} \psi_n \, d\mathbf{r}\\ &\quad - \tfrac{1}{2}\sum_{ij}\left[(p_i+q_i)(p_j+q_j) - \tfrac{1}{3}\delta_{ij}(\mathbf{p}+\mathbf{q})^2\right]\int \psi_p^\dagger (x_i x_j - \tfrac{1}{3}\delta_{ij} r^2)\psi_n \, d\mathbf{r}\\ &\quad - \tfrac{1}{6}(\mathbf{p}+\mathbf{q})^2 \int \psi_p^\dagger r^2 \psi_n \, d\mathbf{r} + \cdots .\end{aligned} \tag{6.4}$$

In this series, the $\bar{L}$th power terms with respect to the lepton momenta come, of course, from the $\bar{L}$th term of the expansion of the exponential in Eq. (6.3). (We reserve the symbol L for later use.) The $\bar{L} = 2$ term consists, however, of two parts, the symmetrical tensor with spur zero and the scalar. After Konopinski and Uhlenbeck [1], we adopt the notation,

$$\begin{aligned}&\int \psi_p^\dagger \psi_n \, d\mathbf{r} = \int 1, \qquad \int \psi_p^\dagger \mathbf{r} \psi_n \, d\mathbf{r} = \int \mathbf{r},\\ &\int \psi_p^\dagger (x_i x_j - \tfrac{1}{3}\delta_{ij} r^2)\psi_n \, d\mathbf{r} = R_{ij}, \qquad \int \psi_p^\dagger r^2 \psi_n \, d\mathbf{r} = \int r^2, \qquad \text{etc.}\end{aligned} \tag{6.5}$$

For the beta decay of complex nuclei, the same symbols are adopted, such that

$$\int \psi_f^\dagger \sum_{i=1}^{A} \mathbf{r}^{(i)} \tau_+^{(i)} \psi_i \, d\mathbf{r}_1 \cdots d\mathbf{r}_A = \int \mathbf{r}. \tag{6.6}$$

The tensorial rank J of the nuclear matrix element is that of the operator which is located between the initial- and final-state wave functions. Therefore, the J values for $\int 1$, $\int \mathbf{r}$, R_{ij}, $\int r^2$ are 0, 1, 2, 0, respectively. These matrix elements vanish unless the triangular condition $\Delta(J_i, J, J_f)$ is satisfied. More explicitly, nonvanishing values are expected only for the transitions listed in (6.7).

$$\begin{aligned}&\Delta J = 0 && \text{for}\quad J = 0\ (\textstyle\int 1, \int r^2, \cdots),\\ &\Delta J = 0, \pm 1 && (\text{except for } 0 \to 0) \text{ for } J = 1\ (\textstyle\int \mathbf{r}, \cdots),\\ &\Delta J = 0, \pm 1 \pm 2 && (\text{except for } 0 \to 0, \tfrac{1}{2} \to \tfrac{1}{2}, 1 \leftrightarrow 0) \text{ for}\\ & && \quad J = 2\ (R_{ij}, \cdots),\\ &\quad \text{etc.}\end{aligned} \tag{6.7}$$

Furthermore, the parity change of nuclear states is

$$\text{No} \quad \text{for } \bar{L} = \text{even}, \quad \text{and} \quad \text{Yes} \quad \text{for } \bar{L} = \text{odd}. \tag{6.8}$$

The nuclear matrix elements in Eq. (6.4) have contributions of the order of

$$(pR)^{\bar{L}} \sim (0.01)^{\bar{L}}. \tag{6.9}$$

In the conventional theory, the nuclear matrix element with $\bar{L}$ is called the $\bar{L}$th forbidden matrix. However, the matrices with $\bar{L} > J$ do not have a role in the $\bar{L}$th forbidden transition, since the triangular condition is not satisfied (see Eq. (6.1)). They may contribute to the lower forbiddenness of the same parity change. For example, $\int r^2$ is a small correction to the allowed transitions. It is neglected unless p is very large. From this reason, the matrices with $J = \bar{L}$ are only taken into consideration in the formulas for beta decay.

Now, let us define L, which is the resultant orbital angular momentum of electron and antineutrino. L values for $\int 1$, $\int r$, R_{ij}, $\int r^2$ are 0, 1, 2, 0, respectively. Two matrices, $\int 1$ and $\int r^2$, are parts of the s-wave matrix $\int j_0(pr)$, $j_0(pr)$ being the spherical Bessel function of order one. This matrix element can be derived if we perform the Rayleigh expansion, Eq. (2.98), of the plane wave. By expanding $j_0(pr)$ in a power series of pr, we have

$$\int j_0(pr) = \int 1 - \tfrac{1}{6}p^2 \int r^2 + \cdots.$$

This sort of matrix in the spherical representation will be adopted in the theory of the muon capture in section 8.4. It is expressed by [0 0 0]. $\int r^2$ will also appear in the product of the p-wave electron and the p-wave neutrino wave functions. The origin of $\int r^2$ in the calculation with the four-component spinors is a little more complicated, but is essentially the same as described here.

In our formula, which will be given in section 6.3, we adopt the spherical tensor of rank two instead of the cartesian tensors $x_i x_j - \frac{1}{3}\delta_{ij} r^2$ in the Konopinski-Uhlenbeck notation, i.e., we write

$$|\textstyle\int R_{ij}|^2 \quad \text{in place of} \quad \displaystyle\sum_{ij} |R_{ij}|^2.$$

The definition of our $\int R_{ij}$ is given, e.g., by

$$\int \psi^\dagger_{J_f M_f} \sum_{k=1}^{A} (x^{(k)} + iy^{(k)})^2 \tau_+^{(k)} \psi_{J_i M_i} \, d\mathbf{r}_1 \cdots d\mathbf{r}_A$$

$$= (J_i 2 M_i 2 \,|\, J_f M_f) \langle J_f \| R_{ij} \| J_i \rangle, \tag{6.10}$$

$$\langle J_f \| R_{ij} \| J_i \rangle = 2\mathcal{M}(R_{ij}), \tag{6.11}$$

and

$$\int R_{ij} = \left(\frac{2J_f + 1}{2J_i + 1} \right)^{1/2} \mathcal{M}(R_{ij}). \tag{6.12}$$

The reduced nuclear matrix element $\mathcal{M}(R_{ij})$ is frequently used in the papers by the present author. It has a relation,

$$\frac{2J_f + 1}{2J_i + 1} |\mathcal{M}(R_{ij})|^2 = \sum_{ij} |R_{ij}|^2. \qquad (6.13)$$

In this book throughout, we use the symbol $\int R_{ij}$ instead of $\mathcal{M}(R_{ij})$, so that

$$|\int R_{ij}|^2 = \sum_{ij} |R_{ij}|^2. \qquad (6.14)$$

A factor 2 in Eq. (6.11) is introduced in conformity with Eqs. (6.13) and (6.
The spectral shape factors for the nuclear matrix elements appearing in Eq. (6.4) are easily obtained from this equation itself. For example, $|\int \mathbf{r}|^2$ has

$$(\mathbf{p} + \mathbf{q})^2 \sim p^2 + q^2 = (E_o^2 - 1) + (E_o - E)^2. \qquad (6.15)$$

The cross term $\mathbf{p} \cdot \mathbf{q}$ vanishes by integrating over the angle between $\mathbf{p}$ and $\mathbf{q}$. As we can see later, this expression of the spectral shape factor is oversimplified, and the Coulomb correction should be taken into account.

Nuclear matrix elements are systematically introduced if we consider the lepton system which consists of the electron with the quantum numbers s (spin), l (orbital), j (total) and the antineutrino s', l', j'. The lepton system has the resultant spin S, resultant orbital angular momentum L, and resultant total angular momentum J. These quantum numbers should satisfy the following six relations among angular momenta:

$$\begin{aligned} &\mathbf{s} + \mathbf{s}' = \mathbf{S}, \qquad \mathbf{l} + \mathbf{l}' = \mathbf{L}, \qquad \mathbf{S} + \mathbf{L} = \mathbf{J}, \\ &\mathbf{s} + \mathbf{l} = \mathbf{j}, \qquad \mathbf{s}' + \mathbf{l}' = \mathbf{j}', \qquad \mathbf{j} + \mathbf{j}' = \mathbf{J}. \end{aligned} \qquad (6.16$$

The nuclear matrix element is defined so that the Jth rank operator is sandwiched in the initial- and final-state wave functions and it is integrated over the whole space. The first term in the vector coupling in Eq. (6.2) or that of Eq. (6.3) corresponds to the spin singlet state of the lepton system. These terms contain the nuclear matrix elements with

$$S = 0 \qquad \text{and} \qquad L = J. \qquad (6.17$$

From this, we imagine that the spin triplet state of the lepton system is also available. This is, in fact, given by the first term of the axial vector coupling in Eq. (6.2),

$$(\psi_p^\dagger \sigma \psi_n)(\psi_e^\dagger \sigma \psi_\nu). \qquad (6.18$$

Since $S = 1$ in this case, there are three possible values of J for a fixed L,

$$J = L - 1, \quad L, \quad L + 1, \qquad \text{for} \quad S = 1. \qquad (6.19$$

A few examples of these matrix elements are as follows.

Allowed Transition

$$L = 0, \quad J = 1, \qquad \int \boldsymbol{\sigma}. \tag{6.20}$$

First Forbidden Transition

$$\begin{aligned} &L = 1, \quad J = 0, \qquad \int \boldsymbol{\sigma} \cdot \mathbf{r}, \\ &L = 1, \quad J = 1, \qquad \int \boldsymbol{\sigma} \times \mathbf{r}, \\ &L = 1, \quad J = 2, \qquad \int B_{ij}. \end{aligned} \tag{6.21}$$

The last term is the spurless symmetrical tensor,

$$B_{ij} \sim (x_i \sigma_j + x_j \sigma_i - \tfrac{2}{3} \delta_{ij} \mathbf{r} \cdot \boldsymbol{\sigma}).$$

Instead of the original definition by Konopinski and Uhlenbeck, we use the spherical representation with a relation

$$\left| \int B_{ij} \right|^2 = \sum_{ij} |B_{ij}|^2. \tag{6.22}$$

Explicitly,

$$\int \psi^{\dagger}_{J_f M_f} \sum_{k=1}^{A} (x^{(k)} + iy^{(k)})(\sigma_x^{(k)} + i\sigma_y^{(k)}) \tau_+^{(k)} \psi_{J_i M_i} \, d\mathbf{r}_1 \cdots d\mathbf{r}_A$$

$$= (J_i 2 M_i 2 \mid J_f M_f) \mathcal{M}(B_{ij}), \tag{6.23}$$

and

$$\int B_{ij} = \left(\frac{2J_f + 1}{2J_i + 1} \right)^{1/2} \mathcal{M}(B_{ij}). \tag{6.24}$$

Here we do not have a factor of 2, as in Eq. (6.11). For the higher rank tensors we use the following symbols.

Second Forbidden Transition

$$\begin{aligned} &L = 2, \quad J = 1, \qquad \int (\boldsymbol{\sigma} \cdot \mathbf{r}) \mathbf{r} \quad \text{and} \quad \int \boldsymbol{\sigma} r^2, \\ &L = 2, \quad J = 2, \qquad \int T_{ij}, \\ &L = 2, \quad J = 3, \qquad \int S_{ijk}. \end{aligned} \tag{6.25}$$

Now, let us proceed to the discussions on the relativistic terms which are the second terms in the vector and axial vector couplings in Eq. (6.2),

$$C_V (\psi_p^{\dagger} \boldsymbol{\alpha} \psi_n)(\psi_e^{\dagger} \boldsymbol{\alpha} \psi_\nu) \qquad \text{and} \qquad C_A (\psi_p^{\dagger} \gamma_5 \psi_n)(\psi_e^{\dagger} \gamma_5 \psi_\nu). \tag{6.26}$$

Replacing the lepton parts by unity, we have the matrix lements $\int \boldsymbol{\alpha}$ and $\int \gamma_5$. If we make nonrelativistic approximation for nucleons, as was done in Eq. (2.57), we have

$$\int \boldsymbol{\alpha} = -\frac{1}{M}\int \mathbf{p} \qquad \text{and} \qquad \int \gamma_5 = -\frac{1}{M}\int \boldsymbol{\sigma}\cdot\mathbf{p}. \tag{6.27}$$

Since the matrix contains the differential operator, we have to consider an additional minus sign for the parity selection rule. The lowest term of the power series expansion of the lepton covariant gives the nuclear matrix elements with the parity change Yes, which belong to the first forbidden transition. The first few examples for these relativistic terms are

$$\begin{aligned} &L = 0, \quad J = 1, \qquad \int \boldsymbol{\alpha}, \\ &L = 1, \quad J = 0, \qquad \int \boldsymbol{\alpha}\cdot\mathbf{r}, \\ &L = 1, \quad J = 1, \qquad \int \boldsymbol{\alpha}\times\mathbf{r}, \\ &L = 1, \quad J = 2, \qquad \int A_{ij}, \quad \text{etc.}; \end{aligned} \tag{6.28}$$

and

$$\begin{aligned} &L = 0, \quad J = 0, \qquad \int \gamma_5, \\ &L = 1, \quad J = 1, \qquad \int \gamma_5\mathbf{r}, \\ &L = 2, \quad J = 2, \qquad \int \gamma_5 R_{ij}, \quad \text{etc.} \end{aligned} \tag{6.29}$$

Here $\int A_{ij}$ is again defined by Eqs. (6.23) and (6.24), by replacing $\boldsymbol{\sigma}$ by $\boldsymbol{\alpha}$. These matrices in Eqs. (6.28) and (6.29) are also called momentum-type matrices, since they have the momentum operator in the nonrelativistic approximation (see Eq. (6.27)). On the other hand, those in Eqs. (6.5), (6.20), and (6.21) are coordinate-type matrices.

The momentum-type matrices have a contribution to the transition amplitude of the order of

$$\frac{v_{\mathrm{N}}}{c}\,(pR)^L, \tag{6.30}$$

while the parity selection rule is

$$(-)^{L+1}. \tag{6.31}$$

Here v_{N} is the nucleon velocity in the nucleus, $v_{\mathrm{N}}/c \sim 0.1$. Therefore, those with L are classified as the $(L+1)$th forbidden matrix elements. Comparing the classification of the forbidden transition, Eq. (6.1), we notice that only those matrices with $L + 1 = J$ are important unless p is extremely high.

Table 6.1 Nuclear Matrix Elements for Beta Decay

Nuclear Matrix Element	Forbiddenness	Angular Momenta of Lepton System			Parity Change
		S	L	J	
$\int 1$	Allowed	0	0	0	No
$\int \boldsymbol{\sigma}$	Allowed	1	0	1	No
$\int \gamma_5$	First	1^a	1^b	0	Yes
$\int \mathbf{r}$	First	0	1	1	Yes
$\int \boldsymbol{\alpha}$	First	0	1^b	1	Yes
$\int \boldsymbol{\sigma}\cdot\mathbf{r}$	First	1	1	0	Yes
$\int \boldsymbol{\sigma}\times\mathbf{r}$	First	1	1	1	Yes
$\int B_{ij}$	First	1	1	2	Yes
$\int R_{ij}$	Second	0	2	2	No
$\int A_{ij}$	Second	0	2^c	2	No
$\int T_{ij}$	Second	1	2	2	No
$\int S_{ijk}$	Second	1	2	3	No

[a] The effective p wave couples with the spin operator so as to make the pseudoscalar.
[b] L is effectively 1. The differential operator changes the s wave into the effective p wave.
[c] L is effectively 2. The differential operator changes the p wave into the effective d wave.

Summarizing discussions, we have Table 6.1 for nuclear matrix elements. The numbers of these matrix elements are two, six, and four for the allowed, first forbidden, and second as well as higher forbidden transitions, respectively. Since the rank of the tensor is J, the triangular condition $\Delta(J_i, J, J_f)$ should be satisfied.

In order to construct the theory of forbidden beta decays with the Coulomb corrections, we have to generalize the theory developed in section 2.7. Such a formalism is most easily made in the spherical representation, keeping in mind the definition of nuclear matrix elements introduced by Konopinski and Uhlenbeck. This is quite similar to that of muon capture in complex nuclei, which will be developed in section 8.4. A very slight difference in section 8.4 is the classification of nuclear matrix elements to each forbiddenness. Here, in the theory of beta decay, we classify the matrix elements to each forbiddenness by the orbital angular momentum L as given, e.g., by Eqs. (6.20), (6.21), and (6.25), in accordance with the magnitude of the matrix elements (see Eqs. (6.9) and (6.30)). On the other hand, in the theory of muon capture, each matrix element is classified by its rank of tensor J. For example, $\int r^2$ is in the allowed transition instead of the second one, so that we will introduce the matrix $\int j_0(qr)$ instead of the Fermi matrix element of the beta decay. In this manner, we can find the formulation for the beta decay from that of section 8.4. We will, therefore, not reproduce it here, and leave it as a reader's exercise. Formulations of the forbidden beta decays are also reviewed beautifully by Konopinski [2] and Weidermüller [3].

6.2 γ_5 TRANSFORMATION, A RELATION OF THE FORMULAS FOR THE nth AND $(n \pm 1)$th FORBIDDEN TRANSITIONS

In this section, we will show a principle by means of which we can introduce the $V - A$ four-fermion interaction. Using this principle, we can demonstrate a similarity between the nth and the $(n \pm 1)$th forbidden transitions in beta decay. With this relation the labor of calculation of the transition probability is always reduced by a factor of 2. The general rule for deriving the relativistic corrections to the allowed transitions is shown. These topics are studied by the present author in [4]. Let us first study the properties of the four-fermion interaction (which is not necessarily the interaction for beta decay) under the γ_5 transformations.

γ_5 SYMMETRIES AND THE $V - A$ INTERACTION

The four-fermion interaction responsible for the weak decays has many interesting symmetry properties with respect to the transformation of the four fields (or wave functions of four fermions in the sense of Chapter 3) involving γ_5, such as the mass reversal transformation $\psi \to \gamma_5 \psi$, $m \to -m$ [5], or the chiral transformation $\psi \to \gamma_5 \psi$ [6]. (To avoid ambiguity, we assume the phase factor η in the transformation $\psi \to \eta\gamma_5 \psi$ to be equal to unity for all fermions.) In this subsection, these properties of the four-fermion interaction are studied. Through these studies we can find a principle for which the four-fermion interaction is $V - A$.

Let us again write down the most general four-fermion interaction as usual,

$$
\begin{aligned}
H_{\text{int}} &= (\tfrac{1}{2})^{1/2} \sum_i (\bar{\psi}_1 O_i \psi_2)[\bar{\psi}_3 (C_i O_i + C_i' O_i \gamma_5)\psi_4] + \text{h.c.} \\
&= aS + bV + cT + dA + eP, \qquad (6.32)
\end{aligned}
$$

satisfying the invariance requirement under proper Lorentz transformations. In the second line of Eq. (6.32), a, b, c, d, and e are certain constants to be determined by further assumptions. The Dirac equation in Eq. (3.7)

$$(i\gamma_\mu p_\mu + m)\psi = 0 \qquad (6.33)$$

and its conjugate are invariant under the transformation

$$\psi \to \gamma_5 \psi, \qquad \bar{\psi} \to -\bar{\psi}\gamma_5, \qquad m \to -m. \qquad (6.34)$$

We apply this transformation to two of the four fields, e.g.,

$$\psi_1 \to \gamma_5 \psi_1 \, (\bar{\psi}_1 \to -\bar{\psi}_1 \gamma_5),\ \psi_2 \to \gamma_5 \psi_2,\ m_1 \to -m_1, \text{ and } m_2 \to -m_2.$$

Then the terms of the interaction Hamiltonian (6.32) divide into two classes [7, 8], namely, *VA* (invariant) and *STP* (pseudoinvariant), the phases and relative magnitudes among the members of each group being arbitrary. The invariant (pseudoinvariant) refers to the upper (lower) sign of the equation

$$H_{\text{int}}(\text{transformed}) = \pm H_{\text{int}}. \tag{6.35}$$

A similar division of the interaction Hamiltonian arises from the transformation for fields 3 and 4. The results obtained by performing the same transformation for each pair of the four fermion fields are given in Table 6.2. Now,

Table 6.2 Invariance Properties of the Four-Fermion Interactions under the Mass-Reversal Transformation Eq. (6.34) for Two Fields Simultaneously [4]

Fields to be Transformed	Interaction[a]	
1 and 2 or 3 and 4	Invariant	$bV + dA$
	Pseudoinvariant	$aS + cT + eP$
1 and 3 or 2 and 4	Invariant	$a(S+P) + cT + b(V-A)$
	Pseudoinvariant	$a(S-P) + b(V+A)$
1 and 4 or 2 and 3	Invariant	$a(S-P) + b(V-A)$
	Pseudoinvariant	$a(S+P) + cT + b(V+A)$
Every pair of the four fields.	Invariant or pseudoinvariant	$V-A$ or $V+A$ or $S-P$ or $S+P+cT$
	Invariant	$V-A$

[a] *a, b, c, d,* and *e* are arbitrary constants.

we postulate that the four-fermion interaction responsible for a weak decay must be invariant under the transformation in Eq. (6.34) for every pair of the four fields. Then the interaction Hamiltonian Eq. (6.32) reduces to $V-A$:

$$\begin{aligned}2(V-A) = {} & [\bar{\psi}_1\gamma_\mu(1+\gamma_5)\psi_2][\bar{\psi}_3\gamma_\mu(1+\gamma_5)\psi_4](C+C') \\ & + [\bar{\psi}_1\gamma_\mu(1-\gamma_5)\psi_2][\psi_3\gamma_\mu(1-\gamma_5)\psi_4](C-C') + \text{h.c.},\end{aligned} \tag{6.36}$$

where

$$C = C_{\text{V}} = -C_{\text{A}} \qquad \text{and} \qquad C' = C'_{\text{V}} = -C'_{\text{A}}.$$

The ratio C'/C is not determined from the present postulate, but needs at least an experimental result, such as the longitudinal polarization of the neutrino, to determine its value. This kind of principle to determine $V-A$ in Eq. (6.36) has a certain importance, although we know several theories [9-11] that can derive a unique form of $V-A$ as described in Chapter 4.

In fact, one of the γ_5 symmetries of H_β is very useful to derive the relativistic corrections to the allowed beta decay, and to give a symmetry of formulas for the nth and the $(n \pm 1)$th forbidden transitions. This will be given in the next subsection.

The unique form of $V - A$ theory is derived by a more stringent condition. For example, if we require the invariance of Eq. (6.32) under the transformation (6.34) for every one of the four fields separately (i.e., not simultaneously) [9], then we have a unique form of four-fermion interaction

$$H_{\text{int}} = C[\bar{\psi}_1 \gamma_\mu (1 + \gamma_5)\psi_2]\,[\bar{\psi}_3 \gamma_\mu (1 + \gamma_5)\psi_4] + \text{h.c.}, \tag{6.37}$$

which is $V - A$ with $C = C'$. The same form of four-fermion interaction has been also obtained by Sudarshan and Marshak by postulating that the wave functions of the four fermions have the same chirality eigenvalue [11], and by Feynman and Gell-Mann, who approached the same problem rather differently [10].

General Rule for Transformation of Formulas

Now we shall consider a special case of beta decay. In particular, we are interested in the transformation for neutral particles. The interaction Hamiltonian of beta decay is

$$\begin{aligned} 2^{1/2}H_\beta = {} & (\psi_\text{p}^\dagger \psi_\text{n})[\psi_\text{e}^\dagger(C_\text{V} + \gamma_5 C_\text{V}')\psi_\nu] - (\psi_\text{p}^\dagger \alpha \psi_\text{n})[\psi_\text{e}^\dagger \alpha (C_\text{V} + \gamma_5 C_\text{V}')\psi_\nu] \\ & + (\psi_\text{p}^\dagger \sigma \psi_\text{n})[\psi_\text{e}^\dagger \sigma (C_\text{A} + \gamma_5 C_\text{A}')\psi_\nu] \\ & - (\psi_\text{p}^\dagger \gamma_5 \psi_\text{n})[\psi_\text{e}^\dagger \gamma_5 (C_\text{A} + \gamma_5 C_\text{A}')\psi_\nu] + \text{h.c.} \end{aligned} \tag{6.38}$$

Here we do not specify the relative magnitudes of the vector and axial vector coupling constants. If we apply the transformation

$$\begin{aligned} &\psi_\text{n} \to \gamma_5 \psi_\text{n}, \qquad m_\text{n} \to -m_\text{n}, \qquad \psi_\nu \to \gamma_5 \psi_\nu, \qquad m_\nu \to -m_\nu, \\ &C_\text{V} \rightleftharpoons -C_\text{A}, \qquad C_\text{V}' \rightleftharpoons -C_\text{A}' \end{aligned} \tag{6.39}$$

to Eq. (6.38), it follows from the invariance of H_β that the coupling constants of the vector and axial vector interactions must have a phase difference of π. Furthermore, we can show a similarity between the nth and the $(n \pm 1)$th forbidden transitions as follows. ($n = 0$ refers to the allowed transition.) Under the same transformation, a term in the transition probability, e.g.,

$$\begin{aligned} &(\psi_\text{p}^\dagger \alpha \psi_\text{n})^* (\psi_\text{p}^\dagger \sigma \times \mathbf{r} \psi_\text{n})[\psi_\text{e}^\dagger(-\alpha)(C_\text{V} + \gamma_5 C_\text{V}')\psi_\nu]^* \\ &\quad \times [\psi_\text{e}^\dagger(\sigma \times \mathbf{P})(C_\text{A} + \gamma_5 C_\text{A}')\psi_\nu], \end{aligned} \tag{6.40}$$

which appears in the first forbidden transition, becomes

$$(\psi_p^\dagger \sigma \psi_n)^*(\psi_p^\dagger \alpha \times \mathbf{r} \psi_n)[\psi_e^\dagger \sigma (C_A + \gamma_5 C_A')\psi_\nu]^* \\ \times [\psi_e^\dagger (-\alpha \times \mathbf{P})(C_V + \gamma_5 C_V')\psi_\nu], \tag{6.41}$$

which appears in the allowed transition. Here **P** is the momentum of the electron–neutrino system. These two terms have the same dependence on the electron energy, which can be proved by using the identity

$$(\sigma \times \mathbf{P})\left[1 - \frac{\alpha \cdot \mathbf{q} - m_\nu \beta}{E_\nu}\right]\alpha = (\alpha \times \mathbf{P})\left[1 - \frac{\alpha \cdot \mathbf{q} + m_\nu \beta}{E_\nu}\right]\sigma, \tag{6.42}$$

where the expression in the square brackets indicates the projection operator for the neutrino which appears in the trace of Eq. (6.40). On the other hand, the trace of Eq. (6.41) contains the same expression as on the right-hand side of Eq. (6.42) except for the sign of the m_ν term. Therefore, the electron-energy dependence is identical for Eqs. (6.40) and (6.41) if m_ν vanishes. In general, all formulas concerning beta decay retain the same energy dependence under the transformation

$$C_V \rightleftharpoons C_A \gamma_5, \qquad C_V \alpha \rightleftharpoons C_A \sigma, \\ C_V' \rightleftharpoons C_A' \gamma_5, \qquad C_V' \alpha \rightleftharpoons C_A' \sigma, \tag{6.43}$$

where the Dirac operators act on nucleon wave functions. In these interchanges of the coupling constants, the minus sign is not required, since the transition probability is a function of the square of the coupling constants. Some examples of the similarity between nuclear matrix elements for the nth and the $(n \pm 1)$th forbidden transitions are listed in Table 6.3. Using this fact, the labor of calculation of the transition probabilities in beta decay is always reduced by half.

An application of this invariance property is given below. A formula with the relativistic corrections for the allowed transition can be obtained from that for the first forbidden transition with the replacement

$$\int \gamma_5 \to \int 1, \qquad \int \sigma \cdot \mathbf{r} \to \int \alpha \cdot \mathbf{r}, \\ \int \alpha \to \int \sigma, \qquad \int \sigma \times \mathbf{r} \to \int \alpha \times \mathbf{r}, \\ \int \mathbf{r} \to \int \gamma_5 \mathbf{r}, \qquad \int B_{ij} \to \int A_{ij} \\ C_V \rightleftharpoons C_A, \qquad C_V' \rightleftharpoons C_A', \tag{6.44}$$

and vice versa. As we have studied in section 6.1 and in Table 6.1, the matrix elements in the left-hand side of Eq. (6.44) are the proper first forbidden matrix elements for which formulas for various beta-decay phenomena are published. (The proper or improper matrix elements refer to those listed in Table 6.1 or not.) From these formulas we can derive the relativistic corrections for the allowed transitions with explicit calculations. The replacement above does not involve the corrections due to the coordinate-type matrix elements of the second forbidden transition, which are smaller than relativistic corrections but not negligible in some cases (see the first paper in [4]). If the formula for the first forbidden transition involved the corrections due to the third forbidden transition, especially $\int \boldsymbol{\alpha} r^2$, $\int (\boldsymbol{\alpha}\cdot\mathbf{r})\mathbf{r}$, and $\int \gamma_5 r^2$ in the fourth column of Table 6.3, we could derive the formula for the allowed transition with relativistic corrections as well as corrections due to the

Table 6.3 Correspondence between Nuclear Matrix Elements[a]

Allowed	First Forbidden	Second Forbidden	Third Forbidden
$C_V \int 1$	$C_A \int \gamma_5$		
$C_A \int \boldsymbol{\sigma}$	$C_V \int \boldsymbol{\alpha}$		
	$C_V \int \mathbf{r}$	$C_A \int \gamma_5 \mathbf{r}$	
	$C_A \int \boldsymbol{\sigma}\cdot\mathbf{r}$	$C_V \int \boldsymbol{\alpha}\cdot\mathbf{r}$	
	$C_A \int \boldsymbol{\sigma}\times\mathbf{r}$	$C_V \int \boldsymbol{\alpha}\times\mathbf{r}$	
	$C_A \int B_{ij}$	$C_V \int A_{ij}$	
		$C_A \int \boldsymbol{\sigma} r^2$	$C_V \int \boldsymbol{\alpha} r^2$
		$C_A \int (\boldsymbol{\sigma}\cdot\mathbf{r})\mathbf{r}$	$C_V \int (\boldsymbol{\alpha}\cdot\mathbf{r})\mathbf{r}$
		$C_A \int T_{ij}$	$C_V \int U_{ij}$
		$C_A \int S_{ijk}$	$C_V \int V_{ijk}$
		$C_V \int r^2$	$C_A \int \gamma_5 r^2$
		$C_V \int R_{ij}$	$C_A \int R_{ij}\gamma_5$
			etc.

[a] By interchanging the nuclear matrix elements in the same row, the formula for the nth forbidden transition is transformed into that for the $(n \pm 1)$th forbidden transition, and vice versa. ($n = 0$ refers to the allowed transition.) A_{ij}, B_{ij}, R_{ij}, $R_{ij}\gamma_5$, T_{ij}, and U_{ij} are the second rank tensors constructed with $\mathbf{r}$ and $\boldsymbol{\alpha}$, $\boldsymbol{\sigma}$, $\mathbf{r}$, $\gamma_5\mathbf{r}$, $\boldsymbol{\sigma}\times\mathbf{r}$, and $\boldsymbol{\alpha}\times\mathbf{r}$, respectively. S_{ijk} (or V_{ijk}) is the third rank tensor constructed with $\mathbf{r}$, $\mathbf{r}$, and $\boldsymbol{\sigma}$(or $\boldsymbol{\alpha}$).

coordinate-type matrix elements, $\int \boldsymbol{\sigma} r^2$, $\int (\boldsymbol{\sigma}\cdot\mathbf{r})\mathbf{r}$ and $\int r^2$, by the same rule (6.43). Unfortunately, this is not the case, and improper third forbidden matrix elements $\int \boldsymbol{\alpha} r^2$, $\int (\boldsymbol{\alpha}\cdot\mathbf{r})\mathbf{r}$, and $\int \gamma_5 r^2$ are usually omitted from the Table 6.1. We have, therefore, to compute explicitly the effect due to $\int \boldsymbol{\sigma} r^2$, $\int (\boldsymbol{\sigma}, \mathbf{r})\mathbf{r}$, and $\int r^2$, and possibly we derive the effect due to $\int \boldsymbol{\alpha} r^2$, $\int (\boldsymbol{\alpha}\cdot\mathbf{r})\mathbf{r}$, and $\int \gamma_5 r^2$ for the third forbidden transition. The formulas with these small terms for the beta-ray spectrum and directional correlation are given in the first paper in [4], while those for parity-nonconserving phenomena are given in [12].

In the next section, we collect formulas for the allowed and first forbidden transitions in various beta-decay phenomena.

6.3 FORMULAS FOR THE ALLOWED AND FIRST FORBIDDEN TRANSITIONS

We have already given a number of formulas for the allowed transitions in Chapters 2 and 3. Since, however, the higher-order corrections are not included in those formulas, we collect here the formulas for the allowed transitions with these corrections. On the other hand, we give the formulas for the first forbidden transitions only with the proper matrix elements in Table 6.1, by neglecting the effect due to the third forbidden matrices in Table 6.3.

Most of the formulas can be expressed in terms of the particle parameters $b_{LL'}^{(n)}$ for beta decays defined in Eqs. (3.62), (3.63), etc., so that we give these parameters in this section. They consist of nuclear matrix elements, beta-decay coupling constants, neutrino energy, and combinations of the electron wave functions g_κ and f_κ. We adopt an approximation where these combinations of the electron wave functions are almost constant in the nucleus, so that they are removed from the integrand and evaluated at the nuclear surface. We adopt the atomic units $\hbar = m = c = 1$ unless we specifically mention other units. The decay scheme is $j \to j_1, j \to j_1 \to j_2$, or $j \to j_1 \to j_2 \to j_3$ (see Fig. 6.1). j and m are the nuclear spin and its magnetic number, respectively. L is the rank of the matrix element for beta decay or the multipolarity of the gamma ray. Appropriate subscripts are added in conformity with the decay scheme. P_n (cos θ) is the Legendre polynomial of the nth order;

$$\begin{aligned}
&P_0(\cos\theta) = 1, \qquad\qquad P_1(\cos\theta) = \cos\theta,\\
&P_2(\cos\theta) = \tfrac{1}{2}(3\cos^2\theta - 1), \qquad P_3(\cos\theta) = \tfrac{1}{2}(5\cos^3\theta - 3\cos\theta),\\
&P_4(\cos\theta) = \tfrac{1}{8}(35\cos^4\theta - 30\cos^2\theta + 3),\\
&P_5(\cos\theta) = \tfrac{1}{8}(63\cos^5\theta - 70\cos^3\theta + 15\cos\theta).
\end{aligned} \tag{6.45}$$

$(j_1 j_2 m_1 m_2 \mid jm)$ and $W(abcd; ef)$ are the 3-j and 6-j symbols, respectively. τ is $+1$ (-1) for the right-handed (left-handed) circular polarization of the gamma rays. δ is equal to 0 (+1) for magnetic (electric) radiation. The reduced nuclear matrix elements for gamma transition are expressed by $\langle j_2 \| L \| j_1 \rangle$, etc.

We have assumed parity conservation for nuclear states and for electromagnetic interactions, although in the last few years certain evidence has been reported for the parity-nonconserving weak nuclear forces (see Chapter 4). Consequently, we have a condition

$$\delta + \delta' + L + L' = \text{even}. \tag{6.46}$$

Without such a condition (e.g., in the case of parity nonconservation in nuclear forces), formulas can easily be derived from those given here. We have assumed time-reversal invariance for strong interactions, so that the products of nuclear matrix elements for beta decay in the brackets are all real. Time-reversal invariance or noninvariance in beta and electromagnetic interactions can be tested by using the beta-gamma angular correlation in oriented nuclei [13] and the beta-gamma-gamma angular correlation in triple cascade transition [14], respectively.

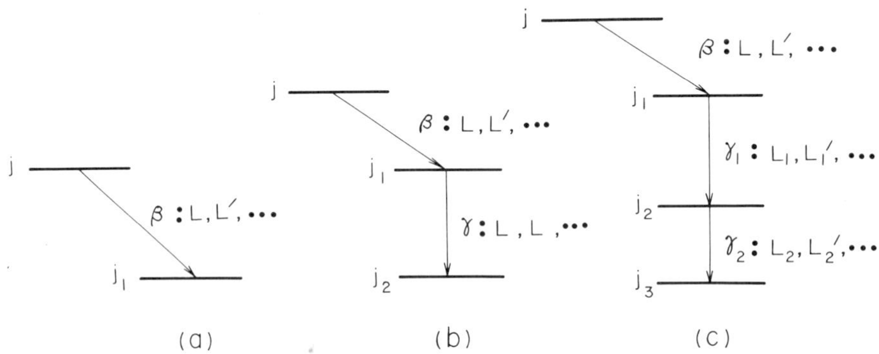

Fig. 6.1 Decay schemes.

We give the formulas for some typical phenomena in beta decay. These are of frequent use. There are many other formulas for different phenomena. We have, however, no space to include those. Instead, we shall cite references at the end of this section. In the following formulas, the constant statistical factors $(2j+1)^{1/2}$, etc., sometimes appear more than once. These are repeated for the purpose of normalization.

Beta-Ray Spectrum

$$P(E)\,dE = \frac{1}{4\pi^3} F_0(\pm Z, E) pEq^2 C(E)\,dE \tag{6.47}$$

with the shape correction factor

$$C(E) = b_{00}^{(0)} - (\tfrac{1}{3})^{1/2} b_{11}^{(0)} + (\tfrac{1}{5})^{1/2} b_{22}^{(0)} \tag{6.48}$$

and the neutrino energy

$$q = E_o - E. \tag{6.49}$$

Here, E and E_o are the electron energy and its maximum value, respectively. The electron momentum is, therefore, given by

$$p = (E^2 + 1)^{1/2}. \tag{6.50}$$

In the denominator of the right-hand side of Eq. (6.47), the factor is $4\pi^3$ instead of $2\pi^3$. This is because the particle parameters include the coupling constant $2\,|C_V|^2 = |C_V|^2 + |C_V'|^2$, etc., and we keep the magnitude of C_V as the old parity-conserving theory. For this purpose we adopt a factor $(\frac{1}{2})^{1/2}$ for the interaction Hamiltonian (see Eqs. (3.1) and (6.32)). The shape correction factor defined in Eq. (6.47) is different from that in Eq. (2.143) with a factor L_0, since we adopt F_0 instead of F as the Fermi function.

$\mathrm{ft}_{1/2}$ Value for a Particular Matrix Element $\int \Omega$

We have to define the corrected integrated Fermi function f_c by

$$f_c = \frac{1}{2\,|C_V|^2 |\int \Omega|^2} \int_1^{E_o} F_0(\pm Z, E) p E q^2 C(E)\, dE \qquad \text{for } \beta^{\mp}. \tag{6.51}$$

The $ft_{1/2}$ value is now given by

$$f_c t_{1/2} = \frac{6146 \pm 10}{|\int \Omega|^2}\ \text{sec} \tag{6.52}$$

where $t_{1/2}$ is in seconds, while all other quantities should be in the atomic units.

Beta-Ray Angular Distribution in Oriented Nuclei

$$W(\theta:\beta) = \sum_n \sum_{L \leqslant L'} \hat{f}_n(j)(-)^{j_1 - j + L + L' + n} \times W(jjLL', nj_1)(2j+1)^{1/2} b_{LL'}^{(n)} P_n(\cos\theta) \tag{6.53}$$

with

$$\hat{f}_n(j) = \sum_m (-)^{j-m} (jjm - m\,|\,n0)(2j+1)^{1/2} a_m$$

and $\Sigma_m a_m = 1$ where the a_m's are the relative populations of the initial magnetic substates (see Appendix A.6). For the nuclear polarization, n can take both even and odd integrals, while for the nuclear alignment n should be even. The angle θ is the direction of the momentum of the beta particle with respect to the nuclear orientation axis.

Gamma-Ray Angular Distribution Following Beta Ray in Oriented Nuclei

(a) The decay scheme is $j \xrightarrow{\beta} j_1 \xrightarrow{\gamma} j_2$, and the beta ray is unobserved.

$$W(\theta:\gamma) = (-)^{j+j_1} \sum_n \left\{ \hat{f}_n(j) \left[\sum_L \int_1^{E_0} F_0(\pm Z, E) pEq^2 (2L+1)^{-1/2} b_{LL}^{(0)}\, dE \right.\right.$$
$$\left. \times (2j+1)^{1/2}(2j_1+1)^{1/2} W(jjj_1j_1, nL) \right]$$
$$\left.\times \left[\sum_{L_1, L_1'} \tau^n \langle j_2 \| L_1 \| j_1 \rangle \langle j_2 \| L_1' \| j_1 \rangle F_n(L_1 L_1' j_2 j_1) \right] \right\} P_n(\cos\theta)$$
$$\text{for } \beta^{\mp}. \qquad (6.54)$$

Here, the geometrical factor F_n for the gamma ray is generally given by

$$F_n(LL'j_aj_b) = F_n(L'Lj_aj_b)$$
$$= (-)^{j_a - j_b - 1}[(2j_b+1)(2L+1)(2L'+1)]^{1/2}$$
$$\times (LL'1 -1 \mid n0) W(j_bj_bLL', nj_a). \qquad (6.55)$$

The angle θ is the direction of the gamma-ray momentum with respect to the orientation axis. The integral n is even and odd if the nucleus is polarized *and* the circular polarization of the gamma ray is observed, otherwise it is even.

(b) Angular distribution of the second gamma ray following beta ray in oriented nuclei. The decay scheme is

$$j \xrightarrow{\beta} j_1 \xrightarrow{\gamma_1} j_2 \xrightarrow{\gamma_2} j_3,$$

and the beta ray and the first gamma ray are unobserved.

$$W(\theta:\gamma_2) = (-)^{j_2 - j} \sum_n \left\{ \hat{f}_n(j) \left[\sum_L \int_1^{E_0} F_0(\pm Z, E) pEq^2 (2L+1)^{-1/2} b_{LL}^{(0)}\, dE \right.\right.$$
$$\left. \times (2j+1)^{1/2}(2j_1+1)^{1/2} W(jjj_1j_1, nj) \right]$$
$$\times \left[\sum_{L_1} (-)^{L_1} \langle j_2 \| L_1 \| j_1 \rangle^2 W(j_1j_1j_2j_2, nL_1)(2j_1+1)^{1/2} \right.$$
$$\left. \times (2j_2+1)^{1/2} \right] \left[\sum_{L_2, L_2'} \tau^n \langle j_3 \| L_2 \| j_2 \rangle \langle j_3 \| L_2' \| j_2 \rangle \right.$$
$$\left.\left. \times F_n(L_2L_2'j_3j_2) \right] \right\} P_n(\cos\theta). \qquad (6.56)$$

The angle θ is the direction of the gamma-ray momentum with respect to the nuclear orientation axis. The integral n is even and odd if the nucleus is polarized *and* the circular polarization of gamma ray is observed, otherwise it is even.

Next, we give the formulas for the beta decay of unoriented nuclei. In these formulas, θ is the angle between the two momenta of the relevant radiations.

Beta-Gamma Circular Polarization Correlation

The decay scheme is $j \xrightarrow{\beta} j_1 \xrightarrow{\gamma} j_2$.

$$W(\theta, \tau:\beta-\gamma) = \sum_n \left\{ \left[\sum_{L \leqslant L'} (-)^{j_1 - j + n} b_{LL'}^{(n)} W(j_1 j_1 LL', nj)(2j_1+1)^{1/2} \right] \times \left[\sum_{L_1, L_1'} \tau^n \langle j_2 \| L_1 \| j_1 \rangle \langle j_2 \| L_1' \| j_1 \rangle F_n(L_1 L_1' j_2 j_1) \right] \right\} \times P_n(\cos\theta). \tag{6.57}$$

Here the summation n runs over even and odd integers. The degree of the circular polarization of the gamma ray in coincidence with beta ray is given by

$$P_\gamma(\theta) = \frac{\sum_{n=\text{odd}} \text{ in Eq. (6.57)}}{\sum_{n=\text{even}} \text{ in Eq. (6.57)}}.$$

Beta-Gamma Directional Correlation

$$W(\theta:\beta-\gamma) = [\text{even } n \text{ terms in Eq. (6.57)}]. \tag{6.58}$$

Beta-Gamma Angular Correlation in the Triple Cascade Transition

The decay scheme is

$$j \xrightarrow{\beta} j_1 \xrightarrow{\gamma_1} j_2 \xrightarrow{\gamma_2} j_3.$$

γ_1 is unobserved.

$$W(\theta, \tau:\beta-\gamma_2) = \sum_n \left\{ \left[\sum_{L \leqslant L'} (-)^{j_1 - j + n} b_{LL'}^{(n)} W(j_1 j_1 LL', nj)(2j_1+1)^{1/2} \right] \times \left[\sum_{L_1} \langle j_2 \| L_1 \| j_1 \rangle^2 W(j_1 n L_1 j_2, j_1 j_2)(2j_1+1)^{1/2} \times (2j_2+1)^{1/2} \right] \left[\sum_{L_2, L_2'} \tau^n \langle j_3 \| L_2 \| j_2 \rangle \langle j_3 \| L_2' \| j_2 \rangle \times F_n(L_2 L_2' j_3 j_2) \right] \right\} P_n(\cos\theta). \tag{6.59}$$

Now we will give the particle parameters $b_{LL'}^{(n)}$. In their expressions, L_i, M_i, N_i, and similar symbols are combinations of the electron wave functions which are given below in Eqs. (6.88)-(6.104).

Particle Parameters

Allowed transitions

$$\begin{aligned} b_{00}^{(0)} &= |\textstyle\int \boldsymbol{\alpha}\cdot\mathbf{r}|^2(|C_V|^2+|C_V'|^2)[\frac{1}{9}q^2L_0-\frac{2}{3}qN_0+M_0] \\ &+|\textstyle\int 1|^2(|C_V|^2+|C_V'|^2)L_0 \\ &+\{i(\textstyle\int\boldsymbol{\alpha}\cdot\mathbf{r})^*(\textstyle\int 1)\}2(|C_V|^2+|C_V'|^2)[\frac{1}{3}qL_0-N_0]. \end{aligned} \tag{6.60}$$

$$\begin{aligned} b_{01}^{(1)} &= \{(\textstyle\int\boldsymbol{\alpha}\cdot\mathbf{r})^*(\textstyle\int\gamma_5\mathbf{r})\}\{2\,\mathrm{Re}\,(C_V^*C_A'+C_V'^*C_A)2[-\frac{1}{9}q^2\Lambda_1 \\ &-\frac{1}{3}q\mathbf{L}_{12}+\frac{1}{3}q\mathbf{N}_{11}+\mathbf{N}_{12}+m_1] \\ &+2\,\mathrm{Im}(C_V^*C_A'+C_V'^*C_A)2[\frac{1}{3}q\mathbf{H}_{12}-\mathbf{J}_{12}]\} \\ &-\{i(\textstyle\int\boldsymbol{\alpha}\cdot\mathbf{r})^*(\textstyle\int\boldsymbol{\sigma})\}\{2\,\mathrm{Re}\,(C_V^*C_A'+C_V'^*C_A)[\frac{2}{3}q\Lambda_1-\mathbf{N}_{11}] \\ &+2\,\mathrm{Im}(C_V^*C_A'+C_V'^*C_A)\mathbf{J}_{11}\} \\ &+\{i(\textstyle\int\boldsymbol{\alpha}\cdot\mathbf{r})^*(\textstyle\int\boldsymbol{\alpha}\times\mathbf{r})\}4\,\mathrm{Re}\,(C_V^*C_V')[-\frac{2}{9}q^2\Lambda_1-\frac{1}{3}q\mathbf{L}_{12}+\mathbf{N}_{12}-2m_1] \\ &-\{i(\textstyle\int 1)^*(\textstyle\int\gamma_5\mathbf{r})\}\{2\,\mathrm{Re}\,(C_V^*C_A'+C_V'^*C_A)[-\frac{2}{3}q\Lambda_1+\mathbf{N}_{11}-2\mathbf{L}_{12}] \\ &+2\,\mathrm{Im}\,(C_V^*C_A'+C_V'^*C_A)(\mathbf{J}_{11}+2\mathbf{H}_{12})\} \\ &-\{(\textstyle\int 1)^*(\textstyle\int\boldsymbol{\sigma})\}2\,\mathrm{Re}\,(C_V^*C_A'+C_V'^*C_A)2\Lambda_1 \\ &-\{(\textstyle\int 1)^*(\textstyle\int\boldsymbol{\alpha}\times\mathbf{r})\}4\,\mathrm{Re}\,(C_V^*C_V')[\frac{2}{3}q\Lambda_1+\mathbf{L}_{12}+\mathbf{N}_{11}]. \end{aligned} \tag{6.61}$$

$$\begin{aligned} b_{11}^{(0)} &= -3^{1/2}\Big[|\textstyle\int\gamma_5\mathbf{r}|^2(|C_A|^2+|C_A'|^2)[\frac{1}{3}q^2L_0-\frac{2}{3}qN_0+2L_1+M_0] \\ &+|\textstyle\int\boldsymbol{\sigma}|^2(|C_A|^2+|C_A'|^2)L_0 \\ &+|\textstyle\int\boldsymbol{\alpha}\times\mathbf{r}|^2(|C_V|^2+|C_V'|^2)[\frac{1}{6}q^2L_0+\frac{2}{3}qN_0+\frac{1}{2}L_1+M_0] \\ &+\{(\textstyle\int\boldsymbol{\sigma})^*(\textstyle\int\boldsymbol{\alpha}\times\mathbf{r})\}2\,\mathrm{Re}\,(C_A^*C_V+C_A'^*C_V')[\frac{1}{3}qL_0+N_0] \\ &+\{i(\textstyle\int\gamma_5\mathbf{r})^*(\textstyle\int\boldsymbol{\sigma})\}2(|C_A|^2+|C_A'|^2)[\frac{1}{3}qL_0-N_0] \\ &+\{i(\textstyle\int\gamma_5\mathbf{r})^*(\textstyle\int\boldsymbol{\alpha}\times\mathbf{r})\}2\,\mathrm{Re}\,(C_A^*C_V+C_A'^*C_V')(L_1-M_0)\Big]. \end{aligned} \tag{6.62}$$

$$\begin{aligned}
b_{11}^{(1)} = 2^{1/2}\Big[&-|\textstyle\int \gamma_5 \mathbf{r}|^2\, 2\,\mathrm{Re}\,(C_A^* C_A')2[\tfrac{1}{3}q\mathbf{L}_{12} + \tfrac{1}{3}q\mathbf{N}_{11} - \mathbf{N}_{12} - \Lambda_2 + m_1] \\
&+ |\textstyle\int \boldsymbol{\sigma}|^2\, 2\,\mathrm{Re}\,(C_A^* C_A')2\Lambda_1 \\
&+ |\textstyle\int \boldsymbol{\alpha}\times\mathbf{r}|^2 2\,\mathrm{Re}\,(C_V^* C_V')[\tfrac{1}{6}q^2\Lambda_1 - \tfrac{1}{3}q\mathbf{L}_{12} + \tfrac{2}{3}q\mathbf{N}_{11} - \mathbf{N}_{12} \\
&\quad + \tfrac{1}{2}\Lambda_2 - 2m_1] \\
&+ \{(\textstyle\int \boldsymbol{\sigma})^*(\int \boldsymbol{\alpha}\times\mathbf{r})\}\{2\,\mathrm{Re}(C_A^* C_V' + C_A'^* C_V)[\tfrac{2}{3}q\Lambda_1 + \mathbf{N}_{11} - \tfrac{1}{2}\mathbf{L}_{12}] \\
&+ 2\,\mathrm{Im}\,(C_A^* C_V' + C_A'^* C_V)[\mathbf{J}_{11} + \tfrac{1}{2}\mathbf{H}_{12}]\} \\
&- \{i(\textstyle\int \gamma_5 \mathbf{r})^*(\int \boldsymbol{\sigma})\}4\,\mathrm{Re}(C_A^* C_A')[-\tfrac{2}{3}q\Lambda_1 + \mathbf{L}_{12} + \mathbf{N}_{11}] \\
&+ \{i(\textstyle\int \gamma_5 \mathbf{r})^*(\int \boldsymbol{\alpha}\times\mathbf{r})\}\{2\,\mathrm{Re}(C_A^* C_V' + C_A'^* C_V)[\tfrac{1}{3}q^2\Lambda_1 \\
&- \tfrac{1}{2}q\mathbf{L}_{12} - \tfrac{1}{2}\mathbf{N}_{12} + \Lambda_2 + 2m_1] \\
&+ 2\,\mathrm{Im}(C_A^* C_V' + C_A'^* C_V)[\tfrac{2}{3}q\mathbf{J}_{11} - \tfrac{1}{6}q\mathbf{H}_{12} - \tfrac{3}{2}\mathbf{J}_{12}]\}\Big].
\end{aligned} \tag{6.63}$$

$$\begin{aligned}
b_{11}^{(2)} = -6^{1/2}\Big[&-|\textstyle\int \gamma_5 \mathbf{r}|^2(|C_A|^2 + |C_A'|^2)[-\tfrac{2}{3}qL_{12} + L_1 + 2N_{12}] \\
&+ |\textstyle\int \boldsymbol{\alpha}\times\mathbf{r}|^2(|C_V|^2 + |C_V'|^2)[\tfrac{1}{3}qL_{12} - \tfrac{1}{4}L_1 + N_{12}] \\
&+ \{(\textstyle\int \boldsymbol{\sigma})^*(\int \boldsymbol{\alpha}\times\mathbf{r})\}[2\,\mathrm{Re}(C_A^* C_V + C_A'^* C_V')\tfrac{1}{2}L_{12} \\
&+ 2\,\mathrm{Im}(C_A^* C_V + C_A'^* C_V')\tfrac{1}{2}H_{12}] \\
&+ \{i(\textstyle\int \gamma_5 \mathbf{r})^*(\int \boldsymbol{\sigma})\}2(|C_A|^2 + |C_A'|^2)L_{12} \\
&+ \{i(\textstyle\int \gamma_5 \mathbf{r})^*(\int \boldsymbol{\alpha}\times\mathbf{r})\}\{2\,\mathrm{Re}(C_A^* C_V + C_A'^* C_V')\tfrac{1}{2}(qL_{12} - L_1 + N_{12}) \\
&- 2\,\mathrm{Im}(C_A^* C_V + C_A'^* C_V')[\tfrac{1}{6}qH_{12} + \tfrac{3}{2}J_{12}]\}\Big].
\end{aligned} \tag{6.64}$$

$$\begin{aligned}
b_{02}^{(2)} = (\tfrac{1}{6})^{1/2}\Big[&\{i(\textstyle\int 1)^*(\int A_{ij})\}2(|C_V|^2 + |C_V'|^2)3L_{12} \\
&+ \{(\textstyle\int \boldsymbol{\alpha}\cdot\mathbf{r})^*(\int A_{ij})\}2(|C_V|^2 + |C_V'|^2)(-qL_{12} + 3N_{12})\Big].
\end{aligned} \tag{6.65}$$

$$\begin{aligned}
b_{12}^{(1)} = -\tfrac{1}{2}(\tfrac{5}{3})^{1/2}\Big[&\{(\textstyle\int \gamma_5 \mathbf{r})^*(\int A_{ij})\}\{2\,\mathrm{Re}(C_A^* C_V' + C_A'^* C_V)[-\tfrac{2}{3}q^2\Lambda_1 \\
&- q\mathbf{L}_{12} + 3\mathbf{N}_{12} - \tfrac{6}{5}\Lambda_2] + 2\,\mathrm{Im}\,(C_A^* C_V' + C_A'^* C_V)(q\mathbf{H}_{12} - 3\mathbf{J}_{12})\} \\
&+ \{i(\textstyle\int \boldsymbol{\sigma})^*(\int A_{ij})\}[2\,\mathrm{Re}(C_A^* C_V' + C_A'^* C_V)3\mathbf{L}_{12} \\
&\quad - 2\,\mathrm{Im}(C_A^* C_V' + C_A'^* C_V)3\mathbf{H}_{12}] \\
&+ \{i(\textstyle\int \boldsymbol{\alpha}\times\mathbf{r})^*(\int A_{ij})\}4\,\mathrm{Re}\,(C_V^* C_V')[-\tfrac{1}{3}q^2\Lambda_1 + q\mathbf{L}_{12} + 3\mathbf{N}_{12} + \tfrac{3}{5}\Lambda_2]\Big].
\end{aligned} \tag{6.66}$$

$$b_{12}^{(2)} = -\tfrac{1}{2}\Big[\{(\textstyle\int \gamma_5 \mathbf{r})^*(\int A_{ij})\}[2\,\mathrm{Re}(C_A^* C_V + C_A'^* C_V')$$
$$\times(-qL_{12} - 3L_1 + 3N_{12})$$
$$+ 2\,\mathrm{Im}(C_A^* C_V + C_A'^* C_V')(-qH_{12} + 3J_{12})]$$
$$+ \{i(\textstyle\int \sigma)^*(\int A_{ij})\}[2\,\mathrm{Re}(C_A^* C_V + C_A'^* C_V')3L_{12}$$
$$+ 2\,\mathrm{Im}(C_A^* C_V + C_A'^* C_V')3H_{12}]$$
$$+ \{i(\textstyle\int \boldsymbol{\alpha}\times\mathbf{r})^*(\int A_{ij})\}2(|C_V|^2 + |C_V'|^2)[qL_{12} + \tfrac{3}{2}L_1 + 3N_{12}]\Big]. \tag{6.67}$$

$$b_{12}^{(3)} = 9(\tfrac{1}{10})^{1/2}\Big[-\{(\textstyle\int \gamma_5 \mathbf{r})^*(\int A_{ij})\}2\,\mathrm{Re}(C_A^* C_V' + C_A'^* C_V)2\Lambda_2$$
$$+ \{i(\textstyle\int \boldsymbol{\alpha}\times\mathbf{r})^*(\int A_{ij})\}4\,\mathrm{Re}(C_V^* C_V')\Lambda_2\Big]. \tag{6.68}$$

$$b_{22}^{(0)} = |\textstyle\int A_{ij}|^2(|C_V|^2 + |C_V'|^2)[\tfrac{1}{12}q^2L_0 + \tfrac{3}{4}L_1]5^{1/2}. \tag{6.69}$$

$$b_{22}^{(1)} = -|\textstyle\int A_{ij}|^2\, 2\,\mathrm{Re}(C_V^* C_V')[\tfrac{1}{12}q^2\Lambda_1 + \tfrac{9}{20}\Lambda_2](10)^{1/2}. \tag{6.70}$$

$$b_{22}^{(2)} = -|\textstyle\int A_{ij}|^2(|C_V|^2 + |C_V'|^2)\tfrac{3}{4}(\tfrac{7}{2})^{1/2}L_1. \tag{6.71}$$

$$b_{22}^{(3)} = |\textstyle\int A_{ij}|^2\, 2\,\mathrm{Re}(C_V^* C_V')9(\tfrac{1}{10})^{1/2}\Lambda_2. \tag{6.72}$$

$$b_{22}^{(4)} = 0. \tag{6.73}$$

First forbidden transitions

$$b_{00}^{(0)} = |\textstyle\int \boldsymbol{\sigma}\cdot\mathbf{r}|^2(|C_A|^2 + |C_A'|^2)[\tfrac{1}{9}q^2L_0 - \tfrac{2}{3}qN_0 + M_0]$$
$$+ |\textstyle\int \gamma_5|^2(|C_A|^2 + |C_A'|^2)L_0$$
$$+ \{i(\textstyle\int \boldsymbol{\sigma}\cdot\mathbf{r})^*(\int \gamma_5)\}2(|C_A|^2 + |C_A'|^2)[\tfrac{1}{3}qL_0 - N_0]. \tag{6.74}$$

$$b_{01}^{(1)} = \{(\textstyle\int \boldsymbol{\sigma}\cdot\mathbf{r})^*(\int \mathbf{r})\}\{2\,\mathrm{Re}(C_A^* C_V' + C_A'^* C_V)2[-\tfrac{1}{9}q^2\Lambda_1 - \tfrac{1}{3}q\mathbf{L}_{12}$$
$$+ \tfrac{1}{3}q\mathbf{N}_{11} + \mathbf{N}_{12} + m_1] + 2\,\mathrm{Im}(C_A^* C_V' + C_A'^* C_V)2[\tfrac{1}{3}q\mathbf{H}_{12} - \mathbf{J}_{12}]\}$$
$$- \{i(\textstyle\int \boldsymbol{\sigma}\cdot\mathbf{r})^*(\int \boldsymbol{\alpha})\}\{2\,\mathrm{Re}(C_A^* C_V' + C_A'^* C_V)[\tfrac{2}{3}q\Lambda_1 - \mathbf{N}_{11}]$$
$$+ 2\,\mathrm{Im}(C_A^* C_V' + C_A'^* C_V)\mathbf{J}_{11}\}$$
$$+ \{i(\textstyle\int \boldsymbol{\sigma}\cdot\mathbf{r})^*(\int \boldsymbol{\sigma}\times\mathbf{r})\}4\,\mathrm{Re}(C_A^* C_A')[-\tfrac{2}{9}q^2\Lambda_1 - \tfrac{1}{3}q\mathbf{L}_{12} + \mathbf{N}_{12} - 2m_1]$$
$$- \{i(\textstyle\int \gamma_5)^*(\int \mathbf{r})\}\{2\,\mathrm{Re}(C_A^* C_V' + C_A'^* C_V)[-\tfrac{2}{3}q\Lambda_1 + \mathbf{N}_{11} - 2\mathbf{L}_{12}]$$
$$+ 2\,\mathrm{Im}(C_A^* C_V' + C_A'^* C_V)(\mathbf{J}_{11} + 2\mathbf{H}_{12})\}$$
$$- \{(\textstyle\int \gamma_5)^*(\int \boldsymbol{\alpha})\}2\,\mathrm{Re}(C_A^* C_V' + C_A'^* C_V)2\Lambda_1$$
$$- \{(\textstyle\int \gamma_5)^*(\int \boldsymbol{\sigma}\times\mathbf{r})\}4\,\mathrm{Re}(C_A^* C_A')[\tfrac{2}{3}q\Lambda_1 + \mathbf{L}_{12} + \mathbf{N}_{11}]. \tag{6.75}$$

$$b_{11}^{(0)} = -3^{1/2}\Big[|\textstyle\int \mathbf{r}|^2(|C_V|^2+|C_V'|^2)[\frac{1}{3}q^2L_0 - \frac{2}{3}qN_0 + 2L_1 + M_0]$$
$$+|\textstyle\int \boldsymbol{\alpha}|^2(|C_V|^2+|C_V'|^2)L_0$$
$$+|\textstyle\int \boldsymbol{\sigma}\times\mathbf{r}|^2(|C_A|^2+|C_A'|^2)[\frac{1}{6}q^2L_0 + \frac{2}{3}qN_0 + \frac{1}{2}L_1 + M_0]$$
$$+\{(\textstyle\int \boldsymbol{\alpha})^*(\int \boldsymbol{\sigma}\times\mathbf{r})\}2\,\mathrm{Re}(C_V^*C_A + C_V'^*C_A')[\frac{1}{3}qL_0 + N_0]$$
$$+\{i(\textstyle\int \mathbf{r})^*(\int \boldsymbol{\alpha})\}2(|C_V|^2+|C_V'|^2)[\frac{1}{3}qL_0 - N_0]$$
$$+\{i(\textstyle\int \mathbf{r})^*(\int \boldsymbol{\sigma}\times\mathbf{r})\}2\,\mathrm{Re}(C_V^*C_A + C_V'^*C_A')(L_1 - M_0)\Big]. \qquad (6.76)$$

$$b_{11}^{(1)} = 2^{1/2}\Big[-|\textstyle\int \mathbf{r}|^2\,2\,\mathrm{Re}(C_V^*C_V')2[\frac{1}{3}q\mathbf{L}_{12} + \frac{1}{3}q\mathbf{N}_{11} - \mathbf{N}_{12} - \Lambda_2 + m_1]$$
$$+|\textstyle\int \boldsymbol{\alpha}|^2\,2\,\mathrm{Re}(C_V^*C_V')2\Lambda_1$$
$$+|\textstyle\int \boldsymbol{\sigma}\times\mathbf{r}|^2\,2\,\mathrm{Re}(C_A^*C_A')[\frac{1}{6}q^2\Lambda_1 - \frac{1}{3}q\mathbf{L}_{12} + \frac{2}{3}q\mathbf{N}_{11} - \mathbf{N}_{12}$$
$$+\tfrac{1}{2}\Lambda_2 - 2m_1]$$
$$+\{(\textstyle\int \boldsymbol{\alpha})^*(\int \boldsymbol{\sigma}\times\mathbf{r})\}\{2\,\mathrm{Re}(C_V^*C_A' + C_V'^*C_A)[\frac{2}{3}q\Lambda_1 + \mathbf{N}_{11} - \frac{1}{2}\mathbf{L}_{12}]$$
$$+2\,\mathrm{Im}(C_V^*C_A' + C_V'^*C_A)[\mathbf{J}_{11} + \tfrac{1}{2}\mathbf{H}_{12}]\}$$
$$-\{i(\textstyle\int \mathbf{r})^*(\int \boldsymbol{\alpha})\}4\,\mathrm{Re}(C_V^*C_V')[-\frac{2}{3}q\Lambda_1 + \mathbf{L}_{12} + \mathbf{N}_{11}]$$
$$+\{i(\textstyle\int \mathbf{r})^*(\int \boldsymbol{\sigma}\times\mathbf{r})\}\{2\,\mathrm{Re}(C_V^*C_A' + C_V'^*C_A)[\frac{1}{3}q^2\Lambda_1 - \frac{1}{2}q\mathbf{L}_{12}$$
$$-\tfrac{1}{2}\mathbf{N}_{12} + \Lambda_2 + 2m_1]$$
$$+2\,\mathrm{Im}(C_V^*C_A' + C_V'^*C_A)[\tfrac{2}{3}q\mathbf{J}_{11} - \tfrac{1}{6}q\mathbf{H}_{12} - \tfrac{3}{2}\mathbf{J}_{12}]\}\Big]. \qquad (6.77)$$

$$b_{11}^{(2)} = -6^{1/2}\Big[-|\textstyle\int \mathbf{r}|^2(|C_V|^2+|C_V'|^2)[-\frac{2}{3}qL_{12} + L_1 + 2N_{12}]$$
$$+|\textstyle\int \boldsymbol{\sigma}\times\mathbf{r}|^2(|C_A|^2+|C_A'|^2)[\frac{1}{3}qL_{12} - \frac{1}{4}L_1 + N_{12}]$$
$$+\{(\textstyle\int \boldsymbol{\alpha})^*(\int \boldsymbol{\sigma}\times\mathbf{r})\}[2\,\mathrm{Re}(C_V^*C_A + C_V'^*C_A')\frac{1}{2}L_{12}$$
$$+2\,\mathrm{Im}(C_V^*C_A + C_V'^*C_A')\tfrac{1}{2}H_{12}]$$
$$+\{i(\textstyle\int \mathbf{r})^*(\int \boldsymbol{\alpha})\}2(|C_V|^2+|C_V'|^2)L_{12}$$
$$+\{i(\textstyle\int \mathbf{r})^*(\int \boldsymbol{\sigma}\times\mathbf{r})\}\{2\,\mathrm{Re}(C_V^*C_A + C_V'^*C_A')\frac{1}{2}(qL_{12} - L_1 + N_{12})$$
$$-2\,\mathrm{Im}(C_V^*C_A + C_V'^*C_A')[\tfrac{1}{6}qH_{12} + \tfrac{3}{2}J_{12}]\}\Big]. \qquad (6.78)$$

$$b_{02}^{(2)} = (\tfrac{1}{6})^{1/2}\Big[\{i(\textstyle\int \gamma_5)^*(\int B_{ij})\}2(|C_A|^2+|C_A'|^2)3L_{12}$$
$$+\{(\textstyle\int \boldsymbol{\sigma}\cdot\mathbf{r})^*(\int B_{ij})\}2(|C_A|^2+|C_A'|^2)(-qL_{12} + 3N_{12})\Big]. \qquad (6.79)$$

$$b_{12}^{(1)} = -\tfrac{1}{2}(\tfrac{5}{3})^{1/2}\Big[\{(\textstyle\int \mathbf{r})^*(\int B_{ij})\}\{2\,\mathrm{Re}(C_V^* C_A' + C_V'^* C_A)[-\tfrac{2}{3}q^2\Lambda_1$$
$$- q\mathbf{L}_{12} + 3\mathbf{N}_{12} - \tfrac{6}{5}\Lambda_2] + 2\,\mathrm{Im}(C_V^* C_A' + C_V'^* C_A)(q\mathbf{H}_{12} - 3\mathbf{J}_{12})\}$$
$$+\{i(\textstyle\int \boldsymbol{\alpha})^*(\int B_{ij})\}[2\,\mathrm{Re}(C_V^* C_A' + C_V'^* C_A)3\mathbf{L}_{12}$$
$$- 2\,\mathrm{Im}(C_V^* C_A' + C_V'^* C_A)3\mathbf{H}_{12}]$$
$$+\{i(\textstyle\int \boldsymbol{\sigma}\times\mathbf{r})^*(\int B_{ij})\}4\,\mathrm{Re}(C_A^* C_A')[-\tfrac{1}{3}q^2\Lambda_1 + q\mathbf{L}_{12} + 3\mathbf{N}_{12} + \tfrac{3}{5}\Lambda_2]\Big]. \quad (6.80)$$

$$b_{12}^{(2)} = -\tfrac{1}{2}\Big[\{(\textstyle\int \mathbf{r})^*(\int B_{ij})\}[2\,\mathrm{Re}(C_V^* C_A + C_V'^* C_A')(-qL_{12} - 3L_1 + 3N_{12})$$
$$+ 2\,\mathrm{Im}(C_V^* C_A + C_V'^* C_A')(-qH_{12} + 3J_{12})]$$
$$+\{i(\textstyle\int \boldsymbol{\alpha})^*(\int B_{ij})\}[2\,\mathrm{Re}(C_V^* C_A + C_V'^* C_A')3L_{12}$$
$$+ 2\,\mathrm{Im}(C_V^* C_A + C_V'^* C_A')3H_{12}]$$
$$+\{i(\textstyle\int \boldsymbol{\sigma}\times\mathbf{r})^*(\int B_{ij})\}2(|C_A|^2 + |C_A'|^2)[qL_{12} + \tfrac{3}{2}L_1 + 3N_{12}]\Big]. \quad (6.81)$$

$$b_{12}^{(3)} = 9(\tfrac{1}{10})^{1/2}\Big[-\{(\textstyle\int \mathbf{r})^*(\int B_{ij})\}2\,\mathrm{Re}(C_V^* C_A' + C_V'^* C_A)2\Lambda_2$$
$$+\{i(\textstyle\int \boldsymbol{\sigma}\times\mathbf{r})^*(\int B_{ij})\}4\,\mathrm{Re}(C_A^* C_A')\Lambda_2\Big]. \quad (6.82)$$

$$b_{22}^{(0)} = |\textstyle\int B_{ij}|^2(|C_A|^2 + |C_A'|^2)[\tfrac{1}{12}q^2L_0 + \tfrac{3}{4}L_1]5^{1/2}. \quad (6.83)$$

$$b_{22}^{(1)} = -|\textstyle\int B_{ij}|^2\,2\,\mathrm{Re}(C_A^* C_A')[\tfrac{1}{12}q^2\Lambda_1 + \tfrac{9}{20}\Lambda_2](10)^{1/2}. \quad (6.84)$$

$$b_{22}^{(2)} = -|\textstyle\int B_{ij}|^2(|C_A|^2 + |C_A'|^2)\tfrac{3}{4}(\tfrac{7}{2})^{1/2}L_1. \quad (6.85)$$

$$b_{22}^{(3)} = |\textstyle\int B_{ij}|^2\,2\,\mathrm{Re}(C_A^* C_A')9(\tfrac{1}{10})^{1/2}\Lambda_2. \quad (6.86)$$

$$b_{22}^{(4)} = 0. \quad (6.87)$$

In the particle parameters, the combinations of the electron wave functions are given below.

$$L_0 = \frac{g_{-1}^2 + f_1^2}{2p^2F_0} \to 1. \quad (6.88)$$

$$L_1 = \frac{g_{-2}^2 + f_2^2}{2p^2F_0R^2} \to \frac{p^2}{9}. \quad (6.89)$$

$$N_0 = \frac{f_{-1}g_{-1} - f_1 g_1}{2p^2 F_0 R} \to -\frac{p^2}{3E} - V. \tag{6.90}$$

$$M_0 = \frac{f_{-1}^2 + g_1^2}{2p^2 F_0 R^2} \to \frac{p^2}{9} + \frac{2p^2}{3E} V + V^2. \tag{6.91}$$

$$L_{12} = \frac{g_{-1} f_2 \cos(\delta_{-1} - \delta_2) - f_1 g_{-2} \cos(\delta_1 - \delta_{-2})}{2p^2 F_0 R} \to -\frac{p^2}{3E}. \tag{6.92}$$

$$N_{12} = \frac{f_{-1} f_2 \cos(\delta_{-1} - \delta_2) + g_1 g_{-2} \cos(\delta_1 - \delta_{-2})}{2p^2 F_0 R^2} \to \frac{p^2}{9} + \frac{p^2}{3E} V. \tag{6.93}$$

$$\mathbf{L}_{12} = \frac{f_1 f_2 \sin(\delta_1 - \delta_2) + g_{-1} g_{-2} \sin(\delta_{-1} - \delta_{-2})}{2p^2 F_0 R} \to \frac{1}{3} p. \tag{6.94}$$

$$\mathbf{N}_{11} = \frac{f_{-1} f_1 - g_{-1} g_1}{2p^2 F_0 R} \sin(\delta_{-1} - \delta_1) \to -\frac{1}{3} p - \frac{p}{E} V. \tag{6.95}$$

$$\mathbf{N}_{12} = \frac{f_{-1} g_{-2} \sin(\delta_{-1} - \delta_{-2}) - g_1 f_2 \sin(\delta_1 - \delta_2)}{2p^2 F_0 R^2} \to -\frac{p^3}{9E} - \frac{1}{3} pV. \tag{6.96}$$

$$\Lambda_1 = \frac{g_{-1} f_1 \sin(\delta_{-1} - \delta_1)}{2p^2 F_0} \to \frac{p}{2E}. \tag{6.97}$$

$$\Lambda_2 = \frac{g_{-2} f_2 \sin(\delta_{-2} - \delta_2)}{2p^2 F_0 R^2} \to \frac{p^3}{18E}. \tag{6.98}$$

$$m_1 = \frac{f_{-1} g_1 \sin(\delta_{-1} - \delta_1)}{2p^2 F_0 R^2} \to -\frac{p^3}{18E} - \frac{1}{3} pV - \frac{p}{2E} V^2. \tag{6.99}$$

$$H_{12} = \frac{g_{-1} f_2 \sin(\delta_{-1} - \delta_2) - f_1 g_{-2} \sin(\delta_1 - \delta_{-2})}{2p^2 F_0 R} \to -\frac{1}{4} p\alpha Z. \tag{6.100}$$

$$J_{12} = \frac{f_{-1} f_2 \sin(\delta_{-1} - \delta_2) + g_1 g_{-2} \sin(\delta_1 - \delta_{-2})}{2p^2 F_0 R^2} \to \left[\frac{p^3}{12E} + \frac{1}{4} pV\right] \alpha Z. \tag{6.101}$$

$$\mathrm{H}_{12} = \frac{f_1 f_2 \cos(\delta_1 - \delta_2) + g_{-1} g_{-2} \cos(\delta_{-1} - \delta_{-2})}{2p^2 F_0 R} \rightarrow -\left[\frac{p^2}{4E} + \frac{1}{3E}\right]\alpha Z. \tag{6.102}$$

$$\mathrm{J}_{11} = \frac{f_{-1} f_1 + g_{-1} g_1}{2p^2 F_0 R} \cos(\delta_{-1} - \delta_1) \rightarrow -\frac{1}{3E}\alpha Z. \tag{6.103}$$

$$\mathrm{J}_{12} = \frac{f_{-1} g_{-2} \cos(\delta_{-1} - \delta_{-2}) - g_1 f_2 \cos(\delta_1 - \delta_2)}{2p^2 F_0 R^2}$$

$$\rightarrow \left[\frac{p^2}{12} + \frac{p^2}{4E} V + \frac{1}{3E} V\right]\alpha Z. \tag{6.104}$$

Here $V = \alpha Z/2R$ is the half of the Coulomb energy at the nuclear surface, R being the nuclear radius. The arrow in each equation indicates the approximation of the point nuclear charge with $(\alpha Z)^2 \ll 1$. This approximation is widely accepted in analysis of experimental data. In the medium and heavy nuclei, the finite nuclear size corrections become important. The combinations of electron wave functions like N_0, M_0, N_{12}, J_{12}, . . ., differ, sometimes as much as 50%, from those in the approximation above. It should be noticed that nuclear matrix elements derived from experimental data on beta-gamma angular correlations vary with the approximation for formulas adopted in numerical analysis, especially in the cases of forbidden transitions. In this connection, the use of numerical tables of electron wave functions given by Bhalla and Rose [15] together with the above formulas is highly recommended to study the nuclear structure. A computer program for numerical solutions, f_κ and g_κ, of the Dirac equation with finite size nucleus is also available [16].

Electron Formulas to Positron Formulas

In this section the $b_{LL'}^{(n)}$'s are given for an electron decay. In order to obtain the corresponding $b_{LL'}^{(n)}$'s for a positron decay, the following substitutions should be performed.

$$Z \rightarrow -Z, \qquad C_\mathrm{A} \rightarrow -C_\mathrm{A}^*, \qquad C'_\mathrm{A} \rightarrow C'^*_\mathrm{A},$$
$$C_\mathrm{V} \rightarrow C_\mathrm{V}^*, \qquad C'_\mathrm{V} \rightarrow -C'^*_\mathrm{V}. \tag{6.105}$$

Beta-Gamma Angular Correlation from Oriented Nuclei

In order to test time-reversal invariance of the beta interaction it was proposed by Morita and Morita [13] and Curtis and Lewis [17] to measure the asymmetry of gamma-ray intensity with respect to the plane made of the nuclear orientation axis and the direction of the beta momentum, in the beta–gamma cascade transitions of the oriented nuclei. The explicit formula for the decay scheme $2 \xrightarrow{\beta} 2 \xrightarrow{\gamma} 0$ is given in the first paper of [13]. It is given in the next equation. The unit vector **J** along the orientation axis of the nucleus is chosen as the z axis. The beta and gamma rays are assumed to be emitted in the directions of **p** and **k** with polar angles $(\theta_1, \phi_1 = 0)$ and (θ_2, ϕ), respectively (see Fig. 6.2). The formula is

$$\begin{aligned} W(\theta_1, \theta_2, \phi, \tau) = & \sum_{L \leqslant L'} \sum_{L_1} \sum_{L_1'} \sum_{\nu} \sum_{n} \sum_{n_1} \hat{f}_\nu(j)[(2j_1+1)(2\nu+1)(2n+1)]^{1/2} \\ & \times (-)^{L'+n+n_1} \tau^{n_1} \begin{pmatrix} j & j_1 & L \\ j & j_1 & L' \\ \nu & n_1 & n \end{pmatrix} \\ & \times [B_{LL'}^{(n)} \sum_{\mu} (-)^\mu (\nu n 0 \mu \mid n_1 \mu) D_{-\mu,0}^{(n)}(0, \theta_1, 0) \\ & \times D_{\mu,0}^{(n_1)}(\phi, \theta_2, 0) + \text{c.c.}] \langle j_2 \| L_1 \| j_1 \rangle \langle j_2 \| L_1' \| j_1 \rangle \\ & \times F_{n_1}(L_1 L_1' j_2 j_1), \end{aligned} \tag{6.106}$$

with the 9-j symbol

$$\begin{pmatrix} a & b & c \\ d & e & f \\ g & h & k \end{pmatrix} = \sum_{\lambda} (2\lambda+1) W(abkf, c\lambda) W(dfhb, e\lambda) W(adkh, g\lambda), \tag{6.107}$$

and the rotation matrix

$$D_{m,0}^{(l)}(\phi, \theta, 0) = \exp(-im\phi) D_{m,0}^{(l)}(0, \theta, 0) = \left(\frac{4\pi}{2l+1}\right)^{1/2} Y_{l,m}^*(\theta, \phi). \tag{6.108}$$

The particle parameters $B_{LL'}^{(n)}$ are related to the parameters $b_{LL'}^{(n)}$ in the following way.

$$\begin{aligned} 2B_{LL'}^{(n)} = & [b_{LL'}^{(n)} \text{ with } \mathrm{Re}(C_i^* C_j) \text{ and } \mathrm{Im}(C_i^* C_j) \text{ replaced by} \\ & C_i^* C_j \text{ and } -iC_i^* C_j, \text{ respectively}]. \end{aligned} \tag{6.109}$$

Here C_i and C_j mean C_V, C_V', C_A, C_A'. Also $i = j$ is included.

The restrictions on the values of the integral numbers ν, n, and n_1 are evident from the algebraic coefficients in Eq. (6.106). They depend on the ranks of the beta-decay matrix elements, the multipolarity of the gamma ray, and the spins of the nuclear states. Furthermore, ν has only even values if nuclei are aligned. n_1 is also restricted to be even, if the circular polarization of gamma ray is not observed. The terms $(\mathbf{J} \cdot \mathbf{p} \times \mathbf{k})^a (\mathbf{J} \cdot \mathbf{k})^b (\mathbf{J} \cdot \mathbf{p})^c (\mathbf{p} \cdot \mathbf{k})^d$, which are pseudoscalar with respect to the time-reversal operation, appear only when $\nu + n + n_1$ is odd (namely, when a is odd). And if $\nu + n + n_1$ is odd, the 9-j coefficient vanishes unless $L \neq L'$. Therefore, the term with which we can test the invariance property of beta interactions under time reversal comes from interferences among beta matrix elements with different ranks. For discussion of the pseudoscalar terms of the form $\alpha Z(\mathbf{J} \cdot \mathbf{p} \times \mathbf{k})^a (\mathbf{J} \cdot \mathbf{k})^b (\mathbf{J} \cdot \mathbf{p})^c (\mathbf{p} \cdot \mathbf{k})^d$, see [13].

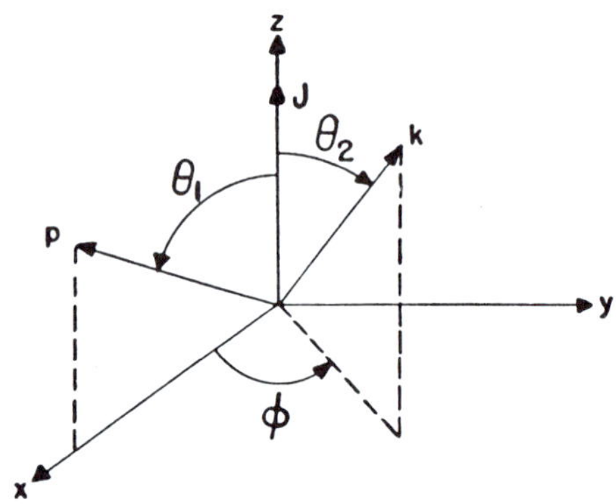

Fig. 6.2 Geometry for beta–gamma angular correlation from oriented nuclei. The unit vector **J** along the orientation axis is chosen as the z axis. The beta and gamma rays are emitted in the directions of **p** and **k** with polar angles $(\theta_1, \phi_1 = 0)$ and (θ_2, ϕ), respectively.

Angular Correlation in Triple Cascade Transitions from Unoriented Nuclei

Here we give the formula for angular correlations of a beta ray and two gamma rays in triple cascade transitions of unoriented nuclei. Three radiations are observed. The decay scheme is

$$j \xrightarrow{\beta} j_1 \xrightarrow{\gamma_1} j_2 \xrightarrow{\gamma_2} j_3.$$

The direction of the momentum **p** of the beta ray is chosen as the z axis. The γ_1 and γ_2 rays are assumed to be emitted in the directions represented by unit vector $\mathbf{k}_1$ and $\mathbf{k}_2$, with polar angles $(\theta_1, \phi_1 = 0)$ and (θ_2, ϕ), respectively (see Fig. 6.3).

$$
\begin{aligned}
&W(\theta_1, \theta_2, \phi, \tau_1, \tau_2: \beta\text{-}\gamma_1\text{-}\gamma_2) \\
&\quad = \sum_{L \leqslant L'} \sum_{L_1} \sum_{L_1'} \sum_{L_2} \sum_{L_2'} \sum_{n} \sum_{n_1} \sum_{n_2} (2j_1 + 1)[(2n + 1)(2n_1 + 1)(2L_1 + 1) \\
&\qquad \times (2L_1' + 1)(2j_2 + 1)]^{1/2} W(j_1 j_1 L L', nj) b_{LL'}^{(n)} \\
&\qquad \times \langle j_2 \| L_1 \| j_1 \rangle \langle j_2 \| L_1' \| j_1 \rangle \\
&\qquad \times \tau_1^{n_1} \tau_2^{n_2} (L_1 L_1' 1 -1 \mid n_1 0) \begin{pmatrix} j_1 & j_2 & L_1 \\ j_1 & j_2 & L_1' \\ n & n_2 & n_1 \end{pmatrix} (-)^{1-j+j_1+L_1+n_1+n_2} \\
&\qquad \times \Big[\sum_{\mu} (-)^{\mu} (n n_1 0 \mu \mid n_2 \mu) D_{-\mu,0}^{(n_1)}(0, \theta_1, 0) D_{\mu,0}^{(n_2)}(\phi, \theta_2, 0)\Big] \\
&\qquad \times \langle j_3 \| L_2 \| j_2 \rangle \langle j_3 \| L_2' \| j_2 \rangle F_{n_2}(L_2 L_2' j_3 j_2),
\end{aligned}
\tag{6.110}
$$

where $n + n_1 + n_2$ is even, and $L + L_i' + \delta_i + \delta_i'$ is even, $i = 1, 2$.

The angular correlation function in triple cascade transitions of unoriented nuclei has terms which are pseudoscalar under time reversal,

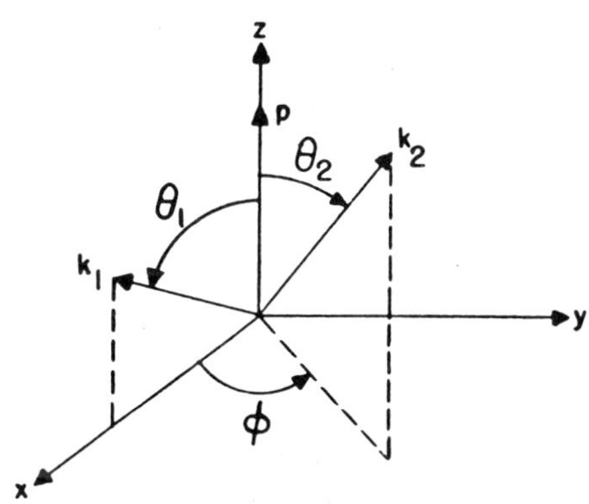

Fig. 6.3 Geometry for beta–gamma–gamma angular correlation from unoriented nuclei. The direction of the momentum **p** of the beta ray is chosen as the z axis. Two gamma rays are assumed to be emitted in the directions of $\mathbf{k}_1$ and $\mathbf{k}_2$, with polar angles $(\theta_1, \phi_1 = 0)$ and (θ_2, ϕ), respectively.

$\alpha Z(\mathbf{p} \cdot \mathbf{k}_1 \times \mathbf{k}_2)^a (\mathbf{p} \cdot \mathbf{k}_1)^b (\mathbf{p} \cdot \mathbf{k}_2)^c (\mathbf{k}_1 \cdot \mathbf{k}_2)^d$ with even a and appropriate values for b, c, d. All the terms $(\mathbf{p} \cdot \mathbf{k}_1 \times \mathbf{k}_2)^a (\mathbf{p} \cdot \mathbf{k}_1)^b (\mathbf{p} \cdot \mathbf{k}_2)^c (\mathbf{k}_1 \cdot \mathbf{k}_2)^d$ with odd a cancel out by assuming the invariance of strong interactions under time reversal.

Approximations for the Higher-Order Matrices in the Particle Parameters of the Allowed Transitions

To perform the higher-order corrections to the formulas for allowed transitions, the following approximations are useful [4]. Assuming uniform density of nuclear matter we have

$$\int r^2 \approx \tfrac{3}{5}R^2 \int 1, \tag{6.111}$$

$$\int \sigma r^2 \approx \tfrac{3}{5}R^2 \int \sigma, \tag{6.112}$$

$$\int (\sigma \cdot \mathbf{r})\mathbf{r} \approx \tfrac{3}{5}R^2 \eta \int \sigma. \tag{6.113}$$

Here η is the nuclear model dependent factor, and is given, e.g., by the single-particle shell model. Note that Eq. (6.111) should not be taken too seriously in the case of the isospin forbidden transition for which $\int 1 = 0$. Equation (6.111) simply expresses the fact that $\int r^2$ is of the order of $\tfrac{3}{5}R^2$. For the momentum-type matrices we adopt nonrelativistic approximation and the empirical mass formula. They are given by

$$i \int \boldsymbol{\alpha} \cdot \mathbf{r} \approx \mp \Lambda \frac{\alpha Z}{4R} \int r^2, \tag{6.114}$$

$$i \int \gamma_5 \mathbf{r} \approx \frac{1}{2M} \int \sigma \mp \Lambda \frac{\alpha Z}{4R} \int (\sigma \cdot \mathbf{r})\mathbf{r}, \tag{6.115}$$

$$\int \alpha \times \mathbf{r} \approx \frac{1}{M} \left[\int \sigma + \int \mathbf{r} \times \mathbf{p}\right], \tag{6.116}$$

with

$$\Lambda = 1 \pm (E_o \mp 2.5)\frac{A^{1/3}}{Z}. \tag{6.117}$$

Here the upper (lower) sign refers to the electron (positron) decay. The first term in Eq. (6.116) comes from the Dirac moment of the proton (while the neutron has a vanishing effect), and the second term the orbital momentum of the nucleons. In the CVC theory, there is a contribution of the anomalous magnetic moments of the nucleons due to the pion cloud, and we have

$$\int \alpha \times \mathbf{r} \approx \frac{1}{M} \left[(1 + \mu_p - \mu_n)\int \sigma + \int \mathbf{r} \times \mathbf{p}\right] \tag{6.118}$$

with

$$1 + \mu_p - \mu_n = 4.70. \tag{6.119}$$

More about Particle Parameters

The particle parameters (of the allowed transitions) with corrections of the coordinate-type matrices are given for use of the beta-ray spectrum and beta-gamma directional correlations in the first paper of [4]. The corrections are as follows.

For $b_{00}^{(0)}$ in Eq. (6.60), add

$$\{(\int 1)^*(\int r^2)\}2(|C_V|^2+|C_V'|^2)(-\tfrac{1}{6}q^2L_0+\tfrac{1}{3}qN_0). \tag{6.120}$$

For $b_{11}^{(0)}$ in Eq. (6.62), add

$$-3^{1/2}2(|C_V|^2+|C_V'|^2)[-\{(\int \sigma)^*(\int \sigma r^2)\}(\tfrac{1}{6}q^2L_0+\tfrac{1}{3}qN_0) + \{(\int \sigma)^*(\int (\sigma\cdot\mathbf{r})\mathbf{r})\}\tfrac{2}{3}qN_0]. \tag{6.121}$$

For $b_{11}^{(2)}$ in Eq. (6.64), add

$$6^{1/2}(|C_A|^2+|C_A'|^2)[\{(\int \sigma)^*(\int \sigma r^2)\}(\tfrac{1}{3}qL_{12}-N_{21}) + \{(\int \sigma)^*(\int (\sigma\cdot\mathbf{r}))\}(\tfrac{1}{3}qL_{12}+3N_{21})]. \tag{6.122}$$

Here we adopt the following symbols.

$$L_0 \to \tfrac{1}{2}(1+\gamma)-\tfrac{5}{3}\alpha ZRE-\tfrac{1}{3}\alpha Z\frac{R}{E}-\tfrac{1}{3}R^2p^2. \tag{6.123}$$

$$N_0 \to -\frac{p^2}{3E}\gamma - V+\tfrac{2}{3}(\alpha Z)^2E-\frac{p^2}{9E}(\alpha Z)^2+\tfrac{11}{18}p^2\alpha ZR. \tag{6.124}$$

$$N_{21}=\frac{g_{-1}g_2\cos(\delta_{-1}-\delta_2)+f_1f_{-2}\cos(\delta_1-\delta_{-2})}{2p^2F_0R^2} \to -\frac{1}{15}p^2-\frac{p^2}{6E}V. \tag{6.125}$$

For L_{12} and H_{12}, we use the expressions in Eqs. (6.92) and (6.100), respectively. Similar corrections for the particle parameters with odd n will be given in [12].

Particle parameters for the second forbidden transitions (with the restriction of even n) are given in [18].

It was found [19] that the formulas for the first forbidden transitions are arranged in quite simplified and organized forms if we adopt:

(a) the Konopinski–Uhlenbeck approximation, which is shown by the right-hand sides of the arrows in Eqs. (6.88)-(6.104);
(b) the following real parameters $u, v, \ldots, z$

$$\int \sigma\times\mathbf{r}=u, \quad \frac{\int \gamma_5}{\int \sigma\times\mathbf{r}}=Vv, \quad \frac{\int \sigma\cdot\mathbf{r}}{\int \sigma\times\mathbf{r}}=iw,$$
$$\frac{C_V\int \mathbf{r}}{C_A\int \sigma\times\mathbf{r}}=-ix, \quad \frac{C_V\int \alpha}{C_A\int \sigma\times\mathbf{r}}=Vy, \quad \frac{\int B_{ij}}{\int \sigma\times\mathbf{r}}=iz. \tag{6.126}$$

The same is true for the second forbidden transition [19]. Since these formulas are very short and easy for numerical calculations, the importance of these parameters was emphasized by Kotani [20]. These parameters are still useful for performing numerical analysis of experimental data with the exact formulas and exact electron wave functions.

Instead of our approach to include the effect of the finite nuclear size by means of numerical solutions of electron wave functions, Yamada and his collaborators developed a theory in which the effective nuclear radius is introduced for each nuclear matrix element. This radius represents effectively the finite size corrections, and it may change from matrix to matrix. The formulas for this theory are given in [21].

Electron-Neutrino Angular Correlation

The electron-neutrino angular correlations are given for the allowed transition in section 2.11, and for the first forbidden transition in [22]. From the formula in the latter reference, we can derive the formula for the allowed transition with higher-order corrections (see the third paper in [4]). The electron-neutrino angular correlation from oriented nuclei was given, e.g., by Jackson, Treiman, and Wyld (see Eqs. (3.116)-(3.118)). Formulas for nuclear fluorescence resonance are given in [23]. The electron-neutrino-gamma angular correlation is given in [24]. A particular case of the electron-neutrino angular correlation is the beta decay of Li^8, the daughter of which splits into two alpha particles. The beta-alpha-alpha angular correlation is given in [25].

Longitudinal Polarization of Beta Particles

This is studied in section 3.9 for the allowed transitions. Formulas for the first forbidden transitions were given by Alder, Stech, and Winther [26] under the assumption of *STP*. The transformation of the formulas for *STP* to those for *VA* interaction, or vice versa, has been studied [27]. The rule is the replacement

$$C_S\beta \to C_V, \qquad C_T\beta\sigma \to C_A\sigma, \qquad C_T\beta\alpha \to C_V\alpha,$$
$$C_P\beta\gamma_5 \to C_A\gamma_5, \tag{6.127}$$

and the same transformation for the primed coupling constants. Here the Dirac matrices operate on nuclear wave functions. For the lepton parts, it is necessary to replace

$$q^m \to (-)^{m+n} q^m, \tag{6.128}$$

where q^m is the mth power of the neutrino energy in the expression of particle parameter $b_{LL'}^{(n)}$. The sign function in Eq. (6.128) appears due to the anticommutability of β and $\boldsymbol{\alpha}$, or γ_5 in the trace calculation. Therefore, the formulas for longitudinal polarization with *VA* is easily derived. The formula for the allowed transition with the higher-order correlation is also derived by successive transformations, Eqs. (6.127) and (6.128), and Eq. (6.44). The result is given in the third paper in [4]. For a special case of the pure Gamow-Teller transition with $C_i = C_i'$, the longitudinal polarization of beta particles is given by

$$P_L = \mp \frac{p}{E}\left[1 + \frac{1}{3E}\left(\pm \alpha ZR \mp 2\frac{C_V \int \boldsymbol{\alpha} \times \mathbf{r}}{C_A \int \sigma} - 2\frac{i\int \gamma_5 \mathbf{r}}{\int \sigma}\right)\right] \qquad \text{for } \beta^{\mp}. \tag{6.129}$$

The formula for the polarization of beta particles in coincidence with gamma rays in the cascade transition is given in [28].

Formulas for the Mössbauer-Type Experiment

In the Mössbauer-type experiment [29], each subtransition, corresponding to the magnetic quantum numbers $m_i \to m_f$, can be observed as a well-separated absorption line if the hyperfine splittings are sufficiently large in the excited and ground states of the nuclei and are appropriately adjusted in the emitter and absorber. In other words, observing the Mössbauer satellites, we can select each magnetic substate, which itself is polarized, while the initial states as a whole are not polarized. The degree of polarization of a magnetic substate, e.g., $j_i m_i$ is unambiguous, and it is m_i/j_i. This sort of nuclear polarization was pointed out by Morita [30], and the beta-satellite gamma correlation was calculated.

Gamma-Gamma Angular Correlation

For the gamma decay, the particle parameters $b_{LL'}^{(n)}$ are given by [31],

$$b_{LL'}^{(n)} = \tau^n (-)^{1+L+L'} \langle j_1 \| L \| j \rangle \langle j_1 \| L' \| j \rangle \times [(2L+1)(2L'+1)]^{1/2} (LL'1-1 \mid n0). \tag{6.130}$$

The gamma-gamma angular correlation is derived from Eqs. (6.57) and (6.130),

$$W(\theta : \gamma - \gamma_1) = \sum_n \Big\{\Big[\sum_L \sum_{L'} \tau^n (-)^{L+L'+n} \langle j_1 \| L \| j \rangle \langle j_1 \| L' \| j \rangle \times F_n(LL'jj_1)\Big]\Big[\sum_{L_1} \sum_{L_1'} \tau_1^n \langle j_2 \| L_1 \| j_1 \rangle \times \langle j_2 \| L_1' \| j_1 \rangle F_n(L_1 L_1' j_2 j_1)\Big]\Big\} P_n(\cos\theta). \tag{6.131}$$

The factor $(-)^{L+L'}$ disappears if we replace the reduced nuclear matrix element

$$\langle j_1 \| L \| j \rangle \qquad \text{by} \qquad (-)^L \langle j \| L \| j_1 \rangle.$$

It is important to keep the sign function in Eq. (6.131) for obtaining the mixing ratio of different multipolarities of the intermediate radiation in the triple gamma cascade transitions, by measuring the γ-γ_1 and γ_1-γ_2 directional correlations [31].

The gamma-ray angular distribution from oriented nuclei is derived from Eqs. (6.53) and (6.130), and so on.

Alpha-Ray Angular Distributions

The particle parameters for the alpha ray are given by [31],

$$\begin{aligned} b_{LL'}^{(n)} &= (-)^{L+L'} \langle j_1 \| L \| j \rangle \langle j_1 \| L' \| j \rangle^* \\ &\quad \times [(2L+1)(2L'+1)]^{1/2} (LL'00 \mid n0). \end{aligned} \tag{6.132}$$

Notice here that the formulas for alpha particles contain only even n terms. Formulas for alpha-ray angular distribution, alpha-gamma directional correlation, etc., are easily derived from Eq. (6.132) and appropriate formulas for beta decay. Since the Coulomb phase shifts are different for different partial waves of the alpha particle, the complex conjugate symbol in Eq. (6.132) still holds, under the assumption of time-reversal invariance for the strong interactions.

Note Added in Proof

A compilation of the formulas used in the treatment of beta decay and electron capture, with associated numerical tables, has been made by Behrens and Jänecke [32].

REFERENCES

1 E. J. Konopinski and G. E. Uhlenbeck, *Phys. Rev.* **60**, 308 (1941).

2 E. J. Konopinski, *The Theory of Beta Radioactivity*, Oxford Univ. Press, London and New York, 1966.

3 H. A. Weidermüller, *Rev. Mod. Phys.* **33**, 574 (1962).

4 M. Morita, *Phys. Rev.* **113**, 1584 (1959); *Phys. Rev.* **114**, 1080 (1959); *Nucl. Phys.* **14**, 106 (1959).

5 D. C. Peaslee, *Phys. Rev.* **91**, 1447 (1953).

6 S. Watanabe, *Phys. Rev.* **106**, 1306 (1957).

7 B. Stech and J. H. D. Jensen, *Z. Phys.* **141**, 175 (1955).

8 J. Tiomno, *Nucl. Phys.* **1**, 226 (1955).

9 J. J. Sakurai, *Nuovo Cimento* **7**, 649 (1958).

10 R. P. Feynman and M. Gell-Mann, *Phys. Rev.* **109,** 193 (1958).
11 E. C. G. Sudarshan and R. E. Marshak, *Phys. Rev.* **109,** 1860 (1958).
12 M. Morita, M. Fuyuki, and S. Tsukada, *Progr. Theoret. Phys.* **47,** 556 (1972).
13 M. Morita and R. S. Morita, *Phys. Rev.* **107,** 1316 (1957); *Phys. Rev.* **110,** 461 (1958).
14 E. M. Henley and B. A. Jacobsohn, *Phys. Rev.* **113,** 225 and 234 (1959); M. Morita, unpublished.
15 C. P. Bhalla and M. E. Rose, *Oak Ridge Natl. Lab. Rept.* ORNL-2954 (1960); ORNL-3207 (1962). A comment on these articles is given by W. Bühring, *Kernforschungszentrum Karlsruhe Report* KFK No. 559 (1967).
16 H. Ohtsubo, unpublished.
17 R. B. Curtis and R. R. Lewis, *Phys. Rev.* **107,** 1381 (1957).
18 M. Yamada and M. Morita, *Progr. Theoret. Phys.* **8,** 431 (1952); Y. Kato and M. Morita, *Progr. Theoret. Phys.* **13,** 276 (1955).
19 M. Morita, *Progr. Theoret. Phys.* **14,** 27 (1955).
20 T. Kotani, *Phys. Rev.* **114,** 795 (1959).
21 Z. Matumoto and M. Yamada, *Progr. Theoret. Phys.* **19,** 285 (1958). Z. Matumoto, *Progr. Theoret. Phys.* **23,** 531 (1960).
22 M. Morita, *Progr. Theoret. Phys.* **9,** 345 (1953); *Phys. Rev.* **90,** 1005 (1953) with erratum **91,** 1580 (1953) (E).
23 M. Goldhaber, L. Grodzins, and A. W. Sunyar, *Phys. Rev.* **109,** 1015 (1958); M. Morita, R. Morita, and M. Yamada, *Phys. Rev.* **111,** 237 (1958); M. Morita and R. Morita, *Phys. Rev.* **111,** 1130 (1958); R. R. Lewis and R. B. Curtis, *Phys. Rev.* **110,** 910 (1958).
24 M. Morita, *Nucl. Phys.* **6,** 132 (1958).
25 M. Morita, *Phys. Rev. Letters* **1,** 112 (1958).
26 K. Alder, B. Stech, and A. Winther, *Phys. Rev.* **107,** 728 (1957).
27 M. Morita and R. Morita, *Phys. Rev.* **109,** 2048 (1958); M. Yamada and M. Morita, *Progr. Theoret. Phys.* **8,** 431 (1952).
28 T. Kotani and M. Ross, *Progr. Theoret. Phys.* **20,** 643 (1958).
29 R. L. Mössbauer, *Z. Phys.* **151,** 124 (1958); *Z. Naturforsch.* **149,** 212 (1959).
30 M. Morita, *Phys. Rev.* **122,** 1525 (1961); *Nucl. Phys.* **51,** 433 (1964).
31 M. Morita, *Progr. Theoret. Phys.* **15,** 445 (1956).
32 H. Behrens and J. Jänecke, *Numerical Data and Functional Relationships in Science and Technology* (K. H. Hellwege and H. Schopper, eds.), Group I, Vol. 4. Springer, Berlin, 1969.

Chapter Seven

NUCLEAR STRUCTURE PROBLEMS STUDIED FROM BETA DECAYS

7.1 SUPERALLOWED AND ALLOWED TRANSITIONS

In the case of allowed transitions, the ft value is defined in Eqs. (2.165) and (2.169),

$$ft_{1/2} = B/|M|^2 \tag{7.1}$$

with

$$|M|^2 = |\textstyle\int 1|^2 + (C_A/C_V)^2 |\textstyle\int \sigma|^2 \tag{7.2}$$

and

$$B = 6146 \pm 10 \text{ sec.} \tag{7.3}$$

The quantity called the $\log ft_{1/2}$ value is the logarithm of Eq. (7.1) to the base ten, where the numerical value of $ft_{1/2}$ is in seconds. Either $ft_{1/2}$ or $\log ft_{1/2}$ gives us a measure of the nuclear matrix elements, which represents the overlapping of the nuclear wave functions of the initial and final states. If the nuclear matrix element $|M|^2$ in Eq. (7.3) is of the order of unity or larger, the $\log ft_{1/2}$ is approximately 3.8 or smaller. This corresponds to almost the full overlapping of the spatial wave functions of the initial and final nuclear states. Actually, there are 33 examples of such beta decays, listed in Table 3.3 in [1]. These examples have the $ft_{1/2}$ values

$ft_{1/2} \approx 800$	for He^6 and Ne^{18} beta decays,
$ft_{1/2} \approx 1000$	for n and H^3 beta decays,
$ft_{1/2} \approx 3000$	for $0^+ \to 0^+$ transitions (Table 2.1),
$ft_{1/2} \approx 4000\text{–}6000$	for most of the mirror transitions.

Here the mirror transition means the beta decay between the mirror nuclei, which have the same odd mass A and $|N - Z| = 1$, N and Z being the neutron and proton numbers, respectively. We call these decays the superallowed (or allowed favored) transitions. There is another group of beta decays which obey the selection rules of allowed transition; $J_i - J_f = 0$ or ± 1, parity change

No. These beta decays have larger values of $\log ft_{1/2}$, which spread from 4 to 9, but have a peak around 5 ± 0.5. These decays are called, simply, allowed transitions or allowed unfavored transitions. The largeness of $\log ft_{1/2}$ and the concomitant smallness of the nuclear matrix element $|M|^2$ are due to the latter's hindrance of the overlapping of the nuclear wave functions. Investigation of this hindrance is a central problem in the study of the nuclear structure in beta decay. We will come back to this point in subsequent sections. First, we will calculate the nuclear matrix elements in the superallowed transitions. To do this, we have to discuss the isospin formalism (it is given in Appendix A.8).

Fermi Matrix Element

The Fermi matrix element is given in Eq. (2.85) by

$$|\textstyle\int 1|^2 = \sum_{M_f} |\int \Psi^\dagger_{J_f M_f} \sum_k \tau_{\pm}^{(k)} \Psi_{J_i M_i} d\mathbf{r}_1 d\mathbf{r}_2 \cdots d\mathbf{r}_A|^2 \quad \text{for } \beta^{\mp}. \tag{7.4}$$

Previously we adopted the symbol ψ for the nuclear wave function which includes both the spatial-spin and isospin parts. The same function is represented by the symbol Ψ here, which is explicitly given by a product $\Psi = \psi\phi$. Now we use ψ for the spatial-spin part and ϕ for the isospin part. Here, the transition operators are equal to $T_{\pm}$,

$$|\textstyle\int 1|^2 = \sum_{M_f} |\int \Psi^\dagger_{J_f M_f} T_{\pm} \Psi_{J_i M_i} d\mathbf{r}_1 d\mathbf{r}_2 \cdots d\mathbf{r}_A|^2 . \tag{7.5}$$

The ordinary spin and spatial parts should be the same in the initial and final states, since the operators acting on these spaces are unit operators. From this and the last two equations in Eq. (A.54), we have the following result,

$$\begin{aligned} |\textstyle\int 1|^2 &= |(\phi_{T_f T_{fz}}, T_{\pm}\phi_{T_i T_{iz}})|^2 \\ &= \delta_{T_f T_i}[T_i(T_i+1) - T_{iz}T_{fz}]. \end{aligned} \tag{7.6}$$

Therefore, the isospin selection rule for the Fermi matrix element is

$$\Delta T = 0 \qquad \text{and} \qquad \Delta T_z = \pm 1. \tag{7.7}$$

In the case of the mirror transitions, $T = \frac{1}{2}$ and $T_z = \frac{1}{2} \leftrightarrow T_z = -\frac{1}{2}$, we have

$$|\textstyle\int 1|^2 = 1. \tag{7.8}$$

In the case of the $0^+ \to 0^+$ transitions, $T = 1$, $T_{iz} = 1$, and $T_{fz} = 0$, we have

$$|\textstyle\int 1|^2 = 2. \tag{7.9}$$

There are certain evidences that the isospin is still a good quantum number even in the medium nuclei (see subsequent sections). For example, measurement of the beta-gamma circular polarization correlation can provide us with information on the ratio

$$R = \frac{C_V \int 1}{C_A \int \sigma}, \tag{7.10}$$

see Table 3.5. In fact, vanishingly small values of R were observed in several cases. The same information is also obtained by measurement of the beta-ray asymmetry from thc polarized nuclei. The latter experiment is being performed by Nagamine [2] in the University of Tokyo, by using a He^3-He^4 dilution refrigerator.

Gamow-Teller Matrix in Mirror Transitions

The Gamow-Teller matrix elements in mirror transitions are given by Mayer and Jensen [3] and Feenberg [4]. Here we follow the former reference. The Gamow-Teller matrix is given for β^+ decay in Eq. (2.86) by

$$|\textstyle\int \sigma|^2 = \sum_{M_f} \sum_m |\int \phi^\dagger_{T_f T_{fz}} \psi^\dagger_{J_f M_f} \sum_k \tau_-^{(k)} \sigma_m^{(k)} \psi_{J_i M_i} \phi_{T_i T_{iz}} d\mathbf{r}_1 d\mathbf{r}_2 \cdots d\mathbf{r}_A|^2. \tag{7.11}$$

Here the isospin functions ϕ are explicitly written. From this, the isospin selection rule for the Gamow-Teller matrix element is generally given by

$$\Delta T = 0, \pm 1, \qquad \Delta T_z = \pm 1. \tag{7.12}$$

The upper sign for ΔT_z refers to the negatron decay, and the lower sign the positron decay. We are now dealing with the mirror transitions, and we put

$$J_i = J_f = J, \qquad T_i = T_f = T = \tfrac{1}{2}, \qquad \text{and} \qquad M_i = M = J, \tag{7.13}$$

and drop T hereafter. The last simplification $M = J$ is because the life of the initial substate M is independent of M. Now, these are three components σ_m with $m = 0, \pm 1$, which are defined in Eq. (2.64),

$$\sigma_0 = \sigma_z, \qquad 2^{1/2} \sigma_{\pm 1} = \mp(\sigma_x \pm i\sigma_y).$$

We have, however, no contribution from the σ_{+1} term, since ψ_{JJ} has its maximum z component. Next, we will transform both the σ_0 and σ_{-1} terms into the form

$$\left\langle \sum_k \sigma_z^{(k)} t_z^{(k)} \right\rangle_{M=J,\ T_z = -\frac{1}{2}},$$

where $\langle\,\rangle$ means the expectation value for the state $\psi_{JJ}\phi_{-1/2}$.

Let us first compute σ_{-1} term. Because of the property of the angular momentum operator (the last equation in (A.54)), we find

$$\psi_{J\,J-1} = \left(\frac{1}{2J}\right)^{1/2} J_- \psi_{JJ}. \tag{7.14}$$

The hermitian conjugate of J_- is J_+, and we have

$$(2J)^{1/2}\psi^\dagger_{JJ-1} = (J_-\psi_{JJ})^\dagger = \psi^\dagger_{JJ}J_+. \tag{7.15}$$

Here **J** is the vector sum of total angular momenta of individual nucleons

$$J_\pm = \sum_k j^{(k)}_\pm,$$

and

$$j_\pm = j_x \pm ij_y. \tag{7.16}$$

Here **j** and **σ** satisfy the commutation relation

$$j^{(k)}_+\sigma^{(k)}_- - \sigma^{(k)}_- j^{(k)}_+ = 2\sigma^{(k)}_z \tag{7.17}$$

with

$$\sigma_\pm = \sigma_x \pm i\sigma_y.$$

The spatial spin part of the Gamow–Teller matrix elements is, therefore, transformed as follows.

$$\begin{aligned}\psi^\dagger_{JJ-1}\sum_k \sigma^{(k)}_{-1}t^{(k)}_-\psi_{JJ} &= \left(\frac{1}{2J}\right)^{1/2}\psi^\dagger_{JJ}J_+\sum_k \sigma^{(k)}_{-1}t^{(k)}_-\psi_{JJ}\\ &= \left(\frac{1}{2J}\right)^{1/2}\psi^\dagger_{JJ}\left[\sum_k \sigma^{(k)}_{-1}t^{(k)}_- J_+ + 2^{1/2}\sum_k \sigma^{(k)}_z t^{(k)}_-\right]\psi_{JJ}\\ &= \left(\frac{1}{J}\right)^{1/2}\psi^\dagger_{JJ}\sum_k \sigma^{(k)}_z t^{(k)}_-\psi_{JJ}.\end{aligned} \tag{7.18}$$

The first term on the right-hand side on the second line vanishes since ψ_{JJ} has its maximum z component. The reduction of the isospin part is similarly performed:

$$\begin{aligned}\phi^\dagger_{-1/2}\sum_k \sigma^{(k)}_z t^{(k)}_-\phi_{1/2} &= \phi^\dagger_{-1/2}\sum_k \sigma^{(k)}_z t^{(k)}_- T_+\phi_{-1/2}\\ &= \phi^\dagger_{-1/2}\left[T_+\sum_k \sigma^{(k)}_z t^{(k)}_- - 2\sum_k \sigma^{(k)}_z t^{(k)}_z\right]\phi_{-1/2}\\ &= -2\phi^\dagger_{-1/2}\sum_k \sigma^{(k)}_z t^{(k)}_z\phi_{-1/2}.\end{aligned} \tag{7.19}$$

Here we have used identities

$$\phi^\dagger_{-1/2}T_+ = (T_-\phi_{-1/2})^\dagger = 0, \tag{7.20}$$

and

$$t^{(k)}_- t^{(k)}_+ - t^{(k)}_+ t^{(k)}_- = -2t^{(k)}_z. \tag{7.21}$$

Combining Eqs. (7.11), (7.18), and (7.19), we have finally

$$\frac{1}{J}\left|2\left\langle\sum_k \sigma^{(k)}_z t^{(k)}_z\right\rangle_{M=J,T_z=-1/2}\right|^2 \tag{7.22}$$

for the σ_{-1} term.

The isospin part for the σ_0 term is also given by Eq. (7.19). The final result for σ_0 term is

$$|2\left\langle \sum_k \phi_z^{(k)} t_z^{(k)} \right\rangle_{M=J,\, T_z=-1/2}|^2. \tag{7.23}$$

From Eqs. (7.11), (7.22), and (7.23), we have the Gamow–Teller matrix element for the mirror transition,

$$|\int \sigma|^2 = \frac{J+1}{J}\, |2\left\langle \sum_k \sigma_z^{(k)} t_z^{(k)} \right\rangle_{M=J,\, T_z=-1/2}|^2. \tag{7.24}$$

This relation is derived by using no other assumption than the charge independence of nuclear forces. It is valid for mirror nuclei only.

If there is only one nucleon outside of the closed shells, the nuclear spin J is equal to the single particle value j. We have to calculate the expectation value of σ_z only. The expectation value for $2t_z$ is simply -1. The single particle wave function with quantum numbers j and μ is given by

$$\psi_{j\mu} = \sum_{mm'} (l\tfrac{1}{2}mm'|j\mu) Y_{lm}(\theta, \phi)\chi_{m'}. \tag{7.25}$$

Here j, l, $\frac{1}{2}$ are the total, orbital, and spin angular momentum, respectively. On the right-hand side the radial wave function is dropped, since it is the same for the initial and the final state, so that the integration over radial variable gives simply unity. The expectation value for σ_z is given by

$$\begin{aligned} \int \psi_{j\mu}^\dagger \sigma_z \psi_{j\mu}\, d\Omega &= \sum_{mm'} \sum_{m_1 m_1'} (l\tfrac{1}{2}mm'|j\mu)(l\tfrac{1}{2}m_1 m_1'|j\mu) \\ &\quad \times \int Y_{lm}^* Y_{lm_1}\, d\Omega\, (\chi_{m'}^\dagger \sigma_z \chi_{m_1'}) \\ &= \sum_{mm'} (l\tfrac{1}{2}mm'|j\mu)^2 (-)^{1/2-m'}. \end{aligned} \tag{7.26}$$

Here the summation includes only two terms for $m' = \pm\frac{1}{2}$. Introducing algebraic forms for the Clebsch–Gordan coefficients in Eq. (7.26) explicitly, we have

$$\begin{aligned} \int \psi_{j\mu}^\dagger \sigma_z \psi_{j\mu}\, d\Omega &= \frac{2\mu}{2l+1} = 1 && \text{for} \quad \mu = j = l + \tfrac{1}{2}, \\ &= -\frac{2\mu}{2l+1} = -\frac{j}{j+1} && \text{for} \quad \mu = j = l - \tfrac{1}{2}, \end{aligned} \tag{7.27}$$

Under the assumption of the single particle shell model, the Gamow–Teller matrix element in mirror transitions is finally given by

$$\begin{aligned} |\int \sigma|^2 &= \frac{j+1}{j} && \text{for} \quad j = l + \tfrac{1}{2} \\ &= \frac{j}{j+1} && \text{for} \quad j = l - \tfrac{1}{2}. \end{aligned} \tag{7.28}$$

For the decay $F^{17} \to O^{17}$, we have the closed core of O^{16}. During the decay, the single proton outside this core is transformed into a single neutron, where the core is assumed to remain inert. Since the single particle orbit is $d_{5/2}$,

$$|\textstyle\int \sigma|^2 = \frac{7}{5} \tag{7.29}$$

from the upper equation in (7.28), while from Eq. (7.8) the Fermi matrix is $|\int 1|^2 = 1$. Using these results and the experimental $ft_{1/2}$ value in this decay

$$(ft_{1/2})_{\mathrm{exp}} = 2380 \text{ sec}, \tag{7.30}$$

we have

$$2380 = 6146\left[1 + \frac{7}{5}\left(\frac{C_A}{C_V}\right)^2\right]^{-1},$$

or

$$|C_A/C_V| = 1.06. \tag{7.31}$$

This shows a marked difference from the ratio $|C_A/C_V| = 1.24$ obtained from the neutron data. This fact that the coupling constant ratio is smaller in complex nuclei than in neutron was reported by Kistner and Rustad [5] in the analysis of $ft_{1/2}$ values for the beta decay of O^{15} and F^{17} together with the $0^+ \to 0^+$ transitions.

For the most of the beta decays with the selection rules of the allowed transitions, the $ft_{1/2}$ values are larger and the matrix elements considerably smaller than those given in Eqs. (7.6) and (7.28). We will come back to this problem later.

7.2 MAGNETIC MOMENTS AND GAMOW-TELLER MATRICES

Equation (7.24) in the last section is extremely useful, since it relates the Gamow–Teller matrix elements with the experimental values of the magnetic moments in mirror nuclei. The magnetic moment of a nucleus can be expressed in a formal way with the help of isospin operators. The expression $\frac{1}{2} + t_z^{(k)}$ is one for the proton and zero for the neutron, while $\frac{1}{2} - t_z^{(k)}$ is zero for the proton and one for the neutron. The magnetic moment is an expectation value for the state $M = J$ which we write here $\langle\,\rangle_J$, and is given in units of the nuclear magneton by

$$\mu = \left\langle \sum_k \left(\tfrac{1}{2} + t_z^{(k)}\right)\left(l_z^{(k)} + \sigma_z^{(k)}(1 + \mu_p)\right) + \left(\tfrac{1}{2} - t_z^{(k)}\right)\sigma_z^{(k)}\mu_n \right\rangle_J. \tag{7.32}$$

Since $j_z = l_z + \frac{1}{2}\sigma_z$, we replace $l_z = j_z - \frac{1}{2}\sigma_z$. The sum of $j_z^{(k)}$ over all k is J, so that

$$\mu = \tfrac{1}{2} J + \tfrac{1}{2}\Big\langle \sum_k \sigma_z^{(k)} (\mu_p + \mu_n + \tfrac{1}{2}) \Big\rangle_J + \Big\langle \sum_k t_z^{(k)} [\, j_z^{(k)} + \sigma_z^{(k)} (\mu_p - \mu_n + \tfrac{1}{2})] \Big\rangle_J ,$$

with $\mu_p = 1.79$, $\mu_n = -1.91$, and $J = \langle \Sigma_k\, j_z^{(k)} \rangle_J$. Hence,

$$\mu = \tfrac{1}{2} J + 0.19 \Big\langle \sum_k \sigma_z^{(k)} \Big\rangle_J + \Big\langle \sum_k t_z^{(k)} (j_z^{(k)} + 4.2\sigma_z^{(k)}) \Big\rangle_J . \tag{7.33}$$

Since we wish to use Eq. (7.33) for the nuclei involved in mirror decays, we shall now assume that all neutrons and protons outside of closed shells, which are the only ones involved in Eq. (7.32), are in a level with the same j. Under these assumptions together with Eq. (7.27), we have

$$\langle j_z^{(k)} \rangle = \langle \sigma_z^{(k)} \rangle \begin{cases} \times j & \text{for} \quad j = l + \tfrac{1}{2}, \\ \times [-(j+1)] & \text{for} \quad j = l - \tfrac{1}{2}. \end{cases} \tag{7.34}$$

We will replace σ_z by j_z in the term not involving t_z, and j_z by σ_z in the term with t_z. The result is

$$\begin{aligned} \mu &= \tfrac{1}{2}(l + 0.88) \frac{J}{j} + (4.7 + l) \Big\langle \sum_k t_z^{(k)} \sigma_z^{(k)} \Big\rangle_J & \text{for} \quad j = l + \tfrac{1}{2}, \\ &= \tfrac{1}{2}(l + 0.12) \frac{J}{j+1} + (3.7 - l) \Big\langle \sum_k t_z^{(k)} \sigma_z^{(k)} \Big\rangle_J & \text{for} \quad j = l - \tfrac{1}{2}. \end{aligned} \tag{7.35}$$

Therefore, we can derive the expectation value

$$\Big\langle \sum_k t_z^{(k)} \sigma_z \Big\rangle_J$$

from the experimental data on the magnetic moment of mirror nuclei. From Eqs. (7.24)-(7.35), we have [3]

$$\begin{aligned} |\textstyle\int \sigma|^2 &= \frac{J+1}{J} \left| \frac{2\mu - (J/j)(l + 0.88)}{4.70 + l} \right|^2 & \text{for } j = l + \tfrac{1}{2}, \\ &= \frac{J+1}{J} \left| \frac{2\mu - [J/(j+1)](l + 0.12)}{3.70 - l} \right|^2 & \text{for } j = l - \tfrac{1}{2}. \end{aligned} \tag{7.36}$$

With these equations, we can find magnitude of the Gamow–Teller matrix elements using the data on the magnetic moments μ of the daughter nuclei. J is the total angular momentum and j is the angular momentum of the level being filled. Furthermore, we can again determine the ratio of the coupling constant from the ft values of the mirror transition (see Eqs. (7.1)-(7.3)),

$$1 + |C_A/C_V|^2 \,|\textstyle\int \sigma|^2 = B/ft_{1/2}.$$

This type of determination for the coupling constant ratio C_A/C_V was performed in the early days. For this purpose, Kofoed-Hansen and Winther derived also the following semi-empirical formula for the Gamow–Teller matrix element [6],

$$|\int \sigma|^2 = 4\frac{J+1}{J}\left|\frac{\mu - g_l J}{g_s - g_l}\right|^2. \tag{7.37}$$

Here μ is the magnetic moment of the daughter nuclei. The gyromagnetic ratios are given by

$$\begin{aligned} g_l &= 1 && \text{for the proton,} \\ &= 0 && \text{for the neutron;} \\ g_s &= 2(1+\mu_p) = 5.5855 && \text{for the proton,} \\ &= 2\mu_n = -3.8256 && \text{for the neutron.} \end{aligned}$$

Several other formulas are also discussed in Feenberg's book [3].

Angular Momentum Distribution in Mirror Nuclei

Recently, experimental values on the magnetic moments of the mirror nuclei became available, together with those on $ft_{1/2}$ values of mirror transitions. Using these data, Sugimoto [7] found some regularities concerning the distributions of the angular momentum between the orbital and spin parts. This analysis is in the framework of the one-body magnetic moment operators, the charge symmetry of nuclear forces, and neglect of the Coulomb interaction. These assumptions are the same as those in the last section. The nuclear moment, (7.32), consists of two parts, called isoscalar and isovector parts:

$$\mu = \mu_S + \mu_V$$

with

$$\mu_S = \tfrac{1}{2}\langle \sum [l_z + (1+\mu_p+\mu_n)\sigma_z]\rangle_J$$

and

$$\mu_V = \tfrac{1}{2}\langle \sum \tau_z [l_z + (1+\mu_p-\mu_n)\sigma_z]\rangle_J. \tag{7.38}$$

We use the short-hand notation, $\langle \sum l_z\rangle_J$, etc., in place of $\langle \sum_k l_z^{(k)}\rangle_J$, etc., in Eq. (7.38) and hereafter. The former is independent of T_z, while the latter changes only in sign for different $T_z = \pm\frac{1}{2}$. Therefore, the sum moment of the mirror states is

$$\begin{aligned} &\mu(T_z = \tfrac{1}{2}) + \mu(T_z = -\tfrac{1}{2}) \\ &\quad = \langle \sum l_z\rangle_J + (1+\mu_p+\mu_n)\langle \sum \sigma_z\rangle_J, \end{aligned} \tag{7.39}$$

where $\langle\,\rangle_J$ is the expectation value for the state with $M = J$ and $T_z = \frac{1}{2}$. Since the spin of the nuclear state is expressed by

$$J = \langle \sum j_z \rangle_J = \langle \sum l_z \rangle_J + \tfrac{1}{2}\langle \sum \sigma_z \rangle_J, \tag{7.40}$$

we can find the quantities $\langle \sum \sigma_z \rangle_J$ and $\langle \sum l_z \rangle_J$ for mirror states separately.

On the other hand, the difference of moments of the mirror states is

$$\mu(T_z = \tfrac{1}{2}) - \mu(T_z = -\tfrac{1}{2}) = 2\langle \sum t_z l_z \rangle_J + (1 + \mu_p - \mu_n)\langle \sum \tau_z \sigma_z \rangle_J, \tag{7.41}$$

while the quantity $\langle \sum \tau_z \sigma_z \rangle_J$ is related to the Gamow–Teller matrix element

$$|\textstyle\int \sigma|^2 = \dfrac{J+1}{J}\,|\langle \sum \tau_z \sigma_z \rangle_J|^2. \tag{7.42}$$

Here we have used $\tau_z = \frac{1}{2} t_z$ and the magnitude of $\langle \sum \tau_z \sigma_z \rangle_J$ is the same for $T_z = \pm\frac{1}{2}$, although the sign is different. Again, we can find the quantities $\langle \sum \tau_z \sigma_z \rangle_J$ and $\langle \sum \tau_z l_z \rangle_J$ separately. The sign of the $\langle \sum \tau_z \sigma_z \rangle_J$ is not determined from $|\int \sigma|^2$, but it is unambigously found from the systematics.

Combining the values thus obtained, Sugimoto calculated the quantities

$$\langle \sum \tfrac{1}{2}(1+\tau_3)\sigma_z \rangle, \qquad \langle \sum \tfrac{1}{2}(1-\tau_3)\sigma_z \rangle,$$
$$\langle \sum \tfrac{1}{2}(1+\tau_3) l_z \rangle, \qquad \langle \sum \tfrac{1}{2}(1-\tau_3) l_z \rangle.$$

In his notation, $\langle \cdots (1+\tau_3) \cdots \rangle$ refers always to the expectation values for the odd-group nucleons, and $\langle \cdots (1-\tau_3) \cdots \rangle$ to those for the even-group nucleons. Numerical values are available for nine examples. And the angular momentum distributions are plotted in Figs. 7.1 and 7.2 as functions of mass number A. The numerical values of $\langle \sum \sigma_z \rangle$ and $\langle \sum \tau_z \sigma_z \rangle$ are almost equal in each case. A marked regularity is seen in the region of the s–d shell ($A = 17 - 40$) as shown in Fig. 7.1a. They are close to the single particle limit for $\frac{5}{2}^+$ at $A = 17$ and decrease approximately linearly with mass number. They are again close to the single particle limit for $\frac{3}{2}^+$ near the end of the s–d shell.

As is shown in Fig. 7.1b, the spin parts of the odd-group nucleons $\langle \sum \frac{1}{2}(1+\tau_3)\sigma_z \rangle$ lie between the single particle limits, and the spin parts of the even-group nucleons $\langle \sum \frac{1}{2}(1-\tau_3)\sigma_z \rangle$ are always approximately zero. This indicates that pairing between like particles plays an important role. The orbital parts $\langle \sum l_z \rangle_J$ deviate naturally from the single particle values such as to compensate the deviations by the spin parts, but the value $\langle \sum \tau_z l_z \rangle_J$ deviates further in the same direction, see Fig. 7.2a. The orbital parts for the odd- or even-group nucleons, $\langle \sum \frac{1}{2}(1+\tau_3)\sigma_z \rangle$ and $\langle \sum \frac{1}{2}(1-\tau_3)\sigma_z \rangle$, deviate from the single particle estimates in a systematic manner, i.e., the individual deviations of the odd-group nucleons from the single particle values are accompanied by the smaller opposite deviations of the even-group values from zero.

Sugimoto inferred from these regularities in the angular momentum distributions that the Gamow–Teller matrix elements are insensitive to the

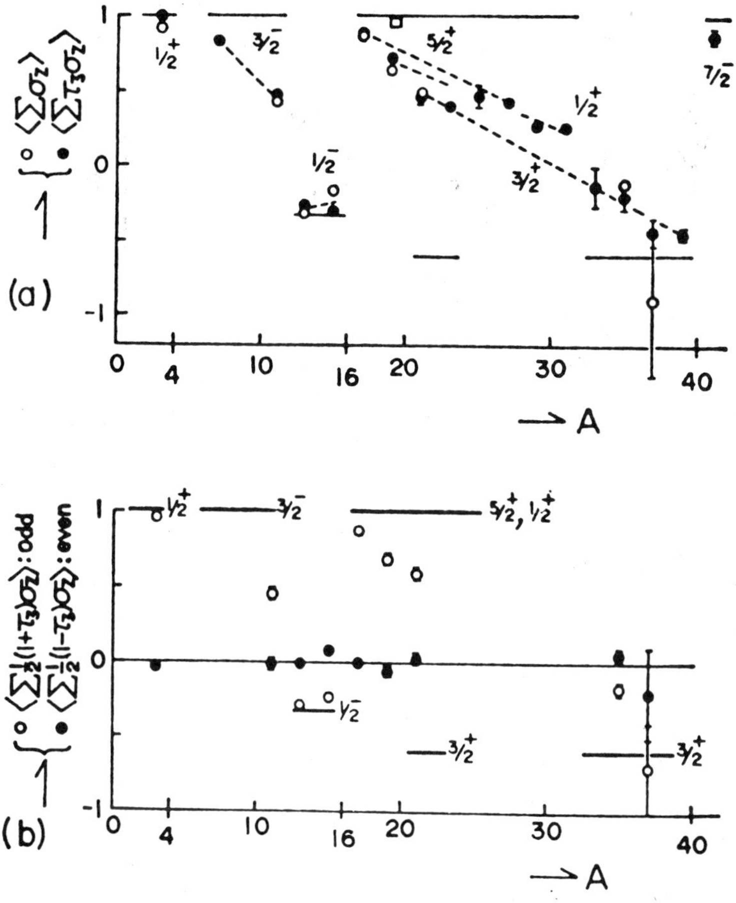

Fig. 7.1 The spin parts of the angular momentum distribution in mirror nuclei [7]. The expectation values are for the state with $T_z = \frac{1}{2}$. The solid lines represent the single particle values. The broken lines connect the points for the same J^π. The square point represents the metastable pair $(\frac{5}{2}^+)$ of $A = 19$. In Fig. 7.1a, τ_3 should read τ_z. In Fig. 7.1b, $\langle \Sigma \frac{1}{2}(1 + \tau_3)\sigma_z \rangle$: odd and $\langle \Sigma \frac{1}{2}(1 - \tau_3)\sigma_z \rangle$: even refer to spin parts of the odd-group and even-group nucleons. In our notation, $\langle \Sigma \frac{1}{2}(1 + \tau_3)\sigma_z \rangle$: odd should read $\langle \Sigma \frac{1}{2}(1 + \tau_z)\sigma_z \rangle_J$ for $P = N + 1$ (or $N = P + 1$) and P = odd (or even), and $\langle \Sigma \frac{1}{2}(1 - \tau_z)\sigma_z \rangle_J$ for $P = N + 1$ (or $N = P + 1$) and P = even (or odd). Similar interpretation holds for $\langle \cdots \rangle$: even.

mesonic effects at least for the cases concerned, but the orbital parts include the possible mesonic contributions.

Leonardi and Rosa-Clot [8] also analyzed the isovector magnetic moments

μ_V and the allowed Gamow–Teller matrix elements in mirror nuclei, finding strong evidence of linear quenching on the average in the atomic number A

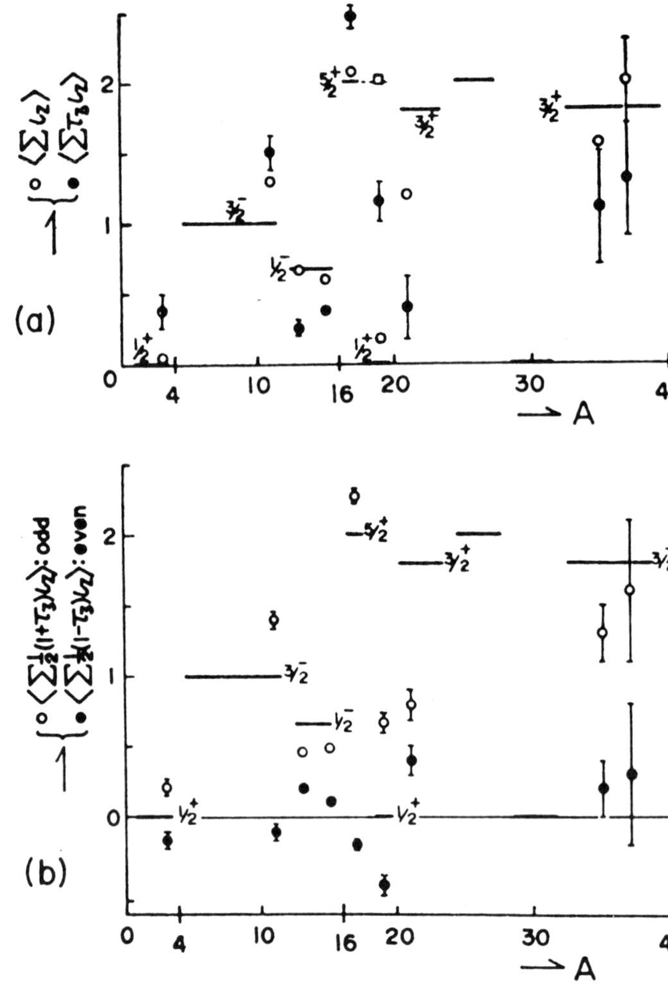

Fig. 7.2 The orbital parts of the angular momentum distribution in mirror nuclei [7]. The expectation value for the state with $T_z = \frac{1}{2}$ is used. In Fig. 7.2a, τ_3 should read τ_z. In 7.2b, $\langle\cdots\rangle$: odd and $\langle\cdots\rangle$: even refer to orbital parts of the odd-group and even-group nucleons (see caption of Fig. 7.1).

(for a given J and parity) of the matrix element $\langle\Sigma\, t_z\sigma_z\rangle_J$. They discussed linear dependence on A for $\langle\Sigma\, \sigma_z\rangle_J$ and $\langle\Sigma\, t_z\sigma_z\rangle_J$ in Fig. 7.1a for each orbit $_{1/2}$, $1p_{3/2}$, $1p_{1/2}$, $1d_{5/2}$, $1d_{3/2}$, and $2s_{1/2}$, separately. This sort of regularity can be satisfactorily explained by the j–j configuration-mixing model [9]. The quantitative results on the corrections

$$\Delta\sigma = \langle \sum \sigma_z \rangle_J^{\text{Schmidt}} - \langle \sum \sigma_z \rangle_J^{j-j\ \text{calc}}$$

obtained by Leonardi and Rosa-Clot are for the case of delta forces, with $V_s = V_t$ (equal singlet and triplet strength of the residual interaction) and assuming the simplest unperturbed configuration (i.e., the mixing between different j with the same l is taken into account). They found

$$\begin{aligned}
&\Delta\sigma = 0 && \text{for } 1s_{1/2},\\
&\Delta\sigma = 0.16(p-1)(V_s/\Delta E_1) \approx 0.22(p-1) && \text{for } 1p_{3/2},\\
&\Delta\sigma = 0 && \text{for } 1p_{1/2},\\
&\Delta\sigma = 0.09(p-1)(V_s/\Delta E_2) \approx 0.2(p-1) && \text{for } 1d_{5/2},\\
&\Delta\sigma = 0.062\, n_1(V_s/\Delta E_2) \approx 0.15 n_1 && \text{for } 2s_{1/2},\\
&\Delta\sigma = -0.063(p-1)(V_s/\Delta E_2) \approx -0.15(p-1) && \text{for } 1d_{3/2}.
\end{aligned} \tag{7.43}$$

Here

$$\Delta E_1 = E(p_{1/2}) - E(p_{3/2}) + \text{pairing}, \quad \Delta E_2 = E(d_{3/2}) - E(d_{5/2}) + \text{pairing},$$

where p is the number of odd nucleons in the shell (odd-hole nucleons in the case of $d_{3/2}$), n_1 is the number of odd nucleons in the adjacent shell $d_{5/2}$, and $E(p_{1/2})$, etc., are the single particle excitation energies. The numerical values are calculated using the parameters in [9].

7.3 EFFECTS OF PAIRING CORRELATIONS ON ALLOWED GAMOW-TELLER TRANSITIONS

As is discussed in section 7.1, there are many beta decays with $\log ft_{1/2} = 4.5 - 5.7$. These are called the hindered (or unfavored) decays. The other beta decays, which have $\log ft_{1/2} \gtrsim 5.7$ and $\Delta l > 1$ (Δl: change of the orbital angular momentum for nucleons), are called the l-forbidden decays. Large hindrance of the nuclear matrix elements is the consequence of wrong overlapping of the initial and final wave functions. A method of explaining hindrance for allowed unfavored decays was proposed by Sakai and Yoshida [10] with the pairing model for the nucleus [11]. In this approach, we start from a shell model and assume that a pairing force acts on nucleons outside a closed shell in such a way that the nucleons couple two by two to spin zero. Since for medium and heavy nuclei protons and neutrons are in different shells, the pairing force between different particles is usually neglected, and only p-p and n-n pairs are taken into account. These pairs behave as quasiparticles which populate the levels. In the j orbit, a pair of nucleons in the states j, m and j, $-m$ has the probability of occupation V_j^2 or of nonoccupation U_j^2 ($V_j^2 + U_j^2 = 1$). The nuclear wave function of the ground state of an even nucleus is given by

$$\Psi_0 = \Pi_{jm}(U_j + V_j(-)^{j-m} a_{jm}^{\dagger} a_{j,-m}^{\dagger})|0\rangle, \tag{7.44}$$

where the product is taken over all states jm of neutrons and protons, and $|0\rangle$ is the vacuum state. Creation and annihilation operators $a^{\dagger}$ and a in the second quantization will be used here without any explanations. (They are, however, discussed briefly in section 9.3.) Numerical values of U_j and V_j are obtained by solving the total Hamiltonian with pairing forces. The odd nucleus with spin $j_1 m_1$ has the wave function

$$\Psi_{j_1 m_1} = a_{j_1 m_1}^{\dagger} \Pi'_{jm}(U_j + V_j(-)^{j-m} a_{jm}^{\dagger} a_{j,-m}^{\dagger})|0\rangle, \tag{7.45}$$

where the prime on the product sign indicates the omission of terms with j_1, $\pm m_1$. The orbit jm is either a neutron or a proton orbit according as the nucleus has odd N or odd Z.

The nucleon part of the beta decay can be written

$$M = \sum_{LM} \langle j_p m_p | LM | j_n m_n \rangle a_{j_p m_p}^{\dagger} a_{j_n m_n}. \tag{7.46}$$

Sakai and Yoshida calculated the matrix elements in the following eight cases: (e, e) → (o, o), (o, o) → (e, e), (o, e) → (e, o), and (e, o) → (o, e) for positron and negatron decays. Here the symbol (e, o) is the nucleus of even proton and odd neutron, etc. They gave the factor ξ in the expression

$$ft_{1/2} = \xi \frac{B}{(2j_f + 1)|\langle j_f \| \sigma \| j_i \rangle|^2}. \tag{7.47}$$

From formula (7.56) and the expression of ξ in Table 7.1, we can derive several relations among $ft_{1/2}$ values. For example, we have

$$\frac{ft_{1/2} \text{ for } (Z+1, N-1) \to (Z, N)}{ft_{1/2} \text{ for } (Z, N-1) \to (Z-1, N)} = \frac{3}{2j_n + 1} \frac{V_p^2(Z)}{U_p^2(Z)},$$

$$\frac{ft_{1/2} \text{ for } (Z, N+1) \to (Z+1, N)}{ft_{1/2} \text{ for } (Z, N) \to (Z+1, N-1)} = \frac{3}{2j_p + 1} \frac{V_n^2(N)}{U_n^2(N)}. \tag{7.48}$$

Therefore, it is possible to derive numerical values of V_p^2, U_p^2, V_n^2, U_n^2, from the foregoing relations with $U^2 + V^2 = 1$.

Sakai and Yoshida [10] determined U^2 and V^2 for the $p_{3/2}$, $d_{5/2}$, and $g_{9/2}$ proton shells, and for the $p_{1/2}$, $d_{3/2}$, and $g_{7/2}$ neutron shells. These examples for g shells are shown in Fig. 7.3, together with those obtained from the data on (d, p) and (d, t) reactions. These values are in good agreement. The theoretical values of U^2 and V^2 which are derived with newly obtained parameters in the pairing theory are also shown there.

Table 7.1 Inverse of the Proportionality Factor ξ in Eq. (7.47)[a]

Decay Mode	Positron Decay	Negatron Decay
(e, e) → (o, o)	$U_{j\mathrm{n}}^2(i)\,V_{j\mathrm{p}}^2(i)$	$U_{j\mathrm{p}}^2(i)\,V_{j\mathrm{n}}^2(i)$
(o, o) → (e, e)	$\frac{1}{3}V_{j\mathrm{n}}^2(f)\,U_{j\mathrm{p}}^2(f)$	$\frac{1}{3}V_{j\mathrm{p}}^2(f)\,U_{j\mathrm{n}}^2(f)$
(o, e) → (e, o)	$\frac{1}{2j_\mathrm{p}+1}U_{j\mathrm{n}}^2(i)\,U_{j\mathrm{p}}^2(f)$	$\frac{1}{2j_\mathrm{p}+1}V_{j\mathrm{p}}^2(f)\,V_{j\mathrm{n}}^2(i)$
(e, o) → (o, e)	$\frac{1}{2j_\mathrm{n}+1}V_{j\mathrm{n}}^2(f)\,V_{j\mathrm{p}}^2(i)$	$\frac{1}{2j_\mathrm{n}+1}U_{j\mathrm{p}}^2(i)\,U_{j\mathrm{n}}^2(f)$

[a] From [10].

Hamamoto [12] applied the pairing theory to investigate the allowed beta transitions of the Gamow–Teller type in odd nuclei. She found that the admixture of seniority-3 states to the unperturbed seniority-1 state has con-

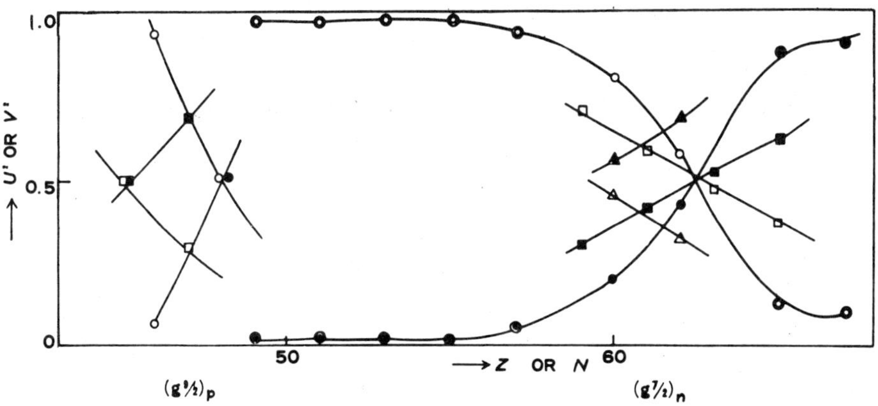

Fig. 7.3 The U^2 and V^2 parameters for the $(g_{9/2})_\mathrm{p}$ and $(g_{7/2})_\mathrm{n}$ orbits deduced from beta decays (circles). The U^2 and V^2 values calculated by using the pairing theory are presented by triangles and squares. Those deduced from (d, p) and (d, t) nuclear reaction studies are presented as double circles [10].

siderable effect on the matrix element. Such an admixture is important in order to calculate the isotope dependence of the $ft_{1/2}$ values. In this theory a parameter which is the strength of the residual interaction is adjusted so as to fit the nuclear magnetic moments. Calculated $ft_{1/2}$ values are found to be in reasonable agreement with the experimental data. The calculation was also performed in the case of the l-forbidden allowed beta decays.

7.4 COMMUTATOR METHOD FOR EVALUATING THE GAMOW-TELLER AND FERMI MATRIX ELEMENTS

One of the older attempts to try to understand nuclear structure by using information on beta-decay matrix elements is a pioneering work on light nuclei ($A \leqslant 40$) performed by Wigner in 1939 [13]. He proposed the supermultiplet theory. We remember that the lowering and raising operators $T_{\pm}$ of the z component in the isospin are the transition operators for the Fermi matrix elements (see Eqs. (7.5)). The Gamow–Teller matrix element includes operators of the type

$$\mathbf{Y} = \sum_{k=1}^{A} \tau_{+}^{(k)} \sigma^{(k)},$$

which raise or lower the quantum numbers for the members of the supermultiplet, which corresponds to an irreducible representation of the group called su_4. As a consequence, the allowed transitions can occur only among members of the same supermultiplet. This statement is valid insofar as the nuclear forces can be regarded approximately as a sum of the Wigner and Majorana forces. Since the appearance of the j-j coupling shell model it has become known that spin-dependent forces like the spin-orbit forces play important roles in nuclei, and the supermultiplet structure of nuclei now is regarded as untenable.

In 1961 Anderson and Wong [14] at Livermore discovered by the (p, n) reactions that there exist isobaric analog states (IAS) not only for light nuclei but also for heavy nuclei ($A > 40$) with small widths of the order 100 keV,

$$\mathrm{p} + |\mathrm{i}\rangle = \mathrm{n} + T_{+}|\mathrm{i}\rangle + Q. \tag{7.49}$$

Here $|\mathrm{i}\rangle$ and $T_{+}|\mathrm{i}\rangle$ stand for the initial state and its isobaric analog. The isospin and its z component for these states are (T, T_z) and $(T, T_z + 1)$, respectively. The Q value is also given by

$$Q = F(\alpha Z/R) \qquad \text{in units of } mc^2 \tag{7.50}$$

with

$$F = 1.1 - 1.2.$$

That is, the Q value given in Eq. (7.50) represents the single particle Coulomb displacement. As a natural extension of the idea of isomultiplet structure for heavy nuclei, persistence of the supermultiplet structure was considered in explaining the hindrance factors of beta transitions. Persistence of the supermultiplet structure of heavy nuclei is also suggested from the fact that the allowed matrix elements for heavy nuclei with $A > 40$ generally take their values

$$|\int 1|^2 \ll |\int \sigma|^2 \ll 1. \tag{7.51}$$

These values are obtained from the beta-gamma circular polarization correlations as a consequence of the parity nonconservation (see Table 3.5). The same kinds of data are also supplied from measurements of the beta-ray distribution from polarized nuclei.

As we have studied in Eq. (7.5), the Fermi transition is induced by the operator T_+. Therefore, if the state $T_+|\mathrm{i}\rangle$ is an eigenstate in the strict sense, the possibility of the Fermi transitions from the initial state $|\mathrm{i}\rangle$ would be exhausted by a particular transition, $|\mathrm{i}\rangle \to T_+|\mathrm{i}\rangle$. This transition is, however, forbidden, since it is energetically impossible. Actually, observed beta decays in the region of heavy elements take place by overlapping of the initial state with the tail of the distribution of the state $T_+|\mathrm{i}\rangle$. As a consequence, the Fermi matrix elements are nonzero but very small, as is shown in Eq. (7.51).

The experimental discovery of the isobaric analog state (IAS) and its usefulness for explaining the hindrance of the Fermi matrix elements led to the conjecture of Fujita and his collaborators [15] that the Gamow–Teller strength for heavy nuclei, expressed as $|\int \sigma|^2$, is concentrated in the several-MeV energy region near the isobaric peak. In other words, Gamow–Teller transitions are observed only in the tail region of a strength distribution, as in the case of the Fermi transitions, and the relation (7.51) can naturally be expected to be valid. This is an extension of the idea of the supermultiplet theory to heavy nuclei. Possible existence of the Gamow–Teller giant resonance was independently pointed out also by Yamada [16], who made an effort to construct a semiempirical formula for $ft_{1/2}$ values. This will be discussed in the next section, while in this section we will discuss a successful theory developed by Fujita, Ikeda, and others [17, 18] for explaining the hindrance of the nuclear matrix elements in the allowed beta decays.

Their idea can be more clearly seen with the example in Fig. 7.4. The 9.7-min isomer of Sn^{125} decays mostly into the 0.325-MeV state of Sb^{125}. According to the single particle shell model, the parent and daughter states are assigned as $2d_{3/2}$ and $2d_{5/2}$, respectively, and the $\log ft_{1/2}$ value should be 3.26. Experimentally, however, the $\log ft_{1/2}$ value is about 5.44, which indicates a hindrance factor of ≈ 150. Here the hindrance factor is the ratio of $ft_{1/2}$ values in the experiment and in the single particle shell model,

$$h = \frac{(ft_{1/2})_{\mathrm{exp}}}{(ft_{1/2})_{\mathrm{sp}}} = \frac{|M|^2_{\mathrm{sp}}}{|M|^2_{\mathrm{exp}}}, \tag{7.52}$$

where the right-hand side is derived from the definition of the $ft_{1/2}$ value, (2.165). That is, the hindrance factor is the ratio of the nuclear matrix elements, as indicated.

Now, the IAS for the 9.7-min state of Sn^{125} lies in Sb^{125} about 12.8 MeV above it, so that the distance between the IAS and the 0.35-MeV state is about 14.8 MeV. The transition operator for the Gamow–Teller matrix element changes a neutron into a proton and may change its spin direction. By the change of a neutron into a proton, the 9.7-min state of Sn^{125} is transferred to its IAS. Further energy shift from this IAS caused by possible spin flip is expected to be about 5 MeV or less judging from the spin-orbit splittings. This energy shift is considerably smaller than the 14.8 MeV which is required to reach the 0.325-MeV state. From this we can imagine that the 0.325-MeV

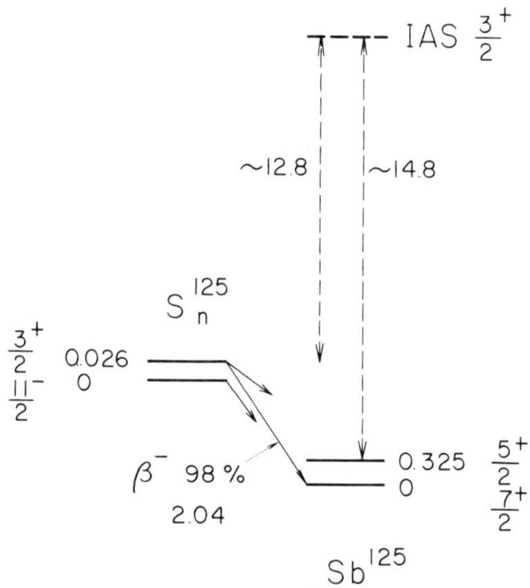

Fig. 7.4 Beta decay of Sb^{125} and isobaric analog state (IAS). The maximum kinetic energy of beta particles in the beta decay of the 0.026-MeV state of Sn^{125} to the ground state of Sb^{125} is 2.04 MeV. The IAS of the initial state is about 14.8 MeV above the ground state of Sb^{125}. The energies are in units of MeV.

state is quite different from the state obtained by the operator of the Gamow–Teller matrix onto the 9.7-min Sn^{125} state. This guess is qualitatively in good agreement with the large hindrance factor.

Fujita *et al.* utilized this persistence of the supermultiplet structure to find hindrance factors with good precision. Their calculation is based on an identity,

$$\langle f|\Omega|i\rangle = \frac{\langle f|[H, \Omega] - \Omega\Delta|i\rangle}{E_f - E_i - \Delta}, \tag{7.53}$$

where $|i\rangle$ and $\langle f|$ stand for the initial and final states with energies E_i and E_f, respectively, Ω is an arbitrary operator, H the total Hamiltonian, Δ a constant.

The square brackets represent the commutator. To the extent that exact wave functions are used, Eq. (7.53) is an identity. But if the wave functions adopted in evaluation of nuclear matrix elements are only approximate ones, the left- and right-hand sides of (7.53) do not necessarily coincide. The prescription made by Fujita and Ikeda is to calculate the right-hand side instead of calculating the left-hand side directly, by putting

$$\Delta = \frac{\langle \mathrm{i} | \Omega^\dagger [H, \Omega] | \mathrm{i} \rangle}{\langle \mathrm{i} | \Omega^\dagger \Omega | \mathrm{i} \rangle} = E_{\Omega_\mathrm{i}} - E_\mathrm{i}. \tag{7.54}$$

Here, the dagger means the hermitian conjugate, and E_{Ω_i} is the energy expectation value for the state $\Omega | \mathrm{i} \rangle$.

Table 7.2 Log $ft_{1/2}$ Values of Unfavoured Transitions.[a]

Region	Parent Nucleus	Daughter Nucleus	Decay Mode	Q value (MeV)	log $ft_{1/2}$ I[b]	II[b]	III[b]
p	${}_{29}\mathrm{Cu}^{62}_{33}$	${}_{28}\mathrm{Ni}^{62}_{34}$	β^+	2.91	5.1	4.6	3.6
	${}_{31}\mathrm{Ga}^{68}_{37}$	${}_{30}\mathrm{Zn}^{68}_{38}$	β^+	1.89	5.2	4.7	4.2
	${}_{35}\mathrm{Br}^{76}_{41}$	${}_{34}\mathrm{Se}^{76}_{42}$	ϵ	0.49	4.9	4.4	3.8
	${}_{35}\mathrm{Br}^{78}_{43}$	${}_{34}\mathrm{Se}^{78}_{44}$	β^+	2.53	4.7	4.5	3.7
	${}_{35}\mathrm{Br}^{80}_{45}$	${}_{34}\mathrm{Se}^{80}_{46}$	β^+	4.1	4.6	4.8	3.6
	${}_{37}\mathrm{Rb}^{80}_{43}$	${}_{36}\mathrm{Kr}^{80}_{44}$	β^+	0.866	4.6	4.6	3.9
	${}_{37}\mathrm{Rb}^{82}_{45}$	${}_{36}\mathrm{Kr}^{82}_{46}$	β^+	3.12	4.4	4.8	3.8
g	${}_{47}\mathrm{Ag}^{106}_{59}$	${}_{46}\mathrm{Pd}^{106}_{60}$	β^+	1.96	5.0	4.3	3.8
	${}_{47}\mathrm{Ag}^{108}_{61}$	${}_{46}\mathrm{Pd}^{108}_{62}$	β^+	0.69	4.4	4.3	3.7
d	${}_{51}\mathrm{Sb}^{118}_{67}$	${}_{50}\mathrm{Sn}^{118}_{68}$	β^+	3.10	4.8	5.0	3.9
	${}_{51}\mathrm{Sb}^{120}_{69}$	${}_{50}\mathrm{Sn}^{120}_{70}$	β^+	1.70	4.5	5.0	3.8
	${}_{55}\mathrm{Ce}^{126}_{71}$	${}_{54}\mathrm{Xe}^{126}_{72}$	β^+	3.8	5.0	4.8	3.7
	${}_{53}\mathrm{I}^{128}_{75}$	${}_{52}\mathrm{Te}^{128}_{76}$	β^+	1.2	5.1	4.9	3.5
	${}_{57}\mathrm{La}^{134}_{77}$	${}_{56}\mathrm{Ba}^{134}_{78}$	β^+	2.7	4.9	4.7	3.4
	${}_{57}\mathrm{La}^{136}_{79}$	${}_{56}\mathrm{Ba}^{136}_{80}$	β^+	1.85	4.6	4.8	3.4

[a] From [18]. In the first column, the region, p, g, or d stands for $2p_{1/2} \rightleftarrows 2p_{3/2}$, $1g_{7/2} \rightleftarrows 1g_{9/2}$, or $2d_{3/2} \rightleftarrows 2d_{5/2}$, respectively. In the fourth column, the decay mode is either positron decay β^+ or electron capture ϵ. The theoretical values (II) log $ft_{1/2}$ are from the theory developed by Fujita *et al.* with the pairing model, whereas entries in III are from the pure pairing model.

[b] Key: I, experimental; II, theoretical; III, pairing.

The adequacy of this prescription for the calculation of beta-decay matrix elements is understood as follows. For the Fermi operator, $[H, \Omega] | \mathrm{i} \rangle$ is almost exactly equal to $\Omega\Delta | \mathrm{i} \rangle$ irrespective of the exactness of the wave function $| \mathrm{i} \rangle$, and the numerator of Eq. (7.54) is very small. On the other hand, the denominator is the energy difference between the final state and IAS, and its absolute magnitude is rather large for most of the medium and heavy nulcei. Consequently, very small values are obtained for Fermi matrix elements in a natural way. For the Gamow–Teller operator the difference between $[H, \Omega]$ and $\Omega\Delta$ is larger when they operate on $|\,i\rangle$, but it is still considerably smaller than the denominator. Thus, right-hand side of Eq. (7.54) naturally yields a rather small Gamow–Teller matrix element if the final state is outside the supermultiplet region.

As an example, the result with the pairing model is shown in Table 7.2, where all transitions are positron decays (or electron capture) of the type (o, o) → (e, e), [18]. The calculated $\log ft_{1/2}$ values are given in comparison with experimental data and the theoretical values in the simple paring model. For the latter, the U and V factors appear in the third row of the second column of Table 7.1. We can see a great improvement for the commutator method from Table 7.2. The commutator method was applied to various beta processes; (a) Gamow–Teller transitions of spherical nuclei [17, 19], (b) Gamow–Teller transitions of deformed nuclei [20], (c) Fermi transitions [21], and (d) first forbidden transitions [22]. It was clarified in the course of studies that the core polarization effect is important for all beta transitions of heavy nuclei, and the commutator method can give qualitative agreement with experimental hindrance factors for a number of cases.

7.5 STATISTICAL THEORY FOR THE TOTAL DECAY RATE

As was mentioned in the last section, Yamada [16] and his collaborators also pointed out the possible existence of the Gamow–Teller giant resonance, and they made an effort to construct a semiempirical formula for the $ft_{1/2}$ values. The nature of their theory is quite statistical, and they call it the gross theory of beta decay. They have introduced a concept of the beta strength function, which they ascertain from the experimental data. Once the strength function is determined, they can estimate the lifetime for the beta decays of the nuclei far from the region of stability lines, and predict the limits of Z and N for exotic nuclei which will be produced by high energy accelerators.

In this section, we will introduce Yamada's gross theory of beta decay in a very simple way. The foundation of his theory and its justification are rather complicated, and it is left to his original papers [16]. We will discuss now the total decay constant λ of the allowed beta decay which generally has several final states. In the usual approximation, λ is written as

$$\lambda = \frac{m^5 c^4}{2\pi^3 \hbar^7} \sum_j \left[|C_V|^2 \left| \int 1_j \right|^2 + |C_A|^2 \left| \int \sigma_j \right|^2 \right] f(E_I - E_j), \tag{7.55}$$

where E_I and E_j are the energies of the initial state and the jth final state, respectively. $\int 1_j$ and $\int \sigma_j$ are the Fermi and the Gamow–Teller matrix elements to the jth final state, and $f(E_I - E_j)$ is the integrated Fermi function, which is dimensionless.

In the gross theory, we assume that the final level density is large enough to replace the summation over j in Eq. (7.55) by an integration as

$$\lambda \approx \frac{m^5c^4}{2\pi^3\hbar^7}\int_{-Q}^{0}[|C_V|^2|M_F(E)|^2+|C_A|^2|\mathbf{M}_{GT}(E)|^2]f(-E)\,dE, \tag{7.56}$$

where the continuous variable E corresponds to $(E_j - E_I)$ in Eq. (7.55), and Q is the ground state Q value. $|M_F(E)|^2$ and $|\mathbf{M}_{GT}(E)|^2$ are the final level density times the squares of the nuclear matrix elements averaged in an appropriate energy interval for the Fermi and the Gamow-Teller transitions, respectively. These quantities are called the Fermi and Gamow-Teller strength functions. The replacement of discrete sums by integrations, such as of Eq. (7.55) by Eq. (7.56), is essential to the gross theory, and Yamada called it the gross approximation or gross treatment. As a consequence, $|M_F(E)|^2$ is the energy distribution of the isobaric analog state $T_+|i\rangle$. As is shown in Fig. 7.5, the Fermi strength function is concentrated in an energy-forbidden transition to the IAS. Similarly, the Gamow-Teller strength is also concentrated in an

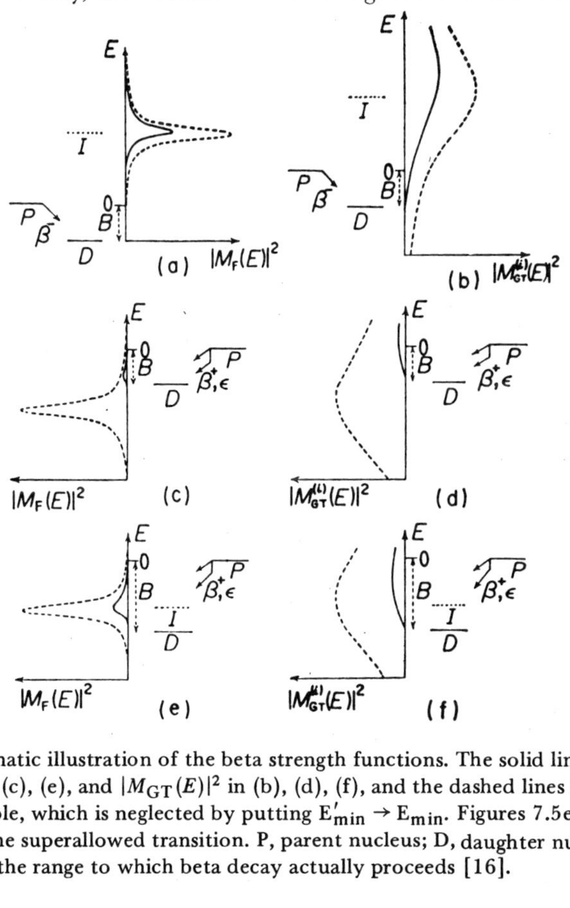

Fig. 7.5 Schematic illustration of the beta strength functions. The solid lines represent $|M_F(E)|^2$ in (a), (c), (e), and $|M_{GT}(E)|^2$ in (b), (d), (f), and the dashed lines correspond to the Pauli principle, which is neglected by putting $E'_{min} \to E_{min}$. Figures 7.5e and 7.5f correspond to the superallowed transition. P, parent nucleus; D, daughter nucleus; I, isobaric analog state; B, the range to which beta decay actually proceeds [16].

energy-forbidden transition to states near the IAS. The beta decay can take to all states which lie between 0 and D in Fig. 7.5 (i.e., in the region marked B). Yamada and Takahashi studied the strength functions by introducing two functions D and dN_1/dE_{sp} in the expression

$$|M_\Omega|^2 = n_\Omega \int_{E_{\min}}^{E_1} D_\Omega(E, E_{sp}) \frac{dN_1}{dE_{sp}} dE_{sp}. \tag{7.57}$$

Here Ω is either F or GT with $n_\Omega = 1$ or $n_\Omega = 3$. The latter comes from the three components of the vector operator $\boldsymbol{\sigma}$. D is the percentage decay rate of the nucleon with the single particle energy E_{sp}. dN_1/dE_{sp} is the number of nucleons in the range E_{sp} to $E_{sp} + dE_{sp}$. $E_{\min}$ is the minimum E_{sp} for which beta decay still takes place. E_1 is the maximum energy of the filled single nucleon states. N_1 is the total number of nucleons which have a possibility of decay, i.e., the total proton number Z for positron decay and the total neutron number N for negatron decay,

$$N_1 = \int_{E_{\min}}^{E_1} \frac{dN_1}{dE_{sp}} dE_{sp} = \begin{cases} Z & \text{for } \beta^+, \\ N & \text{for } \beta^-. \end{cases} \tag{7.58}$$

The Pauli principle can be expressed by the inequality

$$E_{sp} + E > E_i - Q, \tag{7.59}$$

and the lower limit of the integration domain of Eq. (7.57) is written as

$$E'_{\min} = \text{Max}\,(E_{\min}, E_i - Q - E) \tag{7.60}$$

where Max (a, b) denotes the larger one of a and b.

In order to perform numerical estimation of dN_1/dE_{sp} and $E'_{\min}$, the Fermi gas model is the simplest model. We have [16]

$$\frac{dN_1}{dE_{sp}} = \frac{2}{(2\pi\hbar)^3} 4\pi V[2M^{*3}(E_{sp} - E_{\min})]^{1/2} \tag{7.61}$$

and

$$E_{\min} = E_1 - E_F. \tag{7.62}$$

Here, M^* is the effective mass of the nucleon, V is the approximate nuclear volume, and E_F is the Fermi energy given by

$$E_F = \frac{\hbar^2}{2M^*} \left(3\pi^2 \frac{N_1}{V}\right)^{2/3}. \tag{7.63}$$

If we denote the nucleon mass by M and the nuclear radius by $R = r_o A^{1/3}$ fm, where A is the mass number, Eq. (7.63) becomes

$$E_F = \frac{76.52}{(M^*/M)r_o^2}\left(\frac{N_1}{A}\right)^{2/3} \text{ MeV.} \tag{7.64}$$

The following numerical calculations are given by the assumptions

$$r_o = 1.2 \qquad \text{and} \qquad M^*/M = 0.6. \tag{7.65}$$

Here, the effective nucleon mass is given by the nuclear matter problem [23]. The energy E'_{min} is also chosen as

$$E'_{\text{min}} = E_1 - Q - E, \tag{7.66}$$

and the single nucleons with energy $E_{\text{sp}} < E_1 - Q$ do not participate in the actual beta decay.

The distribution function $D_\Omega(E, E_{\text{sp}})$ is simply assumed to be independent of E_{sp}. That is, we assume that every nucleon has the same decay probability no matter what the energy E_{sp} of the nucleon is. Then, with Eqs. (7.57), (7.61), and (7.66), the integration over E_{sp} can be carried out and Eq. (7.56) becomes

$$\lambda = \frac{m^5c^4}{2\pi^3\hbar^7}\int_{-Q}^{0}[|C_V|^2 D_F(E) + 3|C_A|^2 D_{GT}(E)]N_1$$

$$\times\left[1 - \left(1 - \frac{Q+E}{E_F}\right)^{3/2}\right]f(-E)\,dE, \tag{7.67}$$

where the quantity in the square brackets on the second line represents the effect of the Pauli principle. Schematic illustrations of the strength functions are shown in Fig. 7.5 for various cases.

Yamada and Takahashi assumed the following four kinds of distribution functions.

1. *Gaussian type.* For the Fermi transitions,

$$D_F(E) = \frac{1}{(2\pi)^{1/2}\sigma_F}\exp\left[-\frac{(E-\Delta_F)^2}{2\sigma_F^2}\right], \tag{7.68}$$

where the energy center Δ_F and the standard deviation σ_F can be attributed to the Coulomb force

$$\Delta_F = \Delta_C \qquad \text{and} \qquad \sigma_F = \sigma_C. \tag{7.69}$$

Δ_C may be regarded as the single particle Coulomb displacement. For the simple model of a uniformly charged sphere with the radius $1.2A^{1/3}$ fm, we have

$$\Delta_C = \pm(1.44Z_1A^{-1/3} - 0.7825) \text{ MeV} \qquad \text{for } \beta^\mp \tag{7.70}$$

where Z_1 is the proton number of the daughter nucleus for positron decay ai of the parent one for negatron decay. The quantity σ_C as the fluctuation of the Coulomb energy of the single decaying nucleon in the average field,

$$\sigma_C = 0.157 Z_1 A^{-1/3} \text{ MeV}. \tag{7.71}$$

For the Gamow-Teller transition, $D_{GT}(E)$ has the same form as Eq. (7.68) except for replacing Δ_F and σ_F by

$$\Delta_{GT} \approx \Delta_F = \Delta_C \qquad \text{and} \qquad \sigma_{GT}^2 = \sigma_C^2 + \sigma_F^2, \tag{7.72}$$

where σ_N is the energy spread caused by the spin-dependent part of nuclear forces, which will be an adjustable parameter.

2. *Exponential type.* For Fermi transitions,

$$D_F(E) = \frac{1}{2^{1/2}\sigma_F} \exp\left[-\frac{2^{1/2}|E - \Delta_F|}{\sigma_F}\right], \tag{7.73}$$

and a similar form holds for GT transitions.

3. *Modified Lorentz type.* For Fermi transitions,

$$D_F(E) = \frac{(\sigma_F^2 + \gamma^2)}{\pi} \frac{\sigma_F^2}{\gamma}\left[(E - \Delta_F)^2 + \left(\frac{\sigma_F^2}{\gamma}\right)^2\right]^{-1} [(E - \Delta_F)^2 + \gamma^2]^{-1}, \tag{7.74}$$

and a similar form holds for GT transitions. Here γ is an adjustable parameter having the dimension of energy. If we make γ infinity, keeping the quantity σ_Ω^2/γ constant, the distribution function converges to the Lorentz-type function with the half-width $\Gamma_\Omega/2 = \sigma_\Omega^2/\gamma$.

4. *Lorentz type.* For Fermi transitions,

$$D_F(E) = (\Gamma_F/2\pi)[(E - \Delta_F)^2 + (\Gamma_F/2)^2]^{-1}. \tag{7.75}$$

A similar form holds for the GT transitions. It is assumed to be $\Gamma_F/2 = \Gamma_C/2 = \sigma_C^2/\gamma_0$ with $\gamma_0 = 220$ MeV. For the GT transitions, $\Gamma_{GT}^2 = \Gamma_C^2 + \Gamma_N^2$ is adoptec where Γ_N corresponds to σ_N.

Finally, the remaining quantity σ_N, which is the energy spread caused by the spin-dependent part of nuclear forces, is determined empirically. The effects of nuclear forces on a nucleon in the nucleus are fairly independent of Z and A because of the saturation property. Therefore, σ_N is treated as an adjustable parameter neglecting the Z and A dependence. In order to determine σ_N, we take the empirical data [24] on the beta-decay half lives of nuclei that fulfill the following criteria.

A. The total branching ratio of the allowed transitions exceeds or seems exceed 50%.

B. The ground state Q value is large enough to satisfy the inequality $Q \geqslant 10 \times A^{-1/3}$ MeV.

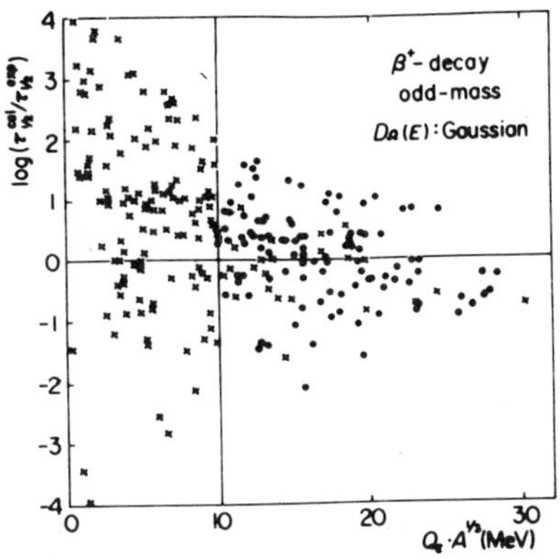

(a)

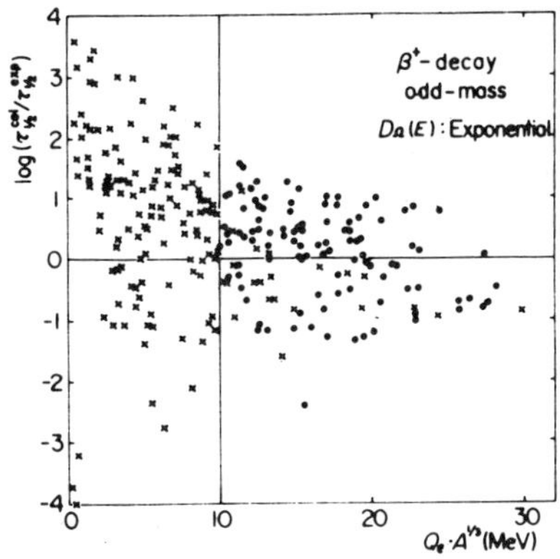

(b)

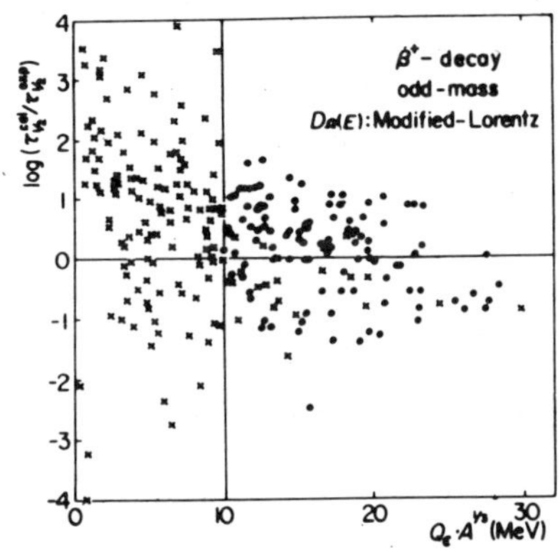

(c)

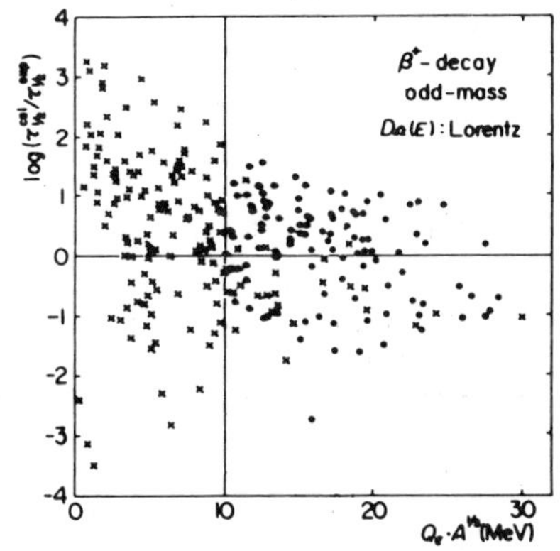

(d)

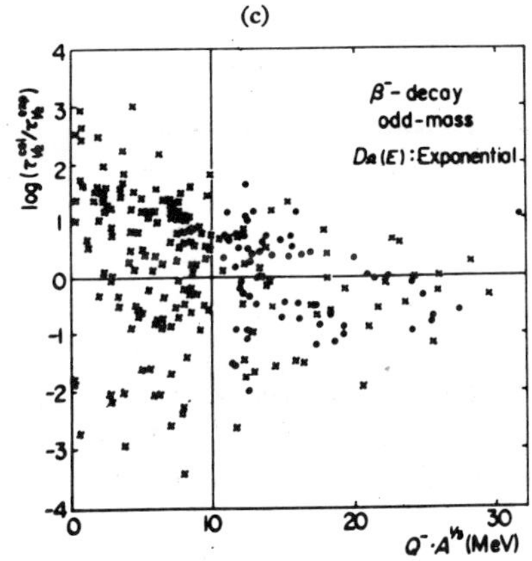

(e)

Fig. 7.6 Comparison of the calculated half-lives $t_{1/2}^{\text{calc}}$ with the experimental data $t_{1/2}^{\text{exp}}$ for odd-mass nuclei in the models (7.68), (7.73)–(7.75), with the best-fit parameters in Table 7.3. The abscissa $QA^{1/3}$ is a sort of measure of the number of final levels, and the second criterion in the text is shown by vertical lines, $QA^{1/3} = 10$. Circles correspond to the nuclei that fulfill criteria A and B, crosses to those that do not [16].

The first criterion is necessary because only the allowed decay is treated, and the second one is imposed to validate the assumption in the gross treatment that the final level density is large. In fact, the larger the Q value, the better the gross approximation (see Fig. 7.6).

The value of the adjustable parameter σ_N is determined so as to minimize the quantity

$$S' = \sum_{n=1} \left[\log_{10}\left(\frac{t_{1/2}^{\text{calc}}(n)}{t_{1/2}^{\text{exp}}(n)}\right)\right]^2, \tag{7.76}$$

which is a measure of the fit between the theory and experiment. The summation is taken over N_0 nuclei that fulfill the criteria above, and $t_{1/2}^{\text{calc}}$ and $t_{1/2}^{\text{exp}}$ are the calculated and experimental values of the half life for beta decay of the nth nucleus. Comparison of the calculated half lives for odd-mass nuclei in comparison with experimental data are shown in Fig. 7.6, while the best-fit values of parameters are listed in Table 7.3, where the mismatch factor is defined as

$$S = 10^x \qquad \text{with} \qquad x = (S'/N_0)^{1/2}. \tag{7.77}$$

Table 7.3 Best-Fit Values of the Width Parameters σ_N and Γ_N and the Mismatch Factors S in Eq. (7.77)[a]

	Positron Decay[b]		Negatron Decay[c]	
$D_{\text{GT}}(E)$	σ_N (MeV)	S	σ_N (MeV)	S
Gaussian	5.0	5.6	6.0	6.7
Exponential	4.4	5.8	5.5	6.4
Modified Lorentz	5.4 ($\gamma = 6$)	6.1	6.8 ($\gamma = 7$)	6.3
Lorentz	$\Gamma_N = 3.4$	6.7	$\Gamma_N = 4.4$	6.4

[a] Only odd-mass nuclei are used. N_0 is the total number of nuclei that satisfy conditions A and B in the text [16].

[b] $N_0 = 105$.

[c] $N_0 = 60$.

7.6 CHARGE DEPENDENCE OF NUCLEAR FORCES STUDIED FROM THE BETA-GAMMA CIRCULAR POLARIZATION CORRELATION OR BETA-RAY ANGULAR DISTRIBUTION

In the last two sections, we discussed the hindrance of the allowed transitions, which led to the discovery of isobaric analog states and the persistent structure of supermultiplet states in heavy nuclei. The energy distribution of these states consistently explains the hindrance factors. The shape of this distribution is, of course, dependent on the properties of nuclear forces and Coulomb energies. Here, we will discuss specifically the charge-dependent properties of nuclear forces studied from beta decays with $\Delta J = 0$ and $\Delta T = \pm 1$.

To do this, we have to know the Fermi matrix element numerically. We can find the ratio of the Fermi and Gamow–Teller matrix elements,

$$R = \frac{C_V \int 1}{C_A \int \boldsymbol{\sigma}}, \tag{7.78}$$

by measuring the beta-gamma circular polarization correlation

$$W(\theta) = 1 + \tau \frac{1}{L+1} \frac{p}{E} A \cos\theta \tag{7.79}$$

where

$$A = \frac{\pm 1 - 2[J(J+1)]^{1/2} R}{J(1+R^2)} \qquad \text{for } \beta^{\mp}. \tag{7.80}$$

The formula is given for the most popular decay scheme, $J \xrightarrow{\beta} J \xrightarrow{\gamma} J - L$, where the gamma ray has a multipolarity of 2^L. The same information is also obtained from the beta-ray angular distribution

$$W(\theta) = 1 + \frac{\langle m \rangle}{J} \frac{p}{E} A \cos\theta \tag{7.81}$$

with

$$A = \frac{\mp 1 - 2[J(J+1)]^{1/2} R}{(J+1)(1+R^2)} \qquad \text{for } \beta^{\mp} \tag{7.82}$$

in the $J \to J$ transition of the polarized nuclei. Here $\langle m \rangle J^{-1}$ represents the degree of nuclear polarization. The magnitude of $\int 1$ and $\int \boldsymbol{\sigma}$ can be derived from the $ft_{1/2}$ value and R, where

$$ft_{1/2} = \frac{6146 \pm 10}{[|\int 1|^2 + (C_A/C_V)^2 |\int \boldsymbol{\sigma}|^2]}. \tag{7.83}$$

Furthermore, if $\int 1 \ll \int \boldsymbol{\sigma}$, we have

$$\int 1 = -R \left(\frac{6146 \pm 10}{ft_{1/2}} \right)^{1/2}. \tag{7.84}$$

Here the sign of $\int \boldsymbol{\sigma}$ is assumed to be plus. Equations (7.80) and (7.82) are derived easily from formulas in section 6.4.

Experimental data on R obtained from the beta-gamma circular polarization correlation are collected by Frauenfelder and Steffen [25]. Recent data on the magnitude of $\int 1$ for 23 nuclides are tabulated in [26]. Similar data from the beta-ray angular distribution in polarized nuclei are being accumulated [2].

In the transitions with $\Delta T = \pm 1$, the Fermi matrix element is zero if the isospin selection rule holds strictly. This is a consequence of the CVC theory, and it is explicitly given in Eq. (7.6). Experimentally, the Fermi matrix elements are of the order of 10^{-3}. Yap obtained these nonzero values theoretically by considering a short-range, charge-dependent internucleon potential in addition to the usual Coulomb potential which introduces a mixing of the different isospin states [27]. To do this, he adopted the charge-dependent nuclear potential H_N which is phenomenologically introduced by Blin-Stoyle and Le Tourneux [28],

$$H_N = \sum_{i<j} V_0 [(\tau_z^{(i)} + \tau_z^{(j)})(p + r\boldsymbol{\sigma}^{(i)} \cdot \boldsymbol{\sigma}^{(j)}) + (\tau_z^{(i)}\tau_z^{(j)} - \tfrac{1}{3}\delta_{ij}\boldsymbol{\tau}^{(i)} \cdot \boldsymbol{\tau}^{(j)})(q + s\boldsymbol{\sigma}^{(i)} \cdot \boldsymbol{\sigma}^{(j)})] e^{-\beta r_{ij}^2}, \tag{7.85}$$

where V_0 is a typical strength for the charge-independent internucleon potential. $\boldsymbol{\sigma}^{(i)}$ and $\boldsymbol{\tau}^{(i)}$ are the spin and isospin matrices for the ith nucleon and the r_{ij} is the internucleon distance. V_0 and β, which is the range parameter, are chosen as follows.

$$V_0 = -51.9 \text{ MeV} \qquad \text{and} \qquad \beta^{-1/2} = 1.73 \text{ fm}.$$

The parameters p and r measure the deviation from charge symmetry, while q and s measure the deviation from charge independence. By charge symmetry, we mean $V_{pp} = V_{nn}$, and by charge independence, we mean the more restrictive conditions $V_{pp} = V_{nn} = V_{pn}$.

Now the Fermi matrix element is written as

$$\int 1 = \langle f | T_\pm | i \rangle. \tag{7.86}$$

For clarity, let us consider a positron decay of a nucleus of spin J and isospin $(T + 1)$ into a nucleus of spin J but isospin T. Owing to the presence of the Coulomb force and the possible charge dependence of the nuclear force, isospin impurities will be introduced. The initial state $|i\rangle$ and the final state $|f\rangle$ will then be perturbed and according to the ordinary perturbation theory; we can write them as follows.

$$\begin{aligned} |i\rangle &= |JM\, T+1\; T_z + 1\rangle + \alpha' |JM\, T+2\; T_z + 1\rangle + \cdots, \\ |f\rangle &= |JMTT_z\rangle + \alpha |JM\, T+1\; T_z\rangle + \cdots. \end{aligned} \tag{7.87}$$

Neglecting second- and higher-order terms in α, we have

$$\langle f | T_\pm | i \rangle = \alpha[(T + T_z + 2)(T - T_z + 1)]^{1/2}, \tag{7.88}$$

where the isospin impurity amplitude α is given by

$$\alpha = -\frac{\langle JMTT_z | H' | JM\, T+1\; T_z \rangle}{E(T+1) - E(T)}, \tag{7.89}$$

where the E's are the unperturbed energies. The charge-dependent perturbation H' is divided two parts

$$H' = H_C + H_N. \tag{7.90}$$

Here, the Coulomb potential H_C is given by

$$H_C = \tfrac{1}{4} e^2 \sum_{i<j} \frac{(1 + \tau_z^{(i)})(1 + \tau_z^{(j)})}{r_{ij}}, \tag{7.91}$$

and the charge-dependent nuclear potential H_N is given in Eq. (7.85). The denominator of Eq. (7.89) is approximated by considering the Coulomb energy difference between members of an isospin multiplet.

The numerical calculation of Yap was performed in the j-j coupling shell model. For example, the beta decay of F^{20} is the transition from the $J = 2^+$, $T = 1$ state of F^{20} to the $J = 2^+$, $T = 0$ state of Ne^{20}, with a maximum kinetic energy of beta particle of 5.40 MeV. Using four-particle configurations $(d_{5/2})^4$ for these states, Yap obtained the result

$$\int 1 = 0.72p - 2.57r - 0.0057 \tag{7.92}$$

where the q and s terms make no contribution, while the constant term comes from the Coulomb potential. The experimental values are

$$\begin{aligned} \int 1 &= -0.0059 \pm 0.0250 \quad [29], \\ &= -0.0088 \pm 0.0250 \quad [30]. \end{aligned} \tag{7.93}$$

From these values we can set the limits for p and r. For the decay of Na^{24}, Blin-Stoyle and Yap [31] obtained, by similar considerations,

$$0.052 > 2r - p > -0.078. \tag{7.94}$$

These results are consistent with the limits

$$0.003 > 3r - p > 0.001 \tag{7.95}$$

obtained by Blin-Stoyle and Yalgin [32] from a comparison of the binding energies of H^3 and He^3. This means that the violation in the charge symmetry of nuclear forces is about 10^{-3} or less.

Yap performed the same analysis for A^{41}, Sc^{44}, Mn^{52}, and Co^{56} [27]. From consistency of parameters s and q (by neglecting r and p) in these beta decays, he obtained the limits

$$2.3\% \leqslant q \leqslant 4\% \qquad \text{and} \qquad 0.5\% \leqslant s \leqslant 1.5\%. \tag{7.96}$$

That is, the violation of the charge independence in nuclear forces is about a few percent.

It should be noted here that there remains an open question about the experimentally obtained magnitudes of the Fermi matrix elements, since there are three Fermi matrix elements which obey the same selection rules,

$$\int 1, \qquad \int \boldsymbol{\alpha}\cdot\mathbf{r}, \qquad \int r^2, \tag{7.97}$$

in the allowed beta decays. The latter two are contributions from the p- and d-wave leptons. They are usually negligibly small compared with $\int 1$. In our case, however, we could not omit these two, since $\int 1$ should be approximately zero. In this sense, experimentally obtained values of $\int 1$ are not exactly equal to the theoretical value of $\int 1$, but a linear combination of the three matrix elements above. Such a formulation has been made [33], and a numerical analysis is being done.

7.7 COMPLETE EXPERIMENTS TO FIND NUCLEAR MATRIX ELEMENTS

Nuclear structure of complex nuclei can be studied also from the forbidden beta decays. To do this, we have to find the magnitudes of the nuclear matrix elements from the data on several kinds of beta-decay phenomena. A set of such experiments is called complete experiments. The number of nuclear matrix elements which contribute to the first forbidden transition varies with the spin change ΔJ. There is only one matrix element

$$\int B_{ij} \qquad \text{for} \quad \Delta J = \pm 2. \tag{7.98}$$

The complete experiments in this case are measurements of the maximum energy of the beta particle and the half life, for determining the $ft_{1/2}$ value,

$$f_c t_{1/2} = \frac{6146 \pm 10}{(1.239 \pm 0.011)^2 \, |\int B_{ij}|^2}. \tag{7.99}$$

Here f_c is the corrected integrated Fermi function

$$f_c = \int_1^{E_o} CF(Z, E)pE(E_o - E)^2 \, dE, \tag{7.100}$$

with

$$C = \tfrac{1}{12} q^2 L_0 + \tfrac{3}{4} L_1.$$

Here C is the spectral shape factor for $|\int B_{ij}|^2$.

Next, there are four matrix elements contributing,

$$\int B_{ij}, \quad \int \boldsymbol{\sigma}\times\mathbf{r}, \quad \int \boldsymbol{\alpha}, \quad \int \mathbf{r} \qquad \text{for} \quad \Delta J = \pm 1. \tag{7.101}$$

Assuming the CVC theory, we can set a relation between $\int \boldsymbol{\alpha}$ and $\int \mathbf{r}$. This reduces the number of parameters from four to three. The complete experiments are, for example, measurements of the $ft_{1/2}$ value, beta-gamma directional correlation, and beta-gamma circular polarization correlation. This is sometimes more than enough, because the beta-gamma angular correlations change with the energy of beta particle, and we can determine the ratios of nuclear matrix elements. We use, however, all available data and determine nuclear matrix elements with chi-square fit to these data.

The most complicated case is the $\Delta J = 0$ transition, where all six matrix elements of the first forbidden transitions are contributing,

$$\int B_{ij}, \quad \int \boldsymbol{\sigma} \times \mathbf{r}, \quad \int \boldsymbol{\alpha}, \quad \int \mathbf{r},$$
$$\int \boldsymbol{\sigma} \cdot \mathbf{r}, \quad \int \gamma_5, \quad \text{for} \quad \Delta J = 0. \tag{7.102}$$

These matrix elements are again determined from the chi-square fit to the available experimental data on the $ft_{1/2}$ value, beta-gamma angular correlations, beta-ray spectrum, etc.

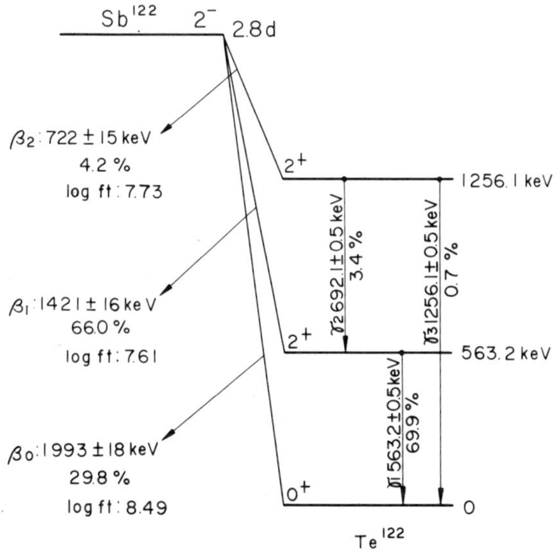

Fig. 7.7 Decay scheme of Sb^{122}.

An example of complete experiments is the case of the beta decay in Sb^{122}. The decay scheme is given in Fig. 7.7. There are three branches of beta decay. The ground state transition is unique. The decays to the first and second 2^+ states are both $\Delta J = 0$ transitions. Manthuruthil and Poirier analyzed theoretically the data on the $ft_{1/2}$ value, beta-gamma directional correlation, beta-gamma circular polarization correlation, and angular distributions of gamma rays from oriented Sb^{122}, and they determined six nuclear matrix elements for the beta decay to the first 2^+ state in Te^{122} [34]. Hayashi and Yamada measured the beta-gamma directional correlations for the beta decays of Sb^{122} to the first and second 2^+ states, and they determined the six nuclear matrix elements for these two beta decays, using their experimental data and those on the $ft_{1/2}$ value [35]. Numerical results for the beta decay to the first 2^+ state coincide in the analyses of these two groups. These values are summarized in Table 7.4, together with those for the beta decay to the second 2^+ state.

Table 7.4 Numerical Values of Nuclear Matrix Elements for the Beta Decays (β_1 and β_2) of Sb^{122} to the First and Second 2^+ States of Te^{122} [a]

	$\frac{\int B_{ij}}{R}$	$\frac{\int \mathbf{r}}{R}$	$\frac{\int i\sigma \times \mathbf{r}}{R}$	$\int i\alpha$	$\frac{\int \sigma \cdot \mathbf{r}}{R}$	$\int i\gamma_5$
β_1	0.254 ± 0.114	0.002 ± 0.015	0.026 ± 0.019	−0.001 ± 0.007	0.627 ± 0.130	−0.116 ± 0.023
β_2	0.278 ± 0.125	0.113 ± 0.146	0.564 ± 0.268	−0.128 ± 0.071	0.454 ± 1.377	−0.079 ± 0.228

[a] From [35]. The ratios of the nuclear matrix elements for β_1 coincide with set II in the analysis of Manthuruthil and Poirier [34]. Numerical values of the nuclear matrix elements are, however, 1.2 times larger than those in this table due to the difference in $fct_{1/2}$ values.

Manthuruthil and Poirier also noticed that the finite size corrections are really important. They emphasized using the formulas in section 6.4, and showed that the nuclear parameters obtained from formulas without such corrections are quite different [34]. The experimental data on the beta-gamma directional

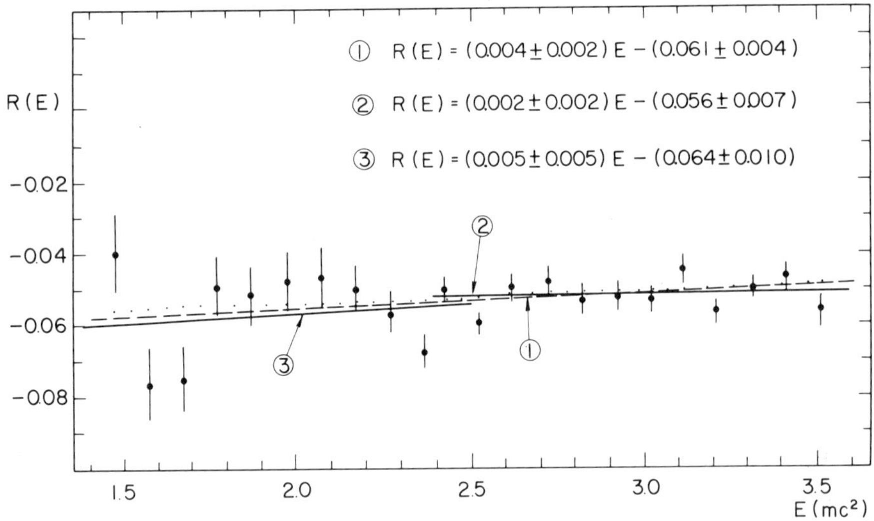

Fig. 7.8 Beta-gamma directional correlation in the decay of Sb^{122} [35]. The correlation coefficient $R(E)$ is plotted for the β_1–γ_1 coincidence in the notation of Fig. 7.7. The coefficient $R(E)$ is defined by $R(E) = A_2^\beta p^{-2}E$, and A_2^β appears in the correlation function $W(\theta) = 1 + A_2^\beta A_2^\gamma P_2(\cos\theta)$.

correlation and beta-ray spectrum given by Hayashi and Yamada [35] are shown in Figs. 7.8 and 7.9.

A great many experimental data on the first forbidden nuclear matrix elements are presently being accumulated in many laboratories. Using these data, theoretical investigations on the nuclear models are being performed. For instance, the first excited 2^+ states in Se^{76}, Sr^{86}, Te^{122}, and Xe^{126} are studied in comparison with available data on beta decays of As^{76}, Rb^{86}, Sb^{122}, and I^{126}, by Matumoto, Yamada, Wang, and Morita, using the rotational model

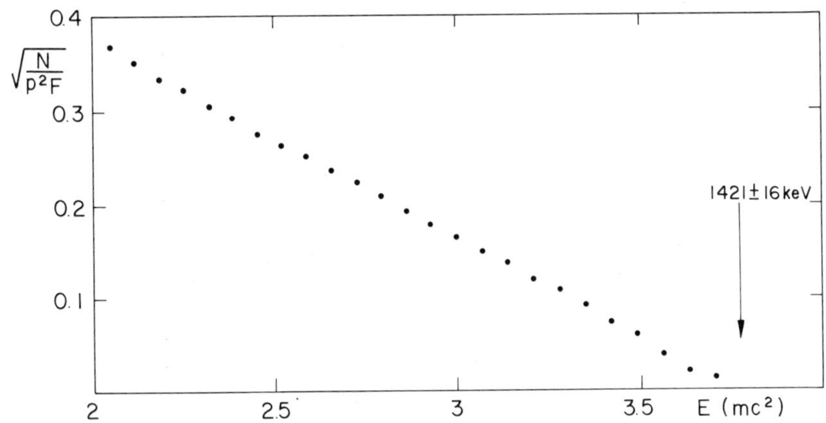

Fig. 7.9 Kurie plot for β_1 of Sb^{122}; see Fig. 7.7 [35].

and *j-j* coupling shell model [36]. A similar work was performed by Kisslinger and Wu [37], using the vibrational model. Recently, Smith and Simms [38] investigated impurities in the isobaric analog state to the ground state of La^{140}, for which the beta-gamma directional as well as circular polarization correlations are measured in detail by Steffen's group and others [39].

7.8 SEARCH FOR HIGHER-ORDER EFFECTS IN ALLOWED TRANSITIONS

In the usual theory of beta decay, only *s*-wave leptons are considered, so that the beta-gamma directional correlation of allowed beta transitions is isotropic and the beta-ray spectrum has the statistical shape. Under a certain condition, however, the contributions from *p*- and *d*-wave leptons may become large enough to cause measurable deviations from the statistical distributions or the beta-gamma directional correlation may display a small anisotropy. These higher-order effects are mainly due to the interference between *s* waves and *p* and *d* waves and are, therefore, characterized by the presence of cross terms of the allowed matrix elements $\int 1$ and $\int \boldsymbol{\sigma}$, and the second forbidden matrix elements $\int i\boldsymbol{\alpha}\cdot\mathbf{r}$, $\int i\gamma_5\mathbf{r}$, $\int \boldsymbol{\alpha}\times\mathbf{r}$, $\int A_{ij}$, $\int r^2$, $\int \boldsymbol{\sigma} r^2$, $\int (\boldsymbol{\sigma}\cdot\mathbf{r})\mathbf{r}$, $\int R_{ij}$, $\int T_{ij}$, and $\int S_{ijk}$. The contributions of the first four matrix elements are of the order of $v_N pR$, where v_N is the average nucleon velocity in the nucleus, p is the momentum of the beta particle, and R is the nuclear radius. (All quantities are again given in units of $\hbar = m = c = 1$). The remaining second forbidden matrix elements give contributions of the order of $(pR)^2$. In the most cases of the beta transitions, pR is about 0.01 or less. Measurable effects are only expected if $p \gg 1$ or if the matrix elements $\int 1$ and $\int \boldsymbol{\sigma}$ arc strongly hindered by some nuclear structure effect which is not operative for at least some type of second forbidden matrix elements. In this connection formulas for beta-ray spectrum and beta–gamma angular correlations are given in section 6.4.

Now, let us first discuss the beta-ray spectrum, which is given by Eq. (2.143)

$$P_{\mp}(E)\,dE = \frac{1}{2\pi^3}\left(|C_V|^2\left|\int 1\right|^2 + |C_A|^2\left|\int \sigma\right|^2\right)F(\pm Z, E)pE(E_o - E)^2 C(E)\,dE$$

for $\beta^{\mp}$,

with the spectral shape factor $C(E)$ (2.144)

$$C(E) = 1 + aE + (b/E) + cE^2$$

in the allowed transition. (The $V - A$ theory with two-component neutrino is assumed.) Here, a, b, and c are energy-independent coefficients, which are linear combinations of the coupling constants and the higher-order matrix elements. Therefore, these coefficients consist of charge-dependent and charge-independent terms. The magnitudes of these coefficients are very small, and they are of the order of 10^{-3} or less, except in the hindered beta decays. Deviation from 1 in the spectral shape factor $C(E)$ was observed in the high-energy beta decay, such as the beta decays of B^{12} and N^{12}. In these two cases, the spectral shape factor is approximately given by

$$C(E) = 1 \pm A^{\mp}E \qquad \text{for } \beta^{\mp} \tag{7.103}$$

with

$$(A^- - A^+) \approx 1\%/\text{MeV},$$

as we have studied it in the experimental proof of the CVC theory.

Several examples of hindered beta decays are tabulated in Table 7.5, with beta-ray spectra [40]. Recent data have also been compiled by Daniel [41].

Table 7.5 Data on Beta-Ray Spectra[s]

Parent Nuclei	Selection Rule	Electron Charge	log $ft_{1/2}$	E_o (mc^2)	Langer *et al.*	Others	
					b	b	a
Na^{22}	Allowed GT	+	7.4	2.1	0.25–0.35[a,b]	−0.016 ± 0.020[c] 0.0008 ± 0.0020[d]	
Na^{24}	Allowed F + GT	−	6.1	3.7		0.072 ± 0.072 or −0.026–0.020[e]	−0.0150 ± 0.0150[c]
P^{32}	Allowed GT	−	7.9	4.3	0.2–0.4[a,f]	0.05–0.093[e]	−0.041 ± 0.013[c] −0.0133 ± 0.0011[g]
Co^{56}	Allowed F + GT	+	8.7	3.9	0.2–0.3[h,i]		
Co^{58}	Allowed F + GT	+	6.6	1.91	0.3[j]		
Ga^{66}	Allowed F	+	7.9	9.1	0.4[k,l]		
Zr^{89}	Allowed F + GT	+	6.1	2.8	0.25–0.45[m,n]		
Y^{90}	First unique	−	8.0	5.5	0.2–0.3[a,f]	Very small[o]	−0.0047 ± 0.0008[g]
In^{114}	Allowed GT	−	4.4	4.9	0.2–0.3[a,f]	0.05 ± 0.02[p]	0.0036 ± 0.0021[g]
Pr^{143}	First	−	7.6	2.8	0.25[q]		
Pm^{147}	First	−	7.6	1.44	0.1–0.3[b]		
Au^{198}	First	−	7.3	2.9	r		

[a] L. M. Langer, *Proc. Rehovoth Conf. Nucl. Structure* (1957), p. 437.
[b] J. H. Hamilton, L. M. Langer, and W. G. Smith, *Phys. Rev.* 112, 2010 (1958).
[c] H. Daniel, *Nucl. Phys.* 8, 191 (1958).
[d] H. Leutz, *Z. Phys.* 164, 78 (1961).
[e] F. T. Porter, F. Wagner, Jr., and M. S. Freedman, *Phys. Rev.* 107, 135 (1957).
[f] O. E. Johnson, R. G. Johnson, and L. M. Langer, *Phys. Rev.* 112, 2004 (1958).
[g] R. T. Nichols, R. E. McAdams, and E. N. Jensen, *Phys. Rev.* 122, 172 (1961).
[h] Or $C = 1 - 0.24E + 0.04E^2$.
[i] J. H. Hamilton, L. M. Langer, and W. G. Smith, *Phys. Rev.* 123, 189 (1961).
[j] J. I. Rhode and O. E. Johnson, *Phys. Rev.* 131, 1227 (1963).
[k] In addition to the second forbidden corrections, $C = 1 + 0.03282\,E + 0.354\,E^{-1} - 0.000419\,E^2$.
[l] D. C. Camp and L. M. Langer, *Phys. Rev.* 129, 1782 (1963).
[m] Or $C = 1 - 0.39E + 0.09E^2$.
[n] J. H. Hamilton, L. M. Langer, and W. G. Smith, *Phys. Rev.* 119, 772 (1960).
[o] T. Yuasa, J. Laberrigue-Frolow, and L. Feuvrais, *J. Phys. Radium* 18, 559 (1957).
[p] H. Daniel and P. H. Panussi, *Z. Phys.* 164, 303 (1961).
[q] J. H. Hamilton, L. M. Langer, R. L. Robinson, and W. G. Smith, *Phys. Rev.* 112, 945 (1958).
[r] $C = 1 - 0.33E + 0.074E^2$. J. H. Hamilton, R. V. Stokendal, D. C. Camp, L. M. Langer, and D. R. Smith, *Nucl. Phys.* 36, 567 (1962).
[s] From [40].

As is shown in Table 7.5, the experimental data are usually given by the spectral shape factor,

$$C(E) = 1 + (b/E) \qquad \text{or} \qquad 1 + aE, \tag{7.104}$$

since there is no significant difference between these two forms in the low energy region. Most of the experimental data are explained with the contribution from the second forbidden transitions. In contrast to this, Langer's group at Indiana University obtained beta-ray spectra which have a marked difference from the data of the other groups. The characteristics of the Indiana group's data are as follows. (1) The b is positive for both electron and positron emissions. (2) The magnitude of b is about 0.2–0.4 (equivalently, a is about -0.05), though it changes from nucleus to nucleus. (3) The effect is observed in both Fermi and Gamow–Teller transitions. (4) The effect is also observed in the unique first forbidden transition of Y^{90}, for which

$$C = \tfrac{1}{12}(p^2 + q^2)\left(1 + \frac{b}{E}\right) \qquad \text{with} \qquad b = 0.2\text{–}0.3. \qquad (7.105)$$

An exhaustive theoretical investigation of this anomaly of beta-ray spectra was performed by Morita and his collaborators, since Yukawa and his group were very much interested in this strange phenomenon, from the point of view of elementary interactions. We havc studied various kinds of strange interactions, such as the Konopinski–Uhlenbeck interaction, which has a derivative, the existence of strange neutrinos [(1) a particle mixture of mass zero and mass nonzero neutrinos, (2) a spin $\frac{3}{2}$ neutrino, (3) emission of a boson neutrino in addition to the usual fermion neutrino, (4) emission of three fermion neutrinos)]; intermediate boson theories (see section 5.11); possible G-parity nonconservation; etc. In these investigations, we found that most of the theories can give the spectral shape factor

$$C(E) = 1 \pm \frac{b}{E} \qquad (\text{or } 1 \pm aE^n \text{ with an integral } n) \qquad \text{for } \beta^{\mp} \qquad (7.106)$$

with different signs of the b (or a) term for different electron charge. Furthermore, the absolute value of b is rather small. If we adopt a six-fermion interaction in addition to the conventional four-fermion interaction, we have the spectral shape factor $[1 + c(E_o - E)^6]$, where c depends on the strength of the six-fermion interaction. If we adjust it to fit $b = 0.2$ or so, then the six-fermion interaction can produce the double-beta decay whose decay rate is comparable to that of the Majorana theory. In this connection, the anomaly in beta-ray spectra is very difficult to understand, although its possible existence stimulates the physicists working on field theory.

On the other hand, the small deviation from the statistical shape is worth measuring, since it gives information on nuclear structure. In fact, we can find magnitudes of nuclear matrix elements from data on this kind of deviation and other available experimental data, as in the case of complete experiments. Among many attempts, such an analysis was performed by Bhalla in the cases of Zr^{89} and In^{114} [42].

Extensive studies of higher-order effects in the beta–gamma directional correlations involving allowed beta transitions were performed by Steffen and his collaborators [43, 44]. They measured the coefficients A_2^β in the beta-gamma directional correlation

$$W(\theta) = 1 + A_2^\beta A_2^\gamma P_2(\cos\theta) \qquad (7.107)$$

with

$$A_2^\gamma = F_2(LLj_2j_1)$$

in the decay scheme $j \xrightarrow{\beta} j_1 \xrightarrow{\gamma} j_2$. Here A_2^γ is a geometrical coefficient depending on the property of gamma transition. Experiments were performed in the cases of Na^{22}, Na^{24}, Sc^{46}, Co^{56}, Co^{60}, Sb^{124}, Cs^{134}, Eu^{152}, Eu^{154}, and the data have been tabulated [44] together with those for F^{20} and Mn^{56} by different groups. In all but three cases, there are measurable effects:

$$A_2^\beta = -0.010 \pm 0.003 \qquad \text{for } F^{20} \quad [45];$$
$$A_2^\beta = 0.0030 \pm 0.0005 \qquad \text{for } Na^{22} \quad [43];$$
$$A_2^\beta = 0.09 \pm 0.03 \qquad \text{for } Tb^{160} \quad [44]. \tag{7.108}$$

The anisotropy factors expected in Na^{24}, Sc^{46}, Co^{60}, Eu^{152}, and Eu^{154} are so small [A_2^β(theory) $\leqslant 10^{-3}$] that it is extremely difficult to observe them experimentally. In some cases, however (e.g., Co^{56}, Mn^{56}, Sb^{124}, Cs^{134}), the large hindrance of the allowed beta matrices is expected from the $ft_{1/2}$ values. This should result in a measurable anisotropy, but Steffen *et al.* [43] observed isotropic correlation in these nuclei

$$A_2^\beta = 0.002 \pm 0.007 \qquad \text{for } CO^{56}. \tag{7.109}$$

They concluded from this fact that the nuclear structure effects which are responsible for the hindrance of the allowed beta decay are also effective in reducing the second forbidden matrix elements. They also pointed out that (1) the small anisotropy for F^{20} may be due to weak magnetism effects caused by CVC, (2) the very small anisotropy for Na^{22} may be due to weak magnetism or to second forbidden nuclear matrix elements, and (3) the rather large anisotropy factor for Tb^{160} is an indication of the K selection rule, which reduces the allowed nuclear matrix elements without affecting the second forbidden ones, since Tb^{160} and Dy^{160} are strongly deformed nuclei.

7.9 NUCLEAR STRUCTURE PROBLEMS STUDIED FROM THE BETA DECAY OF RaE

The beta decay of RaE is famous, since it keeps providing us with information on beta-decay interactions as well as on nuclear structure. In this section, we wish to describe the development of the study of RaE. For a long time it has been known [46] that the spectral shape for the beta decay RaE(Bi^{210}) $\rightarrow$ RaF(Po^{210}) is considerably different from the simple statistical one predicted by the Fermi theory. In Fig. 7.10, we show the Kurie plot given by Daniel [46] and Plassman and Langer [46]. In order to understand the anomalous shape theoretically, great effort was made [47] by a few researchers before the International Conference on Theoretical Physics held in Kyoto 1953, at which the "cancellation theory" of Petschek and Marshak [48] was reported and hot discussions were done between Marshak and Yamada. Marshak *et al.* [48] investigated the possibility of fitting the RaE spectrum with a linear combination of tensor and pseudoscalar interactions, the transition of RaE being assumed to be $0^- \rightarrow 0^+$. On the other hand, Yamada [49]

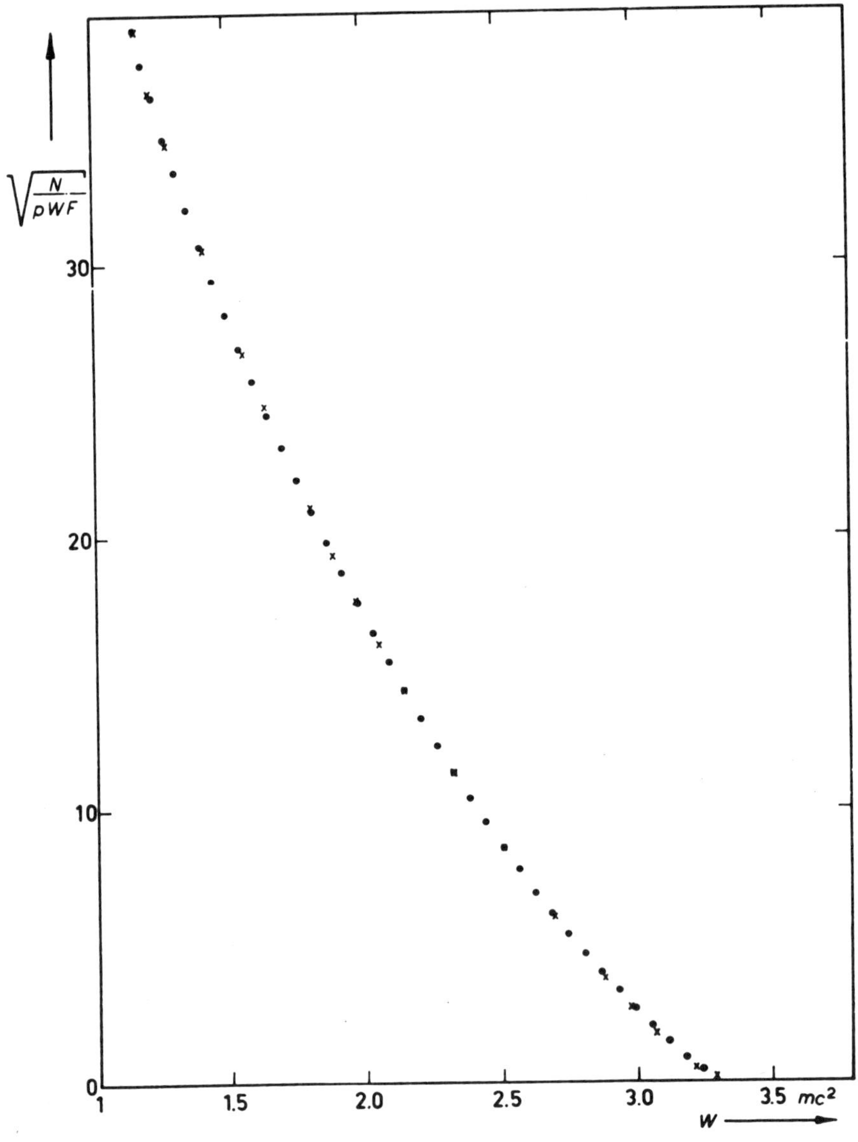

Fig. 7.10 Kurie plot for the beta decay of RaE [46]. Circles are given by Daniel, and crosses are given by Plassman and Langer. The energy *W* should read *E* in our notation.

obtained a good agreement with data by assuming $1^- \rightarrow 0^+$ and either of the combinations *ST* or *VA*. In 1954 a direct spin measurement was made by Smith [50], and showed the spin of Bi^{210} to be one rather than zero.

At the first stage of the cancellation theory, all the nuclear matrix elements, such as $\int \mathbf{r}$, $\int \boldsymbol{\alpha}$ and $\int \boldsymbol{\sigma} \times \mathbf{r}$, were considered to be free parameters. However, in accordance with the success of shell model approach, nuclear matrix elements by themselves have gradually become the object of study. Morita, Fujita, and Yamada [51] analyzed beta spectra of Fe^{59}, Rb^{87}, Tc^{99}, Cs^{137} by assuming that the ratios of nonrelativistic matrix elements can be predicted from the *j-j* coupling shell model, at least in sign, and they were led to the conclusion that $C_F/C_{GT} < 0$, which was confirmed by the later experiments. On the other hand, Lee-Whiting [52] discussed the nuclear structure of RaE and obtained the result, $C_F/C_{GT} > 0$. His basic idea is as follows. (a) The nuclear structure of RaE must be simple, because it has only one neutron and one proton outside the Pb core. (b) In order to explain the fact that the spin of the ground state in Bi^{210} is one instead of zero, the most plausible shell assignment is $(1h_{9/2})_p(1i_{11/2})_n$, for which the predicted ratio is

$$i \int \mathbf{r} / \int \boldsymbol{\sigma} \times \mathbf{r} = +1. \tag{7.110}$$

(c) Combining the result above with the information obtained from the analysis of the beta spectrum [53], he obtained

$$0 \lesssim C_F i \int \mathbf{r} / C_{GT} \int \boldsymbol{\sigma} \times \mathbf{r} \lesssim 1. \tag{7.111}$$

This gives the conclusion $C_F/C_{GT} > 0$, which is opposite to the one obtained from the above-mentioned odd nuclei.

Later, Fujita, Arima, and Yamada [54] examined the argument of Lee-Whiting in detail, and showed that the opposite conclusion can be obtained if the neutron radius is chosen to be 7% larger than the proton radius. And also a question was raised whether the Pb^{208} core is really inert or not. Recently, such a core polarization effect [55, 22] turned out to be generally important.

On the other hand, the shell model studies of Bi^{210} have been pushed further forward. In 1959 Newby and Konopinski [56] gave a very extensive argument on level structure of Bi^{210} and Po^{210}. They showed that the level scheme can be well reproduced by assuming zero-range two-body neutron–proton forces between the neutron and proton outside the inert Pb^{208} double-magic core, and the neutron–proton attraction in the $(h_{9/2}, i_{11/2})_1$ state can be greater by 840 keV than in the $(h_{9/2}, g_{9/2})_0$ state. They also discussed the effect of tensor forces and discovered that they produce attraction in the former and repulsion in the latter state. As a consequence Newby and Konopinski concluded that the ground state most probably has the $(h_{9/2}, i_{11/2})_{1^-}$ configuration and the conclusion of Lee-Whiting can be reconfirmed. However, we already knew at that time that the statement $C_{GT}/C_F = C_A/C_V > 0$ is wrong. Therefore, they conjectured that the pion cloud effect as treated in the conserved vector current theory is important and the conventional theory for forbidden spectra might have to be revised extensively.

The latter point was examined by Fujita [57]. The main point of his argument is the following. Roughly speaking, the conventional theory of the RaE spectrum, in which all the nuclear matrix elements are regarded as free parameters, should not be altered very much even if the CVC theory is adopted, and another evidence was obtained for relation (7.111) to be valid independently of nuclear structure.

More detailed shell model studies on RaE were carried out by Kim and Rasmussen [58]. It was shown that a tensor force with reasonable range and strength can account for the 1^- state of $(h_{9/2}, g_{9/2})$ multiplet being the lowest state of RaE, instead of the 0^- predicted by central forces. Actually it can be shown that almost any resonable attractive central forces brings the spin 0 state lower than the spin 1 state, whereas the former lies 47 keV higher than the latter. However, the high-resolution (d, p) measurement [59] strongly indicated that the ground state of Bi^{210} involves mainly the $g_{9/2}$ neutron state and not the $i_{11/2}$. The result obtained by Kim and Rasmussen is read as

$$\psi(\mathrm{Bi}^{210}, J=1) = 0.969\,|h_{9/2}g_{9/2}> -0.219\,|h_{9/2}i_{11/2}> + 0.005\,|f_{7/2}g_{9/2}> \tag{7.112}$$

and

$$i\int \mathbf{r} / \int \boldsymbol{\sigma} \times \mathbf{r} = -0.91,$$

which is quite in agreement with the empirical range for the ratio given by Eq. (7.111) and $C_A/C_V \approx -1.2$. It was also remarked that the matrix element of a tensor force is not a monotonic function of the force range.

It should also be mentioned that relation (7.111) was first obtained [49] by using the experimental beta spectrum of RaE and the "effective nuclear radius" theory due to Matumoto and Yamada [60]. The purpose of the work by Fujita *et al.* [53] was to examine the possible breakdown of the time-reversal invariance in the weak beta interactions. Suppose that the weak Hamiltonian is not invariant under the time reversal and the ratio of coupling constants C_A/C_V is not real but complex. Then, main term of the expression for the beta spectrum of RaE, in the so-called ξ approximation,

$$|C_V \xi i \int \mathbf{r} - C_V \int \boldsymbol{\alpha} + C_A \xi \int \boldsymbol{\sigma} \times \mathbf{r}|^2 \tag{7.113}$$

with $\xi = \alpha Z/2R$, can produce an additional contribution, such as

$$|\mathrm{Im}\, C_A|^2 \left(\int \boldsymbol{\sigma} \times \mathbf{r}\right)^2, \tag{7.114}$$

which is energy-independent. Even if the quantity (7.114) is very small, it becomes a formidable task to fit the theoretical formula with the experimental spectrum. Later, Telegdi's group [61] carried out the famous measurement of polarized neutron beta decay and found that

$$\phi = -5 \pm 10° \tag{7.115}$$

if ϕ is defined by

$$C_A/C_V \propto \exp[i(\pi + \phi)]. \tag{7.116}$$

Kim and Primakoff [62] pointed out the interesting possibility that a part of the weak Hamiltonian of the axial vector type might be abnormal (even under time reversal) but regular (odd under charge symmetry), and existence or absence of such a part cannot be decided, in principle, from experimental knowledge based on the isodoublet beta transitions such as n → p or $Ne^{19} \to F^{19}$. However, at least two facts seem to favor for the absence of such T-invariance violating interactions :(a) Beta-gamma correlation measurements [63] of polarized Mn^{52}, and (b) the beta spectrum and longitudinal polarization of electrons from Bi^{210} (Ra E), although both facts may be inconclusive.

Nuclear structure studies of RaE are still in progress. Recently, Fujita, Hirata and Shoda [64] argued on the basis of an experiment at Tohoku University [65] that nuclear core polarizations of spin-flip and nonspin-flip types are strongly coupled to each other. If this is true, the effective operator corresponding to $i\int \mathbf{r}$ generally becomes

$$(i\int \mathbf{r})_{\text{eff}} = \alpha i \int \mathbf{r} + \beta \int \boldsymbol{\sigma} \times \mathbf{r},$$

and similarly for $(\int \boldsymbol{\sigma} \times \mathbf{r})_{\text{eff}}$. In order to determine the coefficients α and β, further investigation is required.

REFERENCES

1 C. S. Wu and S. A. Moszkowski, *Beta Decay*, Wiley (Interscience), New York, 1965.
2 K. Nagamine, private communication.
3 M. G. Mayer and J. H. D. Jensen, *Elementary Theory of Nuclear Shell Structure*, Wiley, New York, 1955.
4 E. Feenberg, *Shell Theory of Nucleus*, Princeton Univ. Press, Princeton, N. J., 1955.
5 O. C. Kistner and B. M. Rustad, *Phys. Rev.* **114**, 1329 (1959).
6 O. Kofoed-Hansen and A. Winther, *Phys. Rev.* **86**, 428 (1952); *Danske Mat. Fys. Medd.* **27**, No. 14 (1953); *ibid.* **30**, No. 20 (1956).
7 K. Sugimoto, *Phys. Rev.* **182**, 1051 (1969).
8 R. Leonardi and M. Rosa-Clot, *Phys. Rev. Letters* **21**, 377 (1968).
9 H. Noya, A. Arima, and H. Horie, *Progr. Theoret. Phys. Suppl.* **8**, 33 (1959).
10 M. Sakai and S. Yoshida, *Nucl. Phys.* **50**, 497 (1964).
11 L. S. Kisslinger and R. A. Sorensen, *Rev. Mod. Phys.* **35**, 853 (1963).
12 I. Hamamoto, *Nucl. Phys.* **62**, 49 (1965); **66**, 176 (1965).
13 E. P. Wigner, *Phys. Rev.* **56**, 519 (1939).
14 J. D. Anderson and C. Wong, *Phys. Rev. Letters* **7**, 250 (1961); J. D. Anderson, C. Wong, and J. W. McClure, *Phys. Rev.* **126**, 2170 (1962).
15 K. Ikeda, S. Fujii, and J. Fujita, *Phys. Letters* **3**, 271 (1963); *Phys. Rev.* **133**, B549 (1964).
16 M. Yamada, *Bull. Sci. Eng. Res. Lab. Waseda Univ.* No. **31/32**, 146 (1965); K. Takahashi and M. Yamada, *Progr. Theoret. Phys.* **41**, 1470 (1969).

17 J. Fujita and K. Ikeda, *Nucl. Phys.* **67**, 145 (1965).
18 J. Fujita, Y. Futami, and K. Ikeda, *Progr. Theoret. Phys.* **38**, 107 (1967).
19 J. Fujita and K. Ikeda, *Progr. Theoret. Phys.* **36**, 288 (1966); M. Ichimura, *Progr. Theoret. Phys.* **36**, 853 (1966); H. Ejiri, *J. Phys. Soc. Japan* **22**, 360 (1967); J. Fujita, *Phys. Rev.* **172**, 1047 (1968).
20 J. Fujita, G. T. Emery, and Y. Futami, *Phys. Rev.* **C1**, 2060 (1970).
21 A. Ikeda, *Progr. Theoret. Phys.* **38**, 832 (1967).
22 H. Ejiri, K. Ikeda, and J. Fujita, *Phys. Rev.* **176**, 1277 (1968).
23 K. A. Brueckner and J. L. Gammel, *Phys. Rev.* **109**, 1023 (1958).
24 K. Way *et al.*, *Nuclear Data Sheets*, Nat. Acad. Sci. Nat. Res. Council, Washington, D.C., 1958–1965; Nuclear Data B-1. Academic Press, New York, 1966.
25 H. Frauenfelder and R. M. Steffen, α, β, γ *Ray Spectroscopy* (K. Siegbahn, ed.), p. 1453, North-Holland, Amsterdam, 1966.
26 S. K. Bhattacherjee, S. K. Mitra, and H. C. Padhi, *Nucl. Phys.* **A96**, 81 (1967).
27 C. T. Yap, *Nucl. Phys.* **A100**, 619 (1967).
28 R. J. Blin-Stoyle and J. Le Tourneux, *Ann. Phys.* **18**, 12 (1962).
29 E. Freiberg and V. Soergel, *Z. Phys.* **162**, 114 (1961).
30 V. M. Lobashov and V. A. Nazarenko, *Soviet Phys. JETP* **15**, 257 (1962).
31 R. J. Blin-Stoyle and C. T. Yap, *Nucl. Phys.* **79**, 561 (1966).
32 R. J. Blin-Stoyle and C. Yalgin, *Phys. Letters* **15**, 258 (1965).
33 M. Morita, M. Fuyuki, and S. Tsukada, *Progr. Theoret. Phys.* **47**, 556 (1972).
34 J. C. Manthuruthil and C. P. Poirier, *Nucl. Phys.* **A118**, 657 (1968).
35 S. Yamada and T. Hayashi, *J. Phys. Soc. Japan* **31**, 23 (1971).
36 Z. Matumoto, M. Yamada, I.-T. Wang, and M. Morita, *Phys. Rev.* **129**, 1308 (1963).
37 L. S. Kisslinger and C. Wu, *Phys. Rev.* **136**, B1254 (1964). See also [66]
38 H. S. Smith and P. C. Simms, *Phys. Rev.* **C1**, 1809 (1970).
39 J. Alberghini and R. M. Steffen, *Phys. Letters* **7**, 85 (1963); S. K. Bhattacherjee and S. Mitra, *Phys. Rev.* **131**, 2611 (1963); R. Newsome and H. Fischbeck, *Phys. Rev.* **133**, B273 (1964); I. Estulin and A. Potushkov, *Nucl. Phys.* **36**, 334 (1962); R. M. Singru, P. C. Simms, and R. M. Steffen, *Phys. Rev.* **141**, 1078 (1966); D. E. Ohlms, J. J. Bosken, and P. C. Simms, *Phys. Rev.* **C1**, 1803 (1970).
40 M. Morita, *Progr. Theoret. Phys. Suppl.* **26**, 1 (1963).
41 H. Daniel, *Rev. Mod. Phys.* **40**, 659 (1968).
42 C. P. Bhalla, *Phys. Rev.* **129**, 2130 (1963); **132**, 1177 (1963).
43 R. M. Steffen, *Phys. Rev. Letters* **3**, 277 (1959); Z. W. Garabowski, R. S. Raghavan, and R. M. Steffen, *Phys. Rev.* **139**, B24 (1965).
44 S. Cipolla, Z. W. Garabowski, H. M. Naser, and R. M. Steffen, *Phys. Rev.* **146**, 877 (1966).
45 F. Boehm, V. Soergel, and B. Stech, *Phys. Rev. Letters* **1**, 77 (1958).
46 E. A. Plassman and L. M. Langer, *Phys. Rev.* **96**, 1593 (1954); C. S. Wu, *Beta- and Gamma-Ray Spectroscopy* (K. Siegbahn, ed.), p. 3143. North-Holland, Amsterdam, 1955; *Rev. Mod. Phys.* **22**, 386 (1950); H. Daniel, *Nucl. Phys.* **31**, 293 (1962).
47 S. Nakamura, *Proc. Intern. Conf. Theoret. Phys.* (Kyoto), p. 82, 1953.
48 A. G. Petschek and R. E. Marshak, *Phys. Rev.* **85**, 698 (1952).
49 M. Yamada, *Progr. Theoret. Phys.* **10**, 252 (1953); H. Takebe, S. Nakamura, and M. Taketani, *Progr. Theoret. Phys.* **14**, 317 (1955).
50 K. Smith, referred to by C. S. Wu on p. 343 of the second reference in [46].
51 M. Morita, J. Fujita, and M. Yamada, *Progr. Theoret. Phys.* **10**, 630 (1953).
52 G. E. Lee-Whiting, *Phys. Rev.* **97**, 463 (1955).
53 J. Fujita, M. Yamada, Z. Matumoto, and S. Nakamura, *Progr. Theoret. Phys.* **20**, 287 (1958); J. Sodemann and A. Winther, *Nucl. Phys.* **69**, 369 (1965).

54 J. Fujita, A. Arima, and M. Yamada, *Soryusiron Kenkyu* **8**, 68 (1955).
55 R. M. Spector, *Nucl. Phys.* **40**, 338 (1963); R. M. Spector and R. J. Blin-Stoyle, *Phys. Letters* **1**, 118 (1962).
56 N. Newby and E. J. Konopinski, *Phys. Rev.* **115**, 434 (1959).
57 J. Fujita, *Phys. Rev.* **126**, 202 (1962); J. D. Ullman, *Phys. Letters* **1**, 339 (1962).
58 Y. E. Kim and J. O. Rasmussen, *Nucl. Phys.* **47**, 184 (1963) and **47**, 207 (1963); R. M. Mello and J. Flores, *Nucl. Phys.* **47**, 177 (1963).
59 J. R. Erskine, W. W. Buechner, and H. A. Enge, *Phys. Rev.* **128**, 720 (1962).
60 Z. Matumoto and M. Yamada, *Progr. Theoret. Phys.* **19**, 285 (1958).
61 M. T. Burgy, V. E. Krohn, T. B. Novey, G. R. Ringo, and V. T. Telegdi, *Phys. Rev.* **110**, 1214 (1958).
62 C. W. Kim and H. Primakoff, *Phys. Rev.* **180**, 180 (1969).
63 E. Ambler, R. W. Hayward, D. D. Hoppes, and R. P. Hudson, *Phys. Rev.* **110**, 787 (1958).
64 J. Fujita, M. Hirata, and K. Shoda, *Phys. Letters* **33B**, 550 (1970).
65 K. Shoda, A. Suzuki, M. Sugawara, T. Saito, H. Miyase, and S. Oikawa, *Phys. Rev.* **C3**, 1999 (1971).
66 For collective models of atomic nuclei, see A. Bohr, *Danske Mat. Fys. Medd.* **26** No. 14 (1952); A. Bohr and B. R. Mottelson, *Danske Mat. Fys. Medd.* **27** No. 16 (1953).

Chapter Eight

MUON CAPTURE

8.1 INTRODUCTION

Muon capture in complex nuclei

$$\mu^- + (A, Z) \rightarrow (A, Z - 1) + \nu_\mu \qquad (8.1)$$

is analogous to the orbital electron capture

$$e^- + (A, Z) \rightarrow (A, Z - 1) + \nu_e, \qquad (8.2)$$

which is a counter process of beta decay

$$(A, Z - 1) \rightarrow (A, Z) + e^- + \bar{\nu}_e. \qquad (8.3)$$

Nuclear beta decay has been a useful source of information about nuclear structure, the situation being that the matrix element of the fundamental process

$$n \rightarrow p + e^- + \bar{\nu}_e \qquad \text{or} \qquad p \rightarrow n + e^+ + \nu_e \qquad (8.4)$$

is factored as a product of the nucleon part and the lepton part (the former between the bound nucleons depends largely on the details of the nucleon motion inside the nucleus). By the same token the nuclear muon capture process can be utilized as another tool in the study of nuclear structure. The novelty of this approach lies in the fact that the energy and momentum transferred to the nucleus are large (typically 20–30 MeV), so that more details of the nucleon motion become observable.

Putting it in another way, nuclear structure study through beta decay and muon capture may be regarded as the scanning of the nucleus with the (e, ν_e)-microscope or ν_μ-microscope. The typical nuclear matrix element

$$\int \psi_f^\dagger(\mathbf{r}_1, \mathbf{r}_2, \ldots, \mathbf{r}_A)\left[\sum_{s=1}^{A} \tau_-^{(s)} \exp(i\mathbf{p}\cdot\mathbf{r}_s)\right]\psi_i(\mathbf{r}_1, \mathbf{r}_2, \ldots, \mathbf{r}_A)$$

$$\times\, d\mathbf{r}_1\, d\mathbf{r}_2 \ldots d\mathbf{r}_A \qquad (8.5)$$

can be interpreted as the image or interference patterns of the nucleus under the light of the lepton wave $\exp(i\mathbf{p}\cdot\mathbf{r})$. The wavelength of the light in muon capture is shorter than in beta decay (of the order of 10^{-2}), so that more details of the granular property of the nucleus are likely to be revealed (i.e., the resolving power is higher).

In order to investigate nuclear structure through muon capture, we have to know the interaction which governs this process precisely. Today we know that the muon behaves like a heavy electron, and the hypothesis of muon–electron universality introduces the same interaction with the same coupling constant for both muon and electron. In fact about twenty years ago it had already been pointed out by several authors that muon capture takes place via a weak interaction with a coupling constant which is approximately equal to that of beta decay [1]. (See also [5] in Chapter 4.) This can be studied in detail with the recent data on the muon capture in hydrogen, for which there are no ambiguities due to nuclear structure. We will confirm that the universal *V–A* interaction holds with a greater precision in this case. A similar conclusion can be obtained from systematic study of the total capture rates in various complex nuclei. Once the interaction for the muon capture process is established, we can study nuclear structure from the phenomena of muon capture. This will be done in the next chapter.

8.2 MUON CAPTURE IN HYDROGEN

The simplest nucleus, hydrogen, is available in the laboratory only in the diatomic form of the gaseous or liquid state. The possibility of a strong spin-dependent reaction rate for

$$\mu^- + \mathrm{p} \rightarrow \mathrm{n} + \nu \tag{8.6}$$

makes possible accurate specification of the initial muon–proton state to an interpretation of experimental results. At a low velocity, the negative muon is trapped by the Coulombic proton field and cascades down to the lowest $1s$ state in less than 10^{-9} sec. A neutral mesic atom $(\mathrm{p}\mu^-)$ is formed and this tiny object starts to migrate in the hydrogen medium released from the molecular bond. The mesic atom is formed in the spin triplet ($F = 1$) or singlet ($F = 0$) hyperfine states in the proportion of the statistical weight. The subsequent collision process induces the rapid transition

$$(\mathrm{p}\mu^-)_{F=1} + (\mathrm{pp}) \rightarrow (\mathrm{p}\mu^-)_{F=0} + (\mathrm{pp}) + 0.25 \text{ eV}, \tag{8.7}$$

and the muon capture in gaseous hydrogen takes place from the singlet hyperfine state $(\mathrm{p}\mu^-)_{F=0}$.

In liquid hydrogen a further collision

$$(\mathrm{p}\mu^-)_{F=0} + (\mathrm{pp}) \rightarrow (\mathrm{p}\mu^-\mathrm{p}) + \mathrm{p} + 90 \text{ eV} \tag{8.8}$$

is expected because of the high density of the molecule, and the muon capture in liquid hydrogen takes place from this meso-molecular ion complex. The capture in $(p\mu^- p)$ differs from the capture in $(p\mu^-)$ in two points. First, the muon–proton overlap is different, which should be significant in the local interaction. Second, the muon spin can be either parallel or antiparallel to the proton spin, where two proton spins are parallel to each other in the stable ortho-molecule. The relative spin direction of the muon and proton makes a big difference in the capture rate in the V–A interaction known as the hyperfine effect [2]. We can relate the molecular capture rate W in the $(p\mu^- p)$ complex to the hyperfine atomic capture rate W_1 for $F = 1$ and W_0 for $F = 0$ by

$$W = 2\gamma(\tfrac{3}{4}W_0 + \tfrac{1}{4}W_1), \tag{8.9}$$

where γ is the ratio of the probability of the muon being found at the position of either proton in $(p\mu^- p)$ and the probability of the muon being at the position of the proton in $(p\mu^-)$. The capture rate in gaseous hydrogen is simply W_0.

For singlet hyperfine state $(p\mu^-)_{F=0}$, the rate of the process (8.8) is measured as [3]

$$W = (1.89 \pm 0.20) \times 10^6 \text{ sec}^{-1}. \tag{8.10}$$

If even small amounts of deuterium or other atoms are present, then an irreducible transfer to the deuteron (or heavier atom) is likely. The reduced mass effect makes the $(d\mu^-)$ system more tightly bound by 130 eV. The rate of this mode [3]

$$(p\mu^-) + d \rightarrow (d\mu^-) + p \tag{8.11}$$

is

$$W = c(5.80 \pm 0.30) \times 10^6 \text{ sec}^{-1}, \tag{8.12}$$

where c is the concentration of the deuterium. The muons in the ortho $(p\mu^- p)$ molecular ion, and $(p\mu^-)$ state can also decay naturally with the rate

$$W = \tau_\mu^{-1} = 0.455 \times 10^6 \text{ sec}^{-1}. \tag{8.13}$$

On the other hand, the muon capture in hydrogen is about 1/1000 as likely as natural decay.

The spin-dependent factor in Eq. (8.9) comes from the following considerations. Since the capture rate is scalar, it is given by

$$W = a + b\,\langle \mathbf{S}_p \cdot \mathbf{S}_\mu \rangle \tag{8.14}$$

with spin-independent factors a and b. The expectation value of the relative orientation of the proton and muon spins can be given for the atomic ($p\mu^-$) state

$$\langle \mathbf{S}_p \cdot \mathbf{S}_\mu \rangle = \tfrac{1}{2}F(F+1) - \tfrac{3}{4} = -\tfrac{3}{4} \qquad \text{for the singlet } (F=0),$$
$$= \tfrac{1}{4} \qquad \text{for the triplet } (F=1). \tag{8.15}$$

The atomic capture rates are, therefore, given by

$$W_0 = a - \tfrac{3}{4}b \qquad \text{and} \qquad W_1 = a + \tfrac{1}{4}b. \tag{8.16}$$

From these equations, we have

$$a = \tfrac{1}{4}W_0 + \tfrac{3}{4}W_1 \qquad \text{and} \qquad b = -W_0 + W_1. \tag{8.17}$$

For the molecular ($p\mu^-p$) ion, the expectation value is

$$\langle \mathbf{S}_p \cdot \mathbf{S}_\mu \rangle = -\tfrac{1}{2}, \tag{8.18}$$

since

$$\mathbf{J}^2 = (\mathbf{S}_{pp} + \mathbf{S}_\mu)^2 = \mathbf{S}_{pp}^2 + \mathbf{S}_\mu^2 + 4\langle \mathbf{S}_p \cdot \mathbf{S}_\mu \rangle, \qquad J = S_\mu = \tfrac{1}{2},$$
$$S_{pp} = 1. \tag{8.19}$$

Here, $\mathbf{J}$ and $\mathbf{S}_{pp}$ are the molecular spin and the resultant spin for two protons, respectively. A molecular physics calculation by Halpern [7], shows that the molecule is about 100% in the $J = \frac{1}{2}$ state and almost none in the $J = \frac{3}{2}$ state. From Eqs. (8.14), (8.17), and (8.18) we have

$$W = a - \tfrac{1}{2}b = \tfrac{3}{4}W_0 + \tfrac{1}{4}W_1. \tag{8.20}$$

We now sketch the calculation of the muon capture rate in hydrogen. The numerical coefficient in the expression of the capture rate is the same as that of the K electron capture in Eq. (5.8), except for the lepton masses. N_K is, however, unity, since there is only one bound muon. The result is

$$W = \tfrac{3}{4}W_0 + \tfrac{1}{4}W_1$$
$$= \frac{C_V^2}{2\pi^3}\frac{mc^2}{\hbar}\pi\left(\frac{\alpha m'_\mu}{m}\right)^3\left(\frac{E_\nu}{mc^2}\right)^2\left(\frac{M_n}{M_p + m_\mu}\right)(\tfrac{3}{4}w_0 + \tfrac{1}{4}w_1) \tag{8.21}$$

with

$$E_\nu = [(M_p + m_\mu)^2 - M_n^2]\,[2(M_p + m_\mu)]^{-1}$$

and

$$m'_\mu = m_\mu M_p (M_p + m_\mu)^{-1}.$$

Here, m'_μ is the muon reduced mass and E_ν is the neutrino energy. It should be mentioned that the factor $M_n(M_p + m_\mu)^{-1}$ comes from dp_ν/dE_f in the final state density. The reduced spin-dependent capture rates w_0 and w_1 are also dependent on various coupling constants involved in the interaction Hamiltonian (8.25).

For a simplified *VA* interaction, for which only C_V and C_A can have non-zero values, the transition amplitude is proportional to

$$C_V + C_A\langle \sigma_p \cdot \sigma_\mu \rangle \tag{8.22}$$

where the expectation value in the C_A term is given by

$$\begin{aligned} \langle \sigma_p \cdot \sigma_\mu \rangle &= -3 \qquad \text{for} \quad F = 0, \\ &= 1 \qquad \text{for} \quad F = 1, \end{aligned} \tag{8.23}$$

because the spins are related to σ by $S = \frac{1}{2}\sigma$. Therefore, the reduced capture rates are

$$C_V^2 w_0 = (C_V - 3C_A)^2, \qquad C_V^2 w_1 = (C_V + C_A)^2. \tag{8.24}$$

From this we can easily see that the muon capture rate from the triplet hyperfine state is vanishingly small for the $V - A$ interaction, for which $C_V \approx -C_A$.

In order to perform a detailed calculation of the muon capture rate in hydrogen, we usually adopt the phenomenological $V - A$ theory of the leptonic interaction for the fundamental process

$$\mu^- + p \to n + \nu_\mu.$$

The integrand of the matrix element for this transition is written as

$$H = V_\alpha^\dagger(\bar{\Phi}_\nu \gamma_\alpha \psi_\mu) + A_\alpha^\dagger(\bar{\phi}_\nu i\gamma_\alpha\gamma_5 \psi_\mu) + \text{h.c.}, \tag{8.25}$$

with

$$\phi_\nu = (1 + \gamma_5)2^{-1/2}\psi_\nu.$$

$V_\alpha^\dagger$ and $A_\alpha^\dagger$ are the vector and axial vector nucleon currents:

$$\begin{aligned} V_\alpha^\dagger &= (\bar{\Psi}_n[f_V\gamma_\alpha - ig_V\sigma_{\alpha\beta}p_\beta - ih_V p_\alpha]\psi_p), \\ A_\alpha^\dagger &= (\bar{\Psi}_n[if_A\gamma_\alpha\gamma_5 - g_A p_\alpha\gamma_5 + h_A\sigma_{\alpha\beta}\gamma_5 p_\beta]\psi_p), \end{aligned} \tag{8.26}$$

with

$$p_\beta = [p^{(p)} - p^{(n)}]_\beta \qquad \text{and} \qquad \sigma_{\alpha\beta} = \tfrac{1}{2}(\gamma_\alpha\gamma_\beta - \gamma_\beta\gamma_\alpha).$$

p_β is the difference of four momenta for the proton and neutron. The form factors f, g, and h are functions of p^2, and real if time-reversal invariance holds. In the limit as $p \to 0$, we set

$$f_V(0) = C_V \qquad \text{and} \qquad f_A(0) = C_A. \tag{8.27}$$

The nucleon currents (8.26) are the most general vector and axial vector currents which can be constructed from the Dirac matrices and the nucleon momenta [4]. (In fact, the vector and axial vector which include the sum of the nucleon momenta $[p^{(p)} + p^{(n)}]$ are reducible by using the Dirac equation. This does not hold for the off-mass shell problems.) Assuming G-parity invariance for each weak current, where G is as discussed in section 5.10, we have

$$h_V = h_A = 0. \tag{8.28}$$

This is assumed hereafter. The conserved vector current theory of Feynman and Gell-Mann [5] predicts

$$g_V = (\mu_p - \mu_n)(f_V/2M) \tag{8.29}$$

with $\mu_p - \mu_n = 3.706$. From the isotriplet vector current hypothesis for weak current and the isovector part of the electromagnetic current, we can introduce a momentum transfer dependence of the vector form factor:

$$f_V(p^2) = C_V\,(1 - \tfrac{1}{6}p^2\langle r^2\rangle), \tag{8.30}$$

where

$$\frac{p^2}{M^2} \approx \frac{m_\mu^2}{M^2} = 0.0112,$$

and the root mean square (rms) electromagnetic radius of the nucleon is

$$\langle r^2\rangle^{1/2} = 0.77 \text{ fm}. \tag{8.31}$$

The momentum transfer dependence for the axial vector form factor is unknown. It seems, however, to be very slight. For the sake of reference, we take either

$$f_A(p^2) = C_A \qquad \text{or} \qquad f_A(p^2) = C_A(1 - \tfrac{1}{6}p^2\langle r^2\rangle). \tag{8.32}$$

The g_A term, which is the induced pseudoscalar, is estimated by dispersion-theoretic calculation in a model of one-pion exchange [6]

$$m_\mu g_A \equiv C_P = (7\text{–}8)f_A. \tag{8.33}$$

The coupling constant C_P defined in expression (8.33) is sometimes adopted instead of g_A.

The molecular parameters are still controversial, mainly because we need a precise knowledge of the meson orbit in the molecular ion. We simply adopt the values which most people use [7],

$$2\gamma = 1.00. \tag{8.34}$$

Using the assumptions described earlier, we summarize the predicted capture rates in gaseous and liquid hydrogen in Table 8.1 [8]. For a choice

Table 8.1 Muon Capture Rates in Gaseous and Liquid Hydrogen[a]

	W(gas)		W(liq)	
C_A/C_V	I	II	I	II
−1.18	627	611	473	461
−1.20	643	627	486	473
−1.22	660	643	498	485
−1.24	677	660	511	498
Exp.	640 ± 70[b]		460 ± 42[c]	

[a] Capture rates are given in units of $\sec^{-1}$. We adopt $f_A = C_A$ for case I and $f_A = C_A(1 - \frac{1}{6}p^2\langle r^2\rangle)$ for case II.
[b] A. Alberigi Quaranta *et al. Phys. Letters* **25B**, 429 (1967).
[c] T. E. O. Ericson, *Rep. Intern. Conf. High Energy Physics* (Dubna), 1964.

of beta decay coupling constants

$$C_V M^2 = (1.0003 \pm 0.0017) \times 10^{-5} \qquad \text{and} \qquad C_A/C_V = -1.22$$

and no p^2 dependence for f_A ($f_A = C_A$,), we have

$$W_0 = W(\text{gas}) = 660 \sec^{-1}, \qquad W_1 = 12 \sec^{-1},$$
$$W = W(\text{liq}) = 498 \sec^{-1}. \tag{8.35}$$

The long awaited experiment with gaseous hydrogen was undertaken very recently [9]. The errors are still large, but the first value has been announced as

$$W_{\text{exp}}(\text{gas}) = 640 \pm 70 \sec^{-1}. \tag{8.36}$$

The molecular capture rate has been measured in three laboratories (CERN, Chicago, and Columbia). All the data agree with each other, and the world average reads [10]

$$W_{\text{exp}}(\text{liq}) = 460 \pm 42 \sec^{-1}. \tag{8.37}$$

The atomic capture rate is the primary quantity where the theory and experiments can be compared. Although the experimental errors prevent our discussing the details of the fundamental interaction, we can say that the theory and experiment agree quite well, and the underlying working hypothesis seems to hold completely. The molecular capture rate is subject to the ambiguities of the molecular parameters, but barring its future revision we observe that the predicted value is somewhat higher than the experimental value, though not of statistical significance.

8.3 TOTAL MUON CAPTURE RATE

Primakoff devised a formula for the total muon capture rate $W(A, Z)$ in a nucleus of mass number A and atomic number Z, by using the closure approximation [11]. In the derivation of the formula it is assumed that the states of the daughter nucleus are all excited by the muon capture and that these states consist of a complete set. In this approximation, the total capture rate, which is the sum of the absolute squares of the transition amplitudes over the final states, becomes the expectation value of the square of the interaction operators with respect to the initial state. Primakoff obtained the result

$$W(A, Z)Z_{\text{eff}}^{-4} = D\left(1 - \frac{A-Z}{2A}\,\delta\right) \tag{8.38}$$

with

$$D = \langle\eta\rangle^2 \alpha^3 \left(\frac{m_\mu}{m}\right)^5 \frac{\pi \log 2}{(ft_{1/2})_{\text{neutron}}} \frac{G_V^2 + 3\Gamma_A^2}{C_V^2 + 3C_A^2}.$$

Here W is given in $\sec^{-1}$ and $ft_{1/2}$ in seconds. The average momentum of the emitted neutrino is assumed to be

$$\langle\eta\rangle = \left\langle\frac{q}{m_\mu c}\right\rangle = 0.75\text{–}0.80, \tag{8.39}$$

and the nucleon–nucleon correlation parameter δ is estimated as

$$\delta = 3.0. \tag{8.40}$$

In this connection, the second term in (8.38), $[(A-Z)/2A]\,\delta$, represents the Pauli exclusion principle inhibition factor proportional to the fraction of nucleons which are neutrons. The coupling constants G_V and Γ_A in an effective nonrelativistic interaction are given by

$$G_V = C_V[1 + (q/2M_p)], \qquad \Gamma_A^2 = G_A^2 + \tfrac{1}{3}(G_P^2 - 2G_PG_A), \tag{8.41}$$

with

$$G_A = C_A - C_V(1 + \mu_p - \mu_n)(q/2M_p),$$

$$G_P = [C_P - C_A - C_V(1 + \mu_p - \mu_n)]\,(q/2M_p),$$

and

$$1 + \mu_p - \mu_n = 4.7,$$

which is the difference in the magnetic moments of the proton and neutron.

The coupling constants C_i in the relativistic interactions (8.60) below for muon capture are somewhat different from those for beta decay C_i^β, since the momentum transfers in these reactions are different, as is mentioned in the last section. The effective charge Z_{eff} is somewhat smaller than the atomic number Z of the capturing nucleus. For a nucleus of finite size with a continuous charge distribution, the effective charge density $\langle \rho \rangle$ is defined by

$$\langle \rho \rangle = \frac{\int_0^\infty \rho(\mathbf{r}) \, |\, \psi_\mu(\mathbf{r})\,|^2 \, d\mathbf{r}}{\int_0^\infty |\, \psi_\mu(\mathbf{r})\,|^2 \, d\mathbf{r}}, \tag{8.42}$$

where the $\psi_\mu(\mathbf{r})$ is the muon wave function. (The integration over the entire solid angles is not written explicitly.) The effective charge is defined by

$$Z_{\text{eff}}^4 = \pi a_o^3 \langle \rho \rangle. \tag{8.43}$$

Here a_o is the muon Bohr radius. In particular, the charge distribution is assumed to be uniform, and we have

$$Z_{\text{eff}} = \left[\pi a_o^3 \frac{Z}{\frac{4}{3}\pi R^3} \frac{\int_0^R |\, \psi_\mu(\mathbf{r})\,|^2 \, d\mathbf{r}}{\int_0^\infty |\, \psi_\mu(\mathbf{r})\,|^2 \, d\mathbf{r}} \right]^{1/4} \tag{8.44}$$

where R is the nuclear radius. An extensive numerical calculation for Z_{eff} was performed by Sens [12].

Figure 8.1 shows a plot, for elements with $Z > 8$, of the currently accepted experimental values of $W(A, Z)/Z_{\text{eff}}^4$ versus the fractional neutron excess $(Z - A)/2A$, given by Telegdi [13]. A best fit to the data indicated by the solid circles yields

$$\delta = 3.13, \qquad \text{and} \qquad D = 183 \text{ sec}^{-1}. \tag{8.45}$$

Telegdi gave a theoretical value for D in Eq. (8.38) of

$$D = 169 \text{ sec}^{-1} \tag{8.46}$$

for a universal $V - A$ interaction with $C_{\text{A}}/C_{\text{V}} = -1.21$, conserved vector current, and induced pseudoscalar interaction. The observed value of δ agrees well with Primakoff's estimate $\delta = 3$ in Eq. (8.40), and the general dependence of the reduced capture rate on $(A - Z)/2A$ approximates well the linear behavior implied by Eq. (8.38). With $\langle \eta \rangle = 0.8$, experiments and the UFI (universal Fermi interaction) prediction are in perfect agreement. A prediction for $\langle \eta \rangle = 0.75$ is also indicated for illustrative purposes. The

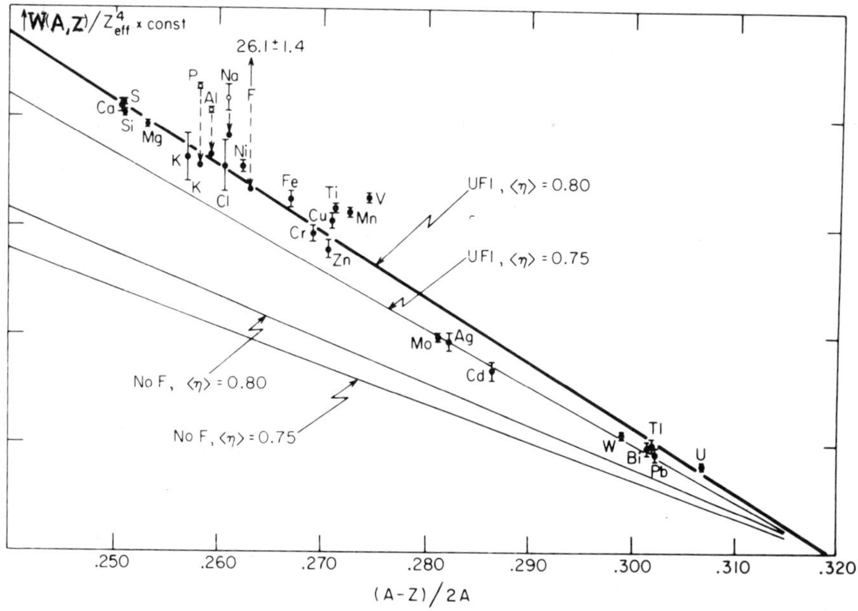

Fig. 8.1 Experimental values of $W(A, Z)/Z_{\text{eff}}^4$ plotted against $(A - Z)/2A$. Ordinate is actually $\pi a_0 W(A, Z)/Z_{\text{eff}}^4$ in units of 10^{-31} cm^3 sec^{-1}. Heavy line through experimental points is best fit with $\delta = 3.13$, D in Eq. (8.38) = 183 sec^{-1}; i.e., corresponds to UFI with $\langle\eta\rangle = 0.80$. Open circles represent experimental points not used in fit because they are affected by hyperfine effects. Downward arrows represent estimated corrections. All lines in the figure correspond to $\delta = 3.13$ (figure plotted by Telegdi [13]).

hypothesis that there is no Fermi interaction in the muon capture interaction [$C_V = 0$] is also examined, as indicated by "no F". From this, we can immediately rule out the hypothesis of no Fermi interaction and can conclude that the interactions for muon capture and beta decay are very similar.

8.4 THEORY OF ALLOWED AND FORBIDDEN TRANSITIONS IN MUON CAPTURE REACTION

We have studied in sections 8.2 and 8.3 that the transition matrix element of the universal $V - A$ interaction, Eqs. (8.25) and (8.26), can explain beautifully experimental data on muon capture rates in hydrogen and data on the total muon capture rates in various complex nuclei.

Now we discuss the possibility of studying nuclear structure by using the muon capture rates in complex nuclei under the assumption that the interactions Eqs. (8.25) and (8.26) are fully determined. The study supplements nuclear structure problems in beta decays and gamma decays where the interactions for these decays are well known. As was pointed out in section 8.1, muon capture is a parallel reaction to the orbital K-electron capture and analogous to positron decay. In this connection, we consider the reaction where the complex nucleus absorbs the muon from its K orbit and a definite spin-parity state of the daughter nucleus is excited. A reaction of this sort is called a partial transition, and the capture rate the partial capture rate. The final state of this reaction is deexcited into the ground state of the daughter nucleus by emitting gamma rays, and the ground state decays successively by emitting a beta ray. If the excitation energy due to the muon capture is large enough, the final state emits neutrons or nuclear fission occurs.

In order to classify the nuclear transition from the definite spin-parity state to the other, we again introduce the concept of forbiddenness for the muon capture reaction, as in the case of beta decay. We call the transition: allowed if the spin change is 0 or 1 and parity change is No; the first forbidden transition if the spin change is 0, 1, or 2 and parity change Yes; the second forbidden transition if the spin change is 2 or 3 and parity change No; and so on. The nuclear matrix elements for the muon capture reaction are classified for each forbiddenness according to their ranks and parities. In fact, we have several kinds of Fermi matrices of rank zero and Gamow–Teller matrices of rank one in the allowed transition, while we have only one Fermi and one Gamow–Teller matrix in the case of beta decay. The reason for this difference becomes clear in the following formulation of muon capture in complex nuclei. If the spin change is $(n + 1)$ for the nth forbidden transition, this is called a unique forbidden transition. (This is because, the beta spectral shape is unique as is, e.g., shown in Eq. (7.100).) The other cases are called nonunique.

Phase Space Volume

The transition rate of the muon capture reaction from a particular nuclear state J_{i} to another state J_{f} is given by the golden rule

$$W = 2\pi < |\,\mathrm{M.E.}\,|^2 >_{\mathrm{av}} q^2 \frac{dq}{dE_{\mathrm{f}}}. \tag{8.47}$$

Here, all quantities are expressed in the atomic units, so that energies and momenta are in units of mc^2 and mc, respectively. We find the neutrino momentum q from energy and momentum conservation:

$$_{Z}M^{A} + m_{\mu} - 1 = q + [q^2 + (_{Z-1}M^{A})^2]^{1/2}. \tag{8.48}$$

The solution of Eq. (8.48) is given by

$$q = (m_\mu - E_o)\left[1 - \frac{m_\mu - E_o}{2({}_ZM^A + m_\mu - 1)}\right]$$
$$\approx (m_\mu - E_o)\left[1 - \frac{m_\mu}{2(m_\mu + AM)}\right], \tag{8.49}$$

with

$$E_o = {}_{Z-1}M^A - {}_ZM^A + 1, \tag{8.50}$$

which is the maximum energy of the electron in the beta decay of the state J_f to the state J_i. Here M is the nucleon mass, as before. We can also find dq/dE_f by differentiating the equation

$$E_f = q + [q^2 + ({}_{Z-1}M^A)^2]^{1/2}. \tag{8.51}$$

The result is

$$\frac{dq}{dE_f} = 1 - \frac{q}{E_f} \approx 1 - \frac{q}{m_\mu + AM}. \tag{8.52}$$

The calculation of the matrix element $\langle |\text{M.E.}|^2 \rangle_{av}$ in Eq. (8.47) was developed by Morita and Fujii [14] in a way quite analogous to the theory of beta decay. We introduce their formalism as follows.

Nonrelativistic Approximation for Nucleon Wave Functions

The integrand of the transition matrix element is again assumed to be Eqs. (8.25) and (8.26). We evaluate the nuclear matrix elements with the nonrelativistic approximation for nucleons, since the velocity of nucleons in the nucleus is about one tenth of the light velocity. This can be done if we transform these operators into equivalent even operators by the Foldy–Wouthuysen transformation as modified by Rose and Osborn [15]. (The equations (8.25) and (8.26) involve odd and even operators. The even operators relate the large component of a nucleon spinor to the large one of another nucleon, and the small to the small, while the odd operators relate the large to the small and small to the large.) In the nonrelativistic approximation the nucleon wave function

$$\psi = \begin{pmatrix} v \\ u \end{pmatrix}$$

is replaced by

$$\psi = \begin{pmatrix} -\dfrac{1}{2M}\,\sigma\cdot \mathbf{p}u \\ u \end{pmatrix}, \tag{8.53}$$

where u and v are the large and small components of the Dirac spinor, corresponding to the eigenvalues -1 and $+1$ for β, respectively. In order to demonstrate the transformation, we choose the first term of the vector part in Eqs. (8.25) and (8.26),

$$(\bar{\psi}_{\rm n}\gamma_\alpha\psi_{\rm p})(\bar{\phi}_\nu\gamma_\alpha\psi_\mu) = (\psi_{\rm n}^\dagger\psi_{\rm p})(\phi_\nu^\dagger\psi_\mu) - (\psi_{\rm n}^\dagger\boldsymbol{\alpha}\psi_{\rm p})(\phi_\nu^\dagger\boldsymbol{\alpha}\psi_\mu). \tag{8.54}$$

As we have studied in the theory of beta decay, the second term in Eqs. (8.54) is the relativistic contribution. It is nonnegligible in the case of muon capture, since the momentum of the neutrino is of the order of the muon mass, and it is about 200 times higher than that for beta decay. We have to take this effect into account. The transform of this term to the even operator form is performed as follows.

$$\begin{aligned}(\psi_{\rm n}^\dagger\boldsymbol{\alpha}\psi_{\rm p})(\phi_\nu^\dagger\boldsymbol{\alpha}\psi_\mu) &= \left(-\frac{1}{2M}(\sigma\cdot\mathbf{p}u_{\rm n})^\dagger,\ u_{\rm n}^\dagger\right)\begin{pmatrix}0 & \sigma\\ \sigma & 0\end{pmatrix}\begin{pmatrix}-\dfrac{1}{2M}\,\sigma\cdot\mathbf{p}u_{\rm p}\\ u_{\rm p}\end{pmatrix}\cdot\mathbf{L}(\boldsymbol{\alpha})\\ &= -\frac{1}{2M}\left[u_{\rm n}^\dagger\sigma\cdot\mathbf{p}\sigma\cdot\mathbf{L}(\boldsymbol{\alpha})u_{\rm p} + u_{\rm n}^\dagger\sigma\cdot\mathbf{L}(\boldsymbol{\alpha})\sigma\cdot\mathbf{p}u_{\rm p}\right].\end{aligned} \tag{8.55}$$

Here, we use the symbol

$$L(\Omega) = \phi_\nu^\dagger\Omega\psi_\mu \qquad \text{with} \qquad \phi_\nu = 2^{-1/2}(1+\gamma_5)\psi_\nu. \tag{8.56}$$

It should be noted that in Eq. (8.55) u and σ are 2×1 and 2×2 matrices. The differential operator $[\mathbf{p} = (\hbar/i)\nabla]$ operates on the lepton covariant $\mathbf{L}(\boldsymbol{\alpha})$ and the proton spinor $u_{\rm p}$ in the first term of Eq. (8.55), while in the second term it operates on $u_{\rm p}$ only. Next, we use a relation

$$\sigma\cdot\mathbf{A}\sigma\cdot\mathbf{B} = \mathbf{A}\cdot\mathbf{B} + i\sigma\cdot(\mathbf{A}\times\mathbf{B}), \tag{8.57}$$

which makes Eq. (8.55)

$$(\psi_{\rm n}^\dagger\boldsymbol{\alpha}\psi_{\rm p})(\phi_\nu^\dagger\boldsymbol{\alpha}\psi_\mu) = -\frac{1}{2M}u_{\rm n}^\dagger[2\mathbf{L}(\boldsymbol{\alpha})\cdot\mathbf{p} + \mathbf{p}\cdot\mathbf{L}(\boldsymbol{\alpha}) + i\sigma\cdot\mathbf{p}\times\mathbf{L}(\boldsymbol{\alpha})]u_{\rm p}. \tag{8.58}$$

Here $\mathbf{p}$ operates on $u_{\rm p}$ in the first term and only on $\mathbf{L}(\boldsymbol{\alpha})$ in the second and third terms.

We perform similar calculations for all terms in Eqs. (8.25) and (8.26), and obtain the integrand of the transition matrix element given by

$$H = u_{\rm n}^\dagger\mathscr{H}u_{\rm p}$$

with the transformed interaction Hamiltonian

$$\mathscr{H} = C_V 1 \cdot L(1) + C_A \sigma \cdot L(\sigma) + \frac{C_V}{2M}\,[2L(\alpha)\cdot \mathbf{p} + \mathbf{p}\cdot L(\alpha)$$

$$+ (1 + \mu_p - \mu_n) i\sigma \cdot \mathbf{p} \times L(\alpha)] + \frac{C_A}{2M}\,[2L(\gamma_5)\sigma\cdot\mathbf{p} + \sigma\cdot\mathbf{p}L(\gamma_5)]$$

$$- \frac{C_P}{2M}\,\sigma\cdot\mathbf{p}L(\beta\gamma_5) - C_S 1\cdot L(\beta) - C_T\sigma\cdot L(\sigma) - \frac{C_T}{E_o}\,\sigma\cdot\mathbf{p}L(\gamma_5). \quad (8.59)$$

Here the coupling constants are introduced by the relations

$$C_V = f_V, \qquad g_V = (\mu_p - \mu_n)\frac{C_V}{2M}, \qquad C_S = -m_\mu h_V,$$

$$C_A = f_A, \qquad C_P = m_\mu g_A, \qquad C_T = -h_A E_o. \quad (8.60)$$

Neutrino and Muon Wave Functions

In order to compute for lepton covariant $L(\Omega)$, we have to know the explicit forms of the wave functions for the neutrino and muon. This is ascertained as follows. The free neutrino state is characterized by its linear momentum $\mathbf{q}$ and the direction of the spin $m = \pm\frac{1}{2}$. Expanding the neutrino plane wave in terms of the normalized spherical spinor $\chi_{\kappa\mu}$, Eq. (2.92), we have

$$\psi_\nu(\mathbf{q}, m; \mathbf{r}) = \sum_{\kappa\mu} i^l (l\tfrac{1}{2}\mu - m m \mid j\mu)\, Y^*_{l\mu-m}(\hat{q})\,\psi^{(\nu)}_{\kappa\mu}, \quad (8.61)$$

with

$$\psi^{(\nu)}_{\kappa\mu} = \begin{pmatrix} -if_\kappa \chi_{-\kappa\mu} \\ g_\kappa \chi_{\kappa\mu} \end{pmatrix}, \quad (8.62)$$

$$g_\kappa = \pi^{-1/2} j_l(qr), \qquad f_\kappa = \pi^{-1/2} S_\kappa j_{\bar{l}}(qr), \quad (8.63)$$

$$\chi_{\kappa\mu} = \sum_{m'} (l\,\tfrac{1}{2}\mu - m' m' \mid j\mu)\, Y_{l\,\mu-m'}(\hat{r})\chi_{m'}. \quad (8.64)$$

Here $\chi_{m'}$ is given by Eq. (A.73). These expressions are obtained from Eqs. (2.101) and (2.105) by making the mass term vanish. (g_κ and f_κ in Eq. (8.63) have no q compared with these terms in Eq. (2.105). However, q^2 appears as the final state density in Eq. (8.47). This holds the consistency.) Remember that the l and $\bar{l}$ of the spherical Bessel functions denote the orbital angular momentum corresponding to κ and its partner $-\kappa$, respectively. $\hat{q}$ and $\hat{r}$ are the unit vectors in the directions of the neutrino momentum and space coordinate, respectively. The normalization of ψ_ν is such that

$$\int \psi^\dagger_\nu(\mathbf{q}', m'; \mathbf{r})\psi_\nu(\mathbf{q}, m; \mathbf{r})\, d\mathbf{r} = \delta_{mm'}\delta(\mathbf{q}' - \mathbf{q}) \quad (8.65)$$

where the integral extends over the whole space.

The bound state wave function for the muon is expressed in the same way as that of the neutrino, but it has only a single value of κ corresponding to a definite orbit. Namely,

$$\psi_\mu(\kappa, \mu; \mathbf{r}) = \psi_{\kappa\mu}^{(\mu)} = \begin{pmatrix} -iF_\kappa \chi_{-\kappa\mu} \\ G_\kappa \chi_{\kappa\mu} \end{pmatrix}, \tag{8.66}$$

G_κ and F_κ being the radial wave functions for the bound state. Explicit forms of these wave functions for a point nucleus are given, e.g., by Bethe and Salpeter [16], for $1s_{1/2}$, $2s_{1/2}$, $2p_{1/2}$, and $2p_{3/2}$. In the lowest state $1s_{1/2}$,

$$\kappa = -1, \qquad \mu = \pm \tfrac{1}{2},$$

$$G_{-1} = \left(\frac{2Z}{a_o}\right)^{3/2} \left[\frac{1+\gamma}{2\Gamma(2\gamma+1)}\right]^{1/2} \exp\left(\frac{-Zr}{a_o}\right)\left(\frac{2Zr}{a_0}\right)^{\gamma-1},$$

$$F_{-1} = -\left(\frac{1-\gamma}{1+\gamma}\right)^{1/2} G_{-1} \qquad \text{with} \qquad \gamma = [1-(\alpha Z)^2]^{1/2}. \tag{8.67}$$

Here a_o is the Bohr radius of the mu mesic atom, Z is the atomic number of the nucleus, and Γ is the gamma function. a_o is explicitly given by

$$a_o = (\alpha m'_\mu)^{-1} \qquad \text{with} \qquad m'_\mu = m_\mu [1 + (m_\mu/AM)]^{-1}, \tag{8.68}$$

m'_μ being the muon reduced mass in the parent mu mesic atom. If we assume αZ (α is the fine structure constant) to be small, G_{-1} and F_{-1} become

$$G_{-1} = 2(\alpha Z m'_\mu)^{3/2} \exp(-\alpha Z m'_\mu r), \qquad F_{-1} = 0. \tag{8.69}$$

The normalization for the muon wave functions ψ_μ is such that

$$\int \psi_\mu^\dagger(\kappa', \mu'; \mathbf{r}) \psi_\mu(\kappa, \mu; \mathbf{r})\, d\mathbf{r} = \delta_{\kappa\kappa'}\delta_{\mu\mu'}. \tag{8.70}$$

Spherical Components of the Matrix Elements

Using these lepton wave functions, we evaluate the spherical component of the interaction Hamiltonian $(\kappa\mu\,|\mathscr{H}|\,\kappa'\mu')$. This is the $\kappa\mu$-$\kappa'\mu'$ component of $\mathscr{H}$ in Eq. (8.59), which is obtained by the replacement of a lepton covariant $L(\Omega)$ by $(\psi_{\kappa\mu}^{(\nu)}, \Omega\psi_{\kappa'\mu'}^{(\mu)})$. The neutrino wave function ϕ_ν should be taken in the case of parity-nonconserving phenomena. Here we are discussing the capture rate, and it is sufficient to use ψ_ν in place of ϕ_ν. Calculation of the spherical component of each term of Eq. (8.59) can be performed by using the Racah algebra which is introduced in Appendix. Finally, we have the following expressions of the spherical components for various terms.

$$\pm 1(\kappa\mu\,|\,\underset{\beta}{1}\,|\,\kappa'\mu') = (-)^{\mu+j'+l'} \sum_{vu} (jj' - \mu\mu'\,|\,u\,\mu'-\mu)$$

$$\times\, \mathbf{Y}_{0vu}^{\mu'-\mu}(\hat{r})[g_\kappa G_{\kappa'} S_{0vu}(\kappa,\kappa') \mp f_\kappa F_{\kappa'} S_{0vu}(-\kappa,-\kappa')]\delta_{vu}, \tag{8.71}$$

with

$$\mathbf{Y}_{0vu}^{\mu'-\mu}(\hat{r}) = (4\pi)^{-1/2} Y_{v\,\mu'-\mu}(\hat{r}). \tag{8.72}$$

Here the upper sign refers to the unit operator and the lower to β.

$$\sigma \cdot (\kappa\mu \,|\, \sigma \,|\, \kappa'\mu') = (-)^{\mu+j'+l'+1} \sum_{vu} (jj' - \mu\mu' \,|\, u\,\mu' - \mu)\mathbf{Y}_{1vu}^{\mu'-\mu}(\hat{r}, \sigma)$$
$$\times [g_\kappa G_{\kappa'} S_{1vu}(\kappa, \kappa') - f_\kappa F_{\kappa'} S_{1vu}(-\kappa, -\kappa')], \tag{8.73}$$

with

$$\mathbf{Y}_{1vu}^{\mu'-\mu}(\hat{r}, \sigma) = \sum_m (1v - m\; m - \mu + \mu' \,|\, u\,\mu' - \mu)$$
$$\times Y_{v\; m+\mu'-\mu}(\hat{r})\mathbf{Y}_{1\,-m}(\sigma). \tag{8.74}$$

Here Y_{lm} and $\mathbf{Y}_{lm}$ are the spherical and solid harmonics, and they have a relation

$$\mathbf{Y}_{lm}(\mathbf{A}) = A^l Y_{lm}(\hat{A}). \tag{8.75}$$

where $\mathbf{A}$, A, and $\hat{A}$ are the arbitrary vector, its magnitude, and the unit vector, respectively. In Eqs. (8.71) and (8.73), S_{0vu} and S_{1vu} are given by

$$S_{kvu}(\kappa, \kappa') = [2(2l+1)(2l'+1)(2j+1)(2j'+1)]^{1/2}$$
$$\times (ll'00 \,|\, v0) \begin{pmatrix} l & l' & v \\ j & j' & u \\ \frac{1}{2} & \frac{1}{2} & k \end{pmatrix} \quad \text{for} \quad k = 0 \text{ and } 1, \tag{8.76}$$

with

$$\begin{pmatrix} l & l' & v \\ j & j' & u \\ \frac{1}{2} & \frac{1}{2} & k \end{pmatrix} = \sum_f (2f+1) W(l'j'\tfrac{1}{2}k, \tfrac{1}{2}f) W(j\tfrac{1}{2}vl', lf) W(jj'vk, uf). \tag{8.77}$$

It is easily seen that in the 9-j symbol, Eq. (8.77), k, v, and u correspond to the resultant spin, resultant orbital, and resultant total angular momentum of the neutrino–muon system, respectively.

$$(\kappa\mu \,|\, \boldsymbol{\alpha} \,|\, \kappa'\mu') \cdot \mathbf{p} = (-)^{\mu+j'+l'+1} \sum_{vu} (jj' - \mu\mu' \,|\, u\,\mu' - \mu)$$
$$\times i[f_\kappa G_{\kappa'} S_{1vu}(-\kappa, \kappa') + g_\kappa F_{\kappa'} S_{1vu}(\kappa, -\kappa')]$$
$$\times \mathbf{Y}_{1vu}^{\mu'-\mu}(\hat{r}, \mathbf{p}). \tag{8.78}$$

Here the definition of $\mathbf{Y}_{1vu}^{\mu'-\mu}(\hat{r}, \mathbf{p})$ is similar to Eqs. (8.74) and (8.75).

$$\mathbf{p}\cdot(\kappa\mu\,|\,\boldsymbol{\alpha}\,|\,\kappa'\mu') = 3^{1/2}(-)^{\mu+j'+l'+1}\sum_{vu}(jj'-\mu\mu'\,|\,u\,\mu'-\mu)$$
$$\times\left[\left(\frac{v+1}{2v+3}\right)^{1/2}\mathbf{Y}_{0\,v+1u}^{\mu'-\mu}(\hat{r})\delta_{v+1\,u}D_{+}-\left(\frac{v}{2v-1}\right)^{1/2}\right.$$
$$\left.\times\,\mathbf{Y}_{0v-1u}^{\mu'-\mu}(\hat{r})\delta_{v-1\,u}D_{-}\right][f_\kappa G_{\kappa'}S_{1vu}(-\kappa,\kappa')$$
$$+\,g_\kappa F_{\kappa'}S_{1vu}(\kappa,-\kappa')], \tag{8.79}$$

with

$$D_{+}=\frac{d}{dr}-\frac{v}{r},\qquad D_{-}=\frac{d}{dr}+\frac{v+1}{r}. \tag{8.80}$$

Here, $D_{+}(D_{-})$ increases (decreases) the order by one unit, when it operates on the spherical Bessel function.

$$i\sigma\cdot\mathbf{p}\times(\kappa\mu\,|\,\alpha\,|\,\kappa'\mu') = 6^{1/2}(-)^{\mu+j'+l'+1}\sum_{vu}(jj'-\mu\mu'\,|\,u\,\mu'-\mu)$$
$$\times[(v+1)^{1/2}W(11uv,1v+1)\mathbf{Y}_{1v+1u}^{\mu'-\mu}(\hat{r},\sigma)D_{+}$$
$$-\,v^{1/2}W(11uv,1v-1)\mathbf{Y}_{1\,v-1u}^{\mu'-\mu}(\hat{r},\sigma)D_{-}]$$
$$\times[f_\kappa G_{\kappa'}S_{1vu}(-\kappa,\kappa')+g_\kappa F_{\kappa'}S_{1vu}(\kappa,-\kappa')]. \tag{8.81}$$

$$(\kappa\mu\,|\,\gamma_5\,|\,\kappa'\mu')\sigma\cdot\mathbf{p} = (-)^{\mu+j'+l'}\sum_{vu}(jj'-\mu\mu'\,|\,u\,\mu'-\mu)$$
$$\times\,i\mathbf{Y}_{0vu}^{\mu'-\mu}(\hat{r})[f_\kappa G_{\kappa'}S_{0vu}(-\kappa,\kappa')$$
$$+\,g_\kappa F_{\kappa'}S_{0vu}(\kappa,-\kappa')]\sigma\cdot\mathbf{p}\delta_{vu}. \tag{8.82}$$

$$\sigma\cdot\mathbf{p}\left(\kappa\mu\,\left|\,\begin{matrix}\gamma_5\\ \beta\gamma_5\end{matrix}\,\right|\,\kappa'\mu'\right) = 3^{-1/2}(-)^{\mu+j'+l'+1}\sum_{vu}(jj'-\mu\mu'\,|\,u\,\mu'-\mu)$$
$$\times\left[\left(\frac{v+1}{2v+1}\right)^{1/2}\mathbf{Y}_{1v+1u}^{\mu'-\mu}(\hat{r},\sigma)D_{+}-\left(\frac{v}{2v+1}\right)^{1/2}\right.$$
$$\left.\times\,\mathbf{Y}_{1v-1u}^{\mu'-\mu}(\hat{r},\sigma)D_{-}\right]$$
$$\times[f_\kappa G_{\kappa'}S_{0vu}(-\kappa,\kappa')\pm g_\kappa F_{\kappa'}S_{0vu}(\kappa,-\kappa')]\delta_{vu}. \tag{8.83}$$

Here the upper sign refers to γ_5, and the lower sign to $\beta\gamma_5$.

Matrix Element

For the purpose of illustrating calculation, we consider the matrix element of the reaction $\mu^- + \mathrm{p} \to \nu + n$, introduced by the main even part of axial vector interaction

$$C_A (u_n^\dagger \sigma u_p) L(\sigma). \tag{8.84}$$

Here we adopt the same convention for the neutrino wave functions as described in Eq. (8.71). From Eqs. (8.61), (8.66), and (8.73), we have

$$\begin{aligned}
\text{M.E.} &= C_A \int (u_n^\dagger \sigma u_p)(\psi_\nu^\dagger \sigma \psi_\mu)\, d\mathbf{r} \\
&= C_A \int \sum_{\kappa\mu} (u_n^\dagger \sigma u_p) i^{-l} (l \tfrac{1}{2}\, \mu - m m \mid j\mu) Y_{l\,\mu-m}(\hat{q}) (\kappa\mu \mid \sigma \mid \kappa'\mu')\, d\mathbf{r} \\
&= C_A \int \sum_{\kappa\mu} \sum_{vu} i^{-l} (l \tfrac{1}{2}\, \mu - m\, m \mid j\mu) Y_{l\,\mu-m}(\hat{q}) (-)^{\mu+j'+l'+1} \\
&\quad \times (jj' - \mu\, \mu' \mid u\, \mu' - \mu) \int u_n^\dagger \mathbf{Y}_{1vu}^{\mu'-\mu}(\hat{r}, \sigma) [g_\kappa G_{\kappa'} S_{1vu}(\kappa, \kappa') \\
&\quad - f_\kappa F_{\kappa'} S_{1vu}(-\kappa, -\kappa')] u_p\, d\mathbf{r}.
\end{aligned} \tag{8.85}$$

The matrix element for the reaction $\mu^- + (A, Z) \to \nu + (A, Z-1)$ is obtained from Eq. (8.85) by assuming that every nucleon in a nucleus does not perturb the others. We introduce the subscript s labeling the nucleons and the operator $\tau_-^{(s)}$ which transforms the sth nucleon from the proton to the neutron state. The nonrelativistic wave functions of the neutron and proton in Eq. (8.85) are replaced by those of the initial and final nuclei, where total and magnetic quantum numbers are replaced by $J_i M_i$ and $J_f M_f$. Thus

$$\text{M.E.} = C_A \int U_{J_f}^{M_f \dagger} \sum_{s=1}^{A} \sigma_s \tau_-^{(s)} \psi_\nu^\dagger \sigma \psi_\mu U_{J_i}^{M_i}\, d\mathbf{r}_1\, d\mathbf{r}_2 \cdots d\mathbf{r}_A. \tag{8.85$'$}$$

Furthermore, we define the reduced nuclear matrix element by using the Wigner–Eckert theory.

$$\begin{aligned}
&\int U_{J_f}^{M_f \dagger} \sum_{s=1}^{A} \{\mathbf{Y}_{1vu}^{\mu'-\mu}(\hat{r}_s, \sigma_s) [g_\kappa G_{\kappa'} S_{1vu}(\kappa, \kappa') \\
&\quad - f_\kappa F_{\kappa'} S_{1vu}(-\kappa, -\kappa')]\} \tau_-^{(s)} U_{J_i}^{M_i}\, d\mathbf{r}_1\, d\mathbf{r}_2 \cdots d\mathbf{r}_A \\
&\quad = (J_i u M_i\, \mu' - \mu \mid J_f M_f) \mathscr{M}_{vu}.
\end{aligned} \tag{8.86}$$

The reduced nuclear matrix element $\mathscr{M}_{vu}$ does not depend on the magnetic quantum number and it is, in principle, calculable if we know the nuclear wave functions of the initial and final states, by the definition Eq. (8.86). Now the nuclear matrix element becomes

$$\begin{aligned}
\text{M.E.} &= \sum_{\kappa\mu} \sum_{vu} i^{-l} (l \tfrac{1}{2} \mu - m\, m \mid j\mu) Y_{l\,\mu-m}(\hat{q}) (-)^{\mu+j'+l'+1} \\
&\quad \times (J_i u M_i\, \mu' - \mu \mid J_f M_f)(jj' - \mu\mu' \mid u\, \mu' - \mu) C_A \mathscr{M}_{vu}.
\end{aligned} \tag{8.87}$$

Next, we generalize the argument above to the Hamiltonian density in Eq. (8.59); the matrix element is obtained by replacing $C_A \mathcal{M}_{vu}$ by

$$\sum_i C^{(i)} \mathcal{M}_{vu}^{(i)}.$$

Here the coupling constants $C^{(i)}$ and the reduced nuclear matrix elements $\mathcal{M}_{vu}^{(i)}$ are as given in Table 8.2. The subscript (i) specifies the individual terms in Eq. (8.86). After summing over all magnetic quantum numbers, integrating over $\hat{q}$, and averaging over the initial states, we have the result for the absolute square of the matrix element.

$$\begin{aligned}\langle |\text{M.E.}|^2 \rangle_{\text{av}} &= \frac{1}{(2j'+1)(2J_i+1)} \int d\Omega_q \sum_{m\mu'} \sum_{M_i M_f} |\text{M.E.}|^2 \\ &= \frac{2J_f+1}{(2j'+1)(2J_i+1)} \sum_{ij} \sum_{\kappa} \sum_{u} C^{(i)*} C^{(j)} \Big[\sum_v \mathcal{M}_{vu}^{(i)*} \Big] \\ &\quad \times \Big[\sum_{v'} \mathcal{M}_{v'u}^{(j)} \Big]. \end{aligned} \tag{8.88}$$

Since the subscript u of the nuclear matrix element indicates the rank of the tensors, we have no cross term between two $\mathcal{M}$'s with different u. The orthonormality of the $\chi_{\kappa\mu}$ brings the sum over κ on the outside of the absolute square. Here κ specifies simultaneously the total and orbital angular momenta j and l of the neutrino. On the other hand, there are several cross terms among $\mathcal{M}$'s with different v. By the definition of $\mathcal{M}_{vu}^{(i)}$, the parity changes of the nuclear states are $(-)^v$ for $i = 1, 2, 3$ and $(-)^{v+1}$ for $i = 4, 5, \ldots, 9$.

Muon Capture Rate from the K Orbit

Now we write down the explicit formula for the muon capture rate from the muonic K orbit, $1s_{1/2}$, by assuming the nuclear charge to be not very large, $\alpha Z \ll 1$. (Of course, we can omit this assumption, in which case the formula becomes more lengthy.) In this approximation, the radial wave functions of the muon are given by

$$G_{-1} = 2(\alpha Z m'_\mu)^{3/2} \exp(-\alpha Z m'_\mu r) \qquad \text{and} \qquad F_{-1} = 0. \tag{8.89}$$

From Eqs. (8.47), (8.49), (8.52), and (8.88), we have the formula for the muon capture rate from the K orbit

$$W = P \frac{2J_f+1}{2J_i+1} \left(1 - \frac{q}{m_\mu + AM} \right) q^2 \tag{8.90}$$

Table 8.2 Coupling Constants $C^{(i)}$ and Reduced Nuclear Matrix Elements $\mathscr{M}_{vu}^{(i)}$ in Eqs. (8.88)[a]

i	$C^{(i)}$	Ξ_s
1	C_V	$\mathbf{Y}^m_{0vu}(\hat{r}_s)[g_\kappa G_{\kappa'} S_{0vu}(\kappa, \kappa') - f_\kappa F_{\kappa'} S_{0vu}(-\kappa, -\kappa')]\, \delta_{vu}$
2	C_S	$\mathbf{Y}^m_{0vu}(\hat{r}_s)[g_\kappa G_{\kappa'} S_{0vu}(\kappa, \kappa') + f_\kappa F_{\kappa'} S_{0vu}(-\kappa, -\kappa')]\, \delta_{vu}$
3	$-C_A + C_T$	$\mathbf{Y}^m_{1vu}(\hat{r}_s, \boldsymbol{\sigma}_s)[g_\kappa G_{\kappa'} S_{1vu}(\kappa, \kappa') - f_\kappa F_{\kappa'} S_{1vu}(-\kappa, -\kappa')]$
4	$-C_V/M$	$i[f_\kappa G_{\kappa'} S_{1vu}(-\kappa, \kappa') + g_\kappa F_{\kappa'} S_{1vu}(\kappa, -\kappa')]\, \mathbf{Y}^m_{1vu}(\hat{r}_s, \mathbf{p}_s)$
5	$-3^{1/2} C_V/(2M)$	$\{[(v+1)/(2v+3)]^{1/2}\mathbf{Y}^m_{0\,v+1\,u}(\hat{r}_s)\delta_{v+1\,u}D_+ - [v/(2v-1)]^{1/2}\mathbf{Y}^m_{0\,v-1\,u}(\hat{r}_s)\delta_{v-1\,u}D_-\}$ $\times\, [f_\kappa G_{\kappa'} S_{1vu}(-\kappa, \kappa') + g_\kappa F_{\kappa'} S_{1vu}(\kappa, -\kappa')]$
6	$-(\tfrac{3}{2})^{1/2} C_V(1 + \mu_p - \mu_n)/M$	$[(v+1)^{1/2} W(11uv, 1\, v+1)\mathbf{Y}^m_{1\,v+1\,u}(\hat{r}_s, \boldsymbol{\sigma}_s)D_+$ $- v^{1/2} W(11uv, 1\, v-1)\mathbf{Y}^m_{1\,v-1\,u}(\hat{r}_s, \boldsymbol{\sigma}_s)D_-]\,[f_\kappa G_{\kappa'} S_{1vu}(-\kappa, \kappa') + g_\kappa F_{\kappa'} S_{1vu}(\kappa, -\kappa')]$
7	C_A/M	$i\mathbf{Y}^m_{0vu}(\hat{r}_s)[f_\kappa G_{\kappa'} S_{0vu}(-\kappa, \kappa') + g_\kappa F_{\kappa'} S_{0vu}(\kappa, -\kappa')]\boldsymbol{\sigma}_s \cdot \mathbf{p}_s$
8	$(2C_T/E_o - C_A/M)/(2 \times 3^{1/2})$	$\{[(v+1)/(2v+1)]^{1/2}\mathbf{Y}^m_{1\,v+1\,u}(\hat{r}_s, \boldsymbol{\sigma}_s)D_+ - [v/(2v+1)]^{1/2}\mathbf{Y}^m_{1\,v-1\,u}(\hat{r}_s, \boldsymbol{\sigma}_s)D_-\}$ $\times\, [f_\kappa G_{\kappa'} S_{0vu}(-\kappa, \kappa') \pm g_\kappa F_{\kappa'} S_{0vu}(\kappa, -\kappa')]\, \delta_{vu}$ (rows 8 and 9)
9	$C_P/(2 \times 3^{1/2})M$	

[a] $\mathscr{M}_{vu}^{(i)}$ is defined by

$$\int U_{J_f}^{M_f\dagger} \sum_{s=1}^{A} \Xi_s \tau_-^{(s)} | U_{J_i}^{M_i}\, d\mathbf{r}_1\, d\mathbf{r}_2 \cdots d\mathbf{r}_A = (J_i u M_i m \mid J_f M_f)\mathscr{M}_{vu}^{(i)}$$

where Ξ_s is given in the table. In the last line, the upper sign refers to $i = 8$ and the lower to $i = 9$.

with

$$
\begin{aligned}
P = \sum_{\kappa}\sum_{u} \Bigg| & (C_V + C_S)[0\,l\,u]S_{0u}(\kappa)\delta_{lu} - (C_A - C_T)[1\,l\,u]S_{1u}(\kappa) \\
& - \frac{C_V}{M}[1\bar{l}\,u\,p]S'_{1u}(-\kappa) + 3^{1/2}\frac{C_V q}{2M}\left\{\left(\frac{\bar{l}+1}{2\bar{l}+3}\right)^{1/2}[0\,\bar{l}+1\,u\,+]\delta_{\bar{l}+1\,u}\right. \\
& \left. + \left(\frac{\bar{l}}{2\bar{l}-1}\right)^{1/2}[0\,\bar{l}-1\,u\,-]\delta_{\bar{l}-1\,u}\right\}S'_{1u}(-\kappa) \\
& + \left(\frac{3}{2}\right)^{1/2}\frac{C_V q}{M}(1+\mu_p-\mu_n)\{(\bar{l}+1)^{1/2}W(11u\bar{l},1\bar{l}+1)[1\,\bar{l}+1\,u\,+] \\
& + \bar{l}^{1/2}W(11u\bar{l},1\bar{l}-1)[1\,\bar{l}-1\,u\,-]\}S'_{1u}(-\kappa) \\
& + \frac{C_A}{M}[0\,\bar{l}\,u\,p]S'_{0u}(-\kappa)\delta_{\bar{l}u} \\
& + \left(\frac{1}{3}\right)^{1/2}\frac{q}{2M}\left(C_A - C_P - \frac{2C_T M}{E_o}\right)\left\{\left(\frac{\bar{l}+1}{2\bar{l}+1}\right)^{1/2}[1\,\bar{l}+1\,u\,+]\right. \\
& \left. + \left(\frac{\bar{l}}{2\bar{l}+1}\right)^{1/2}[1\,\bar{l}-1\,u\,-1]\right\}S'_{0u}(-\kappa)\delta_{\bar{l}u}\Bigg|^2 .
\end{aligned} \tag{8.91}
$$

Here we use the following abbreviations for the geometric factors S_{kvu}:

$$
\begin{aligned}
S_{ku}(\kappa) &= S_{kvu}(\kappa, -1) \\
&= [2(2j+1)]^{1/2}W(\tfrac{1}{2}1jl,\tfrac{1}{2}u)\delta_{lv} \qquad \text{for } k = 1, && (8.92)\\
&= \left(\frac{2j+1}{2l+1}\right)^{1/2}\delta_{lv} \qquad \text{for } k = 0, && (8.93)
\end{aligned}
$$

and

$$
S'_{ku}(-\kappa) = S_\kappa S_{ku}(-\kappa), \qquad S_\kappa = \text{sign of } \kappa. \tag{8.94}
$$

In order to present the property of each nuclear matrix element more clearly, we use the symbols $[k\,w\,u]$, $[k\,w\,u\,\pm]$, and $[k\,w\,u\,p]$ in place of $\mathcal{M}_{vu}$. That is,

$$
\begin{aligned}
&\int U_{J_f}^{M_f\dagger}\sum_{s=1}^{A}\Omega_s\tau_-^{(s)}U_{J_i}^{M_i}\,d\mathbf{r}_1\,d\mathbf{r}_2\cdots d\mathbf{r}_A \\
&\qquad = ([k\,w\,u],\ [k\,w\,u\,\pm],\ \text{or } [k\,w\,u\,p])(J_i u M_i m\,|\,J_f M_f).
\end{aligned} \tag{8.95}
$$

where Ω_s is as listed in Table 8.3.

In the expression for P, the summation over v disappears because the muon has zero orbital angular momentum $l' = 0$, and consequently $v = l$ or $v = \bar{l}$. Summation over u for a fixed κ need be performed only for the second, third, and fifth terms in Eq. (8.91). In fact, this summation consists of only

Table 8.3 Definition of Reduced Nuclear Matrix Elements in Eq. (8.95)[a]

Reduced Nuclear Matrix Element	Operator Ω_s
$[0\ w\ u]$	$G_{-1} j_w(qr_s) \mathbf{Y}^m_{0wu}(\hat{r}_s)\, \delta_{wu}$
$[1\ w\ u]$	$G_{-1} j_w(qr_s) \mathbf{Y}^m_{1wu}(\hat{r}_s, \sigma_s)$
$[0\ w\ u\ \pm]$	$G_{-1} [j_w(qr_s) \pm \alpha Z m'_\mu q^{-1} j_{w\mp 1}(qr_s)]\, \mathbf{Y}^m_{0wu}(\hat{r}_s)\, \delta_{wu}$
$[1\ w\ u\ \pm]$	$G_{-1} [j_w(qr_s) \pm \alpha Z m'_\mu q^{-1} j_{w\mp 1}(qr_s)]\, \mathbf{Y}^m_{1wu}(\hat{r}_s, \sigma_s)$
$[0\ w\ u\ p]$	$iG_{-1} j_w(qr_s) \mathbf{Y}^m_{0wu}(\hat{r}_s) \sigma_s \cdot \mathbf{p}_s\, \delta_{wu}$
$[1\ w\ u\ p]$	$iG_{-1} j_w(qr_s) \mathbf{Y}^m_{1wu}(\hat{r}_s, \mathbf{p}_s)$

[a] $\int U^{M_f \dagger}_{J_f} \sum_{s=1}^{A} \Omega_s \tau^{(s)}_{-} U^{M_i}_{J_i}\, d\mathbf{r}_1\, d\mathbf{r}_2 \cdots d\mathbf{r}_A = [k\ w\ u](J_i u M_i m | J_f M_f)$, etc., where Ω_s is as given in the table.

two terms because of the triangular condition among the arguments of the Racah coefficient in $S_{ku}(\kappa)$ and $S'_{ku}(-\kappa)$. Although Eq. (8.91) looks complicated, the selection rules for the spin and parity reduce the number of the nonvanishing reduced nuclear matrix elements to a manageable size. This will be done for different forbiddenness.

Reduced Nuclear Matrix Elements for Allowed and Forbidden Transitions

In accordance with the classification of the forbiddenness, we have the reduced nuclear matrix elements with rank 0 or 1 for the allowed transitions, those with rank 0, 1, or 2 for the first forbidden transitions, etc. These matrix elements can appear for certain restricted numbers of κ. In Table 8.4, we list

Table 8.4 Relations of κ and u with Forbiddenness[a]

Forbiddenness	κ	u
Allowed	-1	0, 1
	2	1
First	1	0, 1
	-2	1, 2
	3	2
nth ($n \geqslant 2$)	n	n
	$-(n+1)$	$n, n+1$
	$n+2$	$n+1$

[a] We have to sum over κ and u in Eq. (8.91).

the quantum number κ over which we have to sum, and the related rank u. With our approximation, Eq. (8.70), the reduced nuclear matrix elements can be specified by three quantum numbers k, w, u, and an additional symbol +

or −, or p. We list all possible reduced nuclear matrix elements according to forbiddenness of the muon capture reaction in Table 8.5. The definition of the forbiddenness is exactly the same as that in the theory of beta decay. It is given in the second and third rows of Table 8.5 as parity and spin changes

Table 8.5 Reduced Nuclear Matrix Elements in the Allowed and Forbidden Transitions of the Muon Capture Reaction, with the Approximation Eq. (8.89)[a]

Forbiddenness	Allowed	First		nth ($n \geqslant 2$)
Parity change	+	−		$(-)^n$
Spin change $\lvert J_f - J_i \rvert$	0, 1	0, 1, 2		n, $(n+1)$
Matrix elements	[0 0 0]		[0 1 1]	[0 n n]
			[0 1 1 +]	[0 n n +]
	[0 0 0 −]		[0 1 1 −]	[0 n n −]
	[1 2 1]	[1 1 0]	[1 3 2]	[1 $n+2$ $n+1$]
	[1 2 1 +]	[1 1 0 +]	[1 3 2 +]	[1 $n+2$ $n+1$ +]
			[1 1 1]	[1 n n]
			[1 1 1 +]	[1 n n +]
			[1 1 1 −]	[1 n n −]
	[1 0 1]		[1 1 2]	[1 n $n+1$]
	[1 0 1 −]		[1 1 2 −]	[1 n $n+1$ −]
	[0 1 1 p]	[0 0 0 p]	[0 2 2 p]	[0 $n+1$ $n+1$ p]
	[1 1 0 p]		[1 2 1 p]	[1 $n+1$ n p]
	[1 1 1 p]		[1 2 2 p]	[1 $n+1$ $n+1$ p]
			[1 0 1 p]	[1 $n-1$ n p]
No. of matrix elements	9	17		14

[a] They are represented by the three quantum numbers k, w, u, and additional symbols ± or p in each forbiddenness. Here u is the rank of the matrix element. For a capture from the K orbit [1 $u+1$ u −], [1 $u-1$ u +], and [0 0 0 +] do not appear, because of the selection rule for the angular momentum. In the second row, the plus and minus indicate no parity change and parity change, respectively.

between the initial and final nuclear states. On the other hand, the classification for the reduced nuclear matrix elements with respect to forbiddenness is somewhat different from that in the theory of beta decay, and the number of the nuclear matrix elements belonging to a particular forbiddenness is considerably larger. (As a matter of fact, the lower rank matrices which appear in the higher forbidden transitions of beta decay theory as the correction to the lower forbiddenness are classified as the lower forbidden matrices.) This number is 9, 17, and 14 for the allowed, first forbidden, and nth forbidden transitions, respectively. For the allowed transition, there are three Fermi matrices and six Gamow–Teller matrices. The selection rule for the nuclear spin of the matrix element is $J_i + J_f \geqslant u \geqslant |J_i - J_f|$, because the third number u represents the rank of the tensor. The parity change is given by the second number w, namely $(-)^w$ for $[k\ w\ u]$ (and $[k\ w\ u\ \pm]$), and $(-)^{w+1}$ for $[k\ w\ u\ p]$, because w represents the rank of the spherical harmonics with the argument $\hat{r}$ and an additional minus sign comes from the operator **p**, which operates on the nuclear wave function. Furthermore, w is equal to v for $[k\ w\ u]$ or $[k\ w\ u\ p]$ and $v \pm 1$ for $[k\ w\ u\ \pm]$. Here v is the orbital angular momentum of the neutrino. From these considerations, and from Eqs. (8.71), (8.73), (8.78)-(8.83), we can call the three numbers k, w, u the resultant spin, effective resultant orbital angular momentum, and resultant total angular momentum of the muon-neutrino system, respectively. (As the exceptional case, the resultant spin of the lepton system is one for $[0\ w\ u\ \pm]$, and zero for $[1\ w\ u\ \pm]$ in the $(C_A - C_P - 2C_T M E_o^{-1})$ term.)

The classification above of the reduced nuclear matrix elements is based on the tensorial rank of the nuclear matrix elements and the property of the neutrino wave function involved as an integrand, namely, the spherical Bessel function of the order w. The argument of the spherical Bessel function $j_w(qr)$ is, roughly speaking, 1 or 2 at the nuclear radius for light or medium nuclei. Consulting the values of $j_w(qr)$'s in the numerical tables of spherical Bessel functions [17], we conclude that

$$j_{w+2}(qr) \lesssim (10^{-1}\text{-}10^{-2}) \times j_w(qr). \tag{8.96}$$

This leads to

$$[k\ w+2\ u'] \lesssim 3 \times 10^{-2}[k\ w\ u], \qquad [k\ w+2\ u'\ \pm] \lesssim 3 \times 10^{-2}[k\ w\ u\ \pm],$$
$$[k\ w+2\ u'\ p] \lesssim 3 \times 10^{-2}[k\ w\ u\ p]. \tag{8.97}$$

Here u' may or may not be equal to u. Approximation (8.97) may not hold in heavy nuclei because of the large nuclear radius. Also, the atomic number of the nucleus becomes larger, and consequently, the neglect of F_{-1} is not justified. We have to compute the muon capture rate using Eq. (8.88) and Table 8.2. For the nuclear matrix elements involving p, $[k\ w\ u\ p]$, which represents the relativistic correction, we have a relation

$$\frac{1}{M}[k\ w\ u\ p] \approx \frac{p}{M}[k\ w\ u], \tag{8.98}$$

with $p/M \sim v/c \sim 10^{-1}$, where the p and v are the nucleon momentum and its velocity in the nucleus. Taking these relative magnitudes of nuclear matrix elements into account, we obtain the classification shown in Table 8.5. Another relation for the nuclear matrix elements is

$$[k\ w\ u\ -] \approx [k\ w\ u], \tag{8.99}$$

because the second term in the integrand is very small compared with the first (see Table 8.3). On the other hand, $[k\ w\ u\ +]$ is different from $[k\ w\ u]$

because two terms in the integrand have almost the same magnitude. Since we have no cross term between two nuclear matrix elements with different u's, the contribution of the $(n + 2)$th forbiddenness to the nth forbiddenness in our representation of Table 8.5 is very small, of the order of 0.1% or less. (There is, however, no contribution if $J_i = 0$.) This means that the daughter nucleus of the muon capture reaction goes mostly to states which have $|J_i - J_f| = 0, 1$ without parity change, or $|J_i - J_f| = 0, 1, 2$ with parity change in the low Z region.

Explicit Formulas for the Muon Capture Rate

Each forbiddenness has an explicit formula for the muon capture rate in [14, 18]. For $J_i \neq 0$, the results depend on the assumption that the hyperfine states of the mu mesic atom are statistically populated. This effect of the hyperfine splitting is omitted from [14, 18]. A method to include this effect is discussed in [19]. We have included the terms up to the order of p/M in the transformed interaction Hamiltonian, Eq. (8.59). The effect of the terms of $(p/M)^2$ is considered by Ohtsubo [20], Friar [21], and Oziewicz [22].

The muon capture rate including the small component F_{-1} of the muon wave function is obtained from Eq. (8.88) and Table 8.2. Classification of the reduced nuclear matrix elements and explicit formulas for the muon capture rate are obtained by Hirooka and Morita [23]. The muon wave functions G_{-1} and F_{-1} are not restricted to the solutions in the pure Coulomb field of the nucleus, and the effect of the finite-size nucleus can be properly taken into account by solving the Dirac equation numerically. As an example, we give an explicit formula for the pure Gamow–Teller transition later in Eq. (9.32). To save space, however, we omit the explicit formulas for the other transitions. We prefer to compute the muon capture rate directly from Eq. (8.88) and Table 8.2 by feeding the appropriate nuclear wave functions and muon wave functions numerically. For this purpose, Table 8.4 gives us the limits for summations over κ and u.

REFERENCES

1 J. Tiomno and J. A. Wheeler, *Rev. Mod. Phys.* **21**, 153 (1949).
2 J. Bernstein, T. D. Lee, C. N. Yang, and H. Primakoff, *Phys. Rev.* **111**, 313 (1958).
3 E. J. Bleser, E. W. Anderson, L. M. Lederman, S. L. Meyer, J. L. Rosen, J. E. Rothberg, and I-T. Wang, *Phys. Rev.* **132**, 2679 (1963).
4 S. Weinberg, *Phys. Rev.* **112**, 1375 (1958).
5 R. P. Feynman and M. Gell-Mann, *Phys. Rev.* **109**, 193 (1958).
6 M. L. Goldberger and S. B. Treiman, *Phys. Rev.* **111**, 355 (1958).
7 W. R. Wessel and P. Philipson, *Phys. Rev. Letters* **13**, 23 (1964); A. Halpern, *Phys. Rev. Letters* **13**, 660 (1964); P. K. Kabir, *Phys. Rev. Letters* **14**, 257 (1965); P. L. Becher and M. R. Patterson, *Bull. Amer. Phys. Soc.* (Ser. II) **12**, 595 (1967).

8 A. Fujii, M. Morita, and H. Ohtsubo, *Progr. Theoret. Phys. Suppl.* Extra No. 303 (1968).

9 A. Alberigi Quaranta, A. Bertin, G. Matone, F. Palmonari, A. Placci, P. Dalpiaz, G. Torelli, and E. Zavattini, *Phys. Letters* **25B**, 429 (1967).

10 T. E. O. Ericson, *Rept. Intern. Conf. High Energy Physics* (Dubna), Atomidat, Moscow 1964.

11 H. Primakoff, *Rev. Mod. Phys.* **31**, 802 (1959).

12 J. C. Sens, *Phys. Rev.* **113**, 679 (1959).

13 V. L. Telegdi, *Phys. Rev. Letters* **8**, 327 (1962).

14 M. Morita and A. Fujii, *Phys. Rev.* **118**, 606 (1960).

15 M. E. Rose and R. K. Osborn, *Phys. Rev.* **93**, 1315 (1954).

16 H. Bethe and M. E. Salpeter, *Quantum Mechanics of One- and Two-Electron Atoms*, p. 69. Academic Press, New York, 1959.

17 E. g., *see* Nat. Bur. of Standards, *Tables of Spherical Bessel Functions*, Columbia Univ. Press, New York, 1947.

18 M. Morita and R. Morita, *J. Phys. Soc. Japan* **19**, 1759 (1964).

19 V. V. Balashov and R. A. Eramzyan, *Atomic Energy Rev.* **5**, 3 (1967).

20 H. Ohtsubo, *Phys. Letters* **22**, 480 (1966).

21 J. L. Friar, *Nucl. Phys.* **87**, 407 (1966).

22 Z. Oziewicz, *Rept. Inst. Theoret. Phys. Wroclaw* No. 208 (1970).

23 H. Hirooka and M. Morita, unpublished (1967); H. Hirooka, Ph.D. thesis, Osaka Univ. (1969).

Chapter Nine

NUCLEAR STRUCTURE PROBLEMS STUDIED FROM MUON CAPTURE REACTIONS

9.1 MUON CAPTURE IN C^{12} (EXPERIMENTS AND SIMPLE THEORIES)

As we saw in Chapter 8, understanding muon capture reactions involves two main problems. One concerns the law of weak interactions itself, and the other is the problem of nuclear structure. Since the former is relatively well established, we now concentrate on nuclear structure problems. We adopt the transition matrix element of the universal $V - A$ types, Eqs. (8.25) and (8.26), which can beautifully explain experimental data on muon capture rates in hydrogen and on the total muon capture rates in various complex nuclei, too. We have also obtained the explicit formulas (8.90) and (8.91) for the partial muon capture rate in complex nuclei, which are dependent on several nuclear matrix elements defined in Table 8.3. Throughout this chapter, we assume G-parity conservation, and we set the coupling constants C_S and C_T to be zero.

A typical example of muon capture is the transition in which C^{12} in the ground state absorbs a muon, yielding the ground state of B^{12}

$$\mu^- + C^{12} \rightarrow B^{12} + \nu.$$

The spins and parities for these nuclear levels are $0^+ \rightarrow 1^+$. Measurements of this partial transition rate have been repeatedly performed (see Table 9.1), and the latest datum of the Carnegie group [1] is

$$W = \left(6.75 \begin{array}{l} +0.30 \\ -0.75 \end{array}\right) \times 10^3 \text{ sec}^{-1}.$$

Theoretical work on this problem was first performed by Fujii and Primakoff [2] using a single particle j-j coupling shell model with Primakoff's original theory of muon capture [3]. A similar calculation was also made by Wolfenstein [4]. Later, the theory of muon capture was formulated in a way similar to the theory of allowed and forbidden transitions in beta decay [5]. In this theory, which was introduced in Chapter 8, the relativistic terms are correctly taken into account, while the matrix elements $[S\ L\ J\ p]$ are not considered in the former theories. Although these relativistic terms are only corrections to the main terms in the allowed or unique forbidden transitions, they make an essential contribution to the capture rate in the case of the nonunique transitions. Morita and Fujii [5] incorporated these terms and calculated the partial muon capture rate in C^{12} (which is classified as an allowed transition).

Table 9.1 Experimental Data on Muon Capture Rate in C^{12}

Year	Muon Capture Rate (10^3 sec^{-1})	Reference
1953	5.9 ± 1.5	*a*
1958	9.18 ± 0.5	*b*
1959	9.05 ± 0.95	*c*
1959	6.8 ± 1.5	*d*
1960	5.8 ± 1.3	*e*
1960	6.8 ± 1.1	*f*
1961	6.31 ± 0.24	*g*
1963	6.6 ± 0.9	*h*
1964	$6.75^{+0.30}_{-0.75}$	*i*

[a] T. N. K. Godfrey, *Phys. Rev.* 92, 512 (1953).
[b] J. O. Burgman, J. Fischer, B. Leontic, A. Lundby, R. Meunier, J. P. Stroot, and J. D. Teja, *Phys. Rev. Letters* 1, 469 (1958).
[c] H. V. Argo, F. B. Harrison, H. W. Kruse, and A. D. McGuire, *Phys. Rev.* 114, 626 (1959).
[d] W. Love, S. Marder, I. Nadelhalft, R. Siegel, and A. E. Taylor, *Bull. Amer. Phys. Soc.* 4, 81 (1959).
[e] B. L. Bloch, Thesis, Carnegie Inst. of Technol. Rept. NYO-9280 (1960).
[f] J. G. Fetkovich, T. H. Fields, and R. L. McIlwain, *Phys. Rev.* 118, 319 (1960).
[g] E. J. Maier, B. L. Bloch, R. M. Edelstein, and R. T. Siegel, *Phys. Rev. Letters* 6, 417 (1961).
[h] G. T. Reynolds, D. B. Scarl, R. A. Swanson, J. R. Waters, and R. A. Zdanis, *Phys. Rev.* 129, 1790 (1963).
[i] E. J. Maier, R. M. Edelstein, and R. T. Siegel, *Phys. Rev.* 133, B663 (1964).

They found that the relativistic corrections change the capture rate by about 10%. Their theory was again based on the single particle *j-j* coupling shell model. In these theories, the calculated muon capture rate and the beta-decay rate of the ground state of B^{12} are about five times larger than the experimental data. Therefore, a normalization has been adopted so that the calculated beta-decay rate coincides with the experimental one. Flamand and Ford considered muon capture in a slightly different manner [6]. They evaluated nuclear matrix elements in the intermediate coupling shell model, where the strength parameter a/K was adjusted so as to give the correct beta-decay rate.

They included the effect of the nuclear finite size on the muon wave functions in the K orbit. The effect was about −6%. In another approach Foldy and Walecka [7] wrote the Gamow–Teller matrix [1 0 1] in terms of the $ft_{1/2}$ value of the ground state of B^{12} and of the transverse magnetic dipole operator in inelastic electron scattering [8] leading to the 15.1-MeV 1^+ state of C^{12}. This level of C^{12} is an isobaric analog of the ground state of B^{12}. The contribution of the smaller nuclear matrix elements was evaluated in the *j-j* and *L-S* coupling shell model. The muon capture rate in their theory is therefore nuclear model independent.

An entirely new approach to calculating the muon capture rate was developed by Kim and Primakoff [9], who treated the nuclei of B^{12} and C^{12} as elementary particles. This theory is based on the hypothesis of the conserved vector hadron weak current and the hypothesis of the partially conserved axial vector hadron weak current. The theory does not use the impulse approximation, which was adopted by all previous authors. Instead, the momentum transfer dependence of the weak nuclear form factor is found from an analysis of suitable empirical nuclear structure data.

In the rest of this section, we demonstrate the theoretical calculations of the partial muon capture rate in C^{12}, using the single particle *j-j* coupling shell model. Although we adopt a simple model, the calculated capture rate is still strongly dependent on the detailed behavior of the nuclear wave functions. This indicates that it may be possible to study nuclear structure from the partial muon capture rate. On the other hand, the nuclear properties are smeared out in the total muon capture rate (see section 8.3).

The muon capture rate is given by Eqs. (8.90) and (8.91), if we neglect the small component F_{-1} of radial wave functions, as is indicated in Eq. (8.89). In our case, the transition is $0^+ \to 1^+$, and the relevant matrix elements are six: [1 0 1]; [1 0 1 −]; [1 2 1]; [1 2 1 +]; [0 1 1 p]; [1 1 1 p]: From Eq. (8.95), these matrix elements are given by

$$([1\ 0\ 1], \text{etc.})(0\ 1\ 0\ M|1\ M) = \int U_{1,1}^{M,-1\dagger}(B^{12}) \sum_{s=1}^{8} \Omega_s \tau_-^{(s)} U_{0,0}^{0,0}(C^{12})\, d\mathbf{r}_1\, d\mathbf{r}_2 \cdots d\mathbf{r}_8. \tag{9.1}$$

Here Ω_s is as given in Table 8.3. We assume the *j-j* coupling shell model and harmonic oscillator wave functions. The four nucleons in the 1*s* shell are assumed to be inert during the ground state to ground state transition. The nuclear wave function is affixed with four labels, $U_{J,T}^{M,T_z}$. Here the two subscripts J and T are the nuclear spin and isospin, while the two superscripts M and T_z are their magnetic quantum numbers, respectively. The wave function for the ground state of C^{12}, which has $J = T = 0$, is given by

$$U_{0,0}^{0,0}(C^{12}) = (\tfrac{1}{2})^{1/2} \sum_m \sum_{t_z=\pm\frac{1}{2}} (\tfrac{3}{2}\tfrac{3}{2} m\ -m\,|\,0\ 0) \times 2t_z \psi_{\frac{3}{2},\frac{1}{2}}^{m,t_z}(2, 3, \ldots, 8)\psi_{\frac{3}{2}}^{-m}(1)\phi_{\frac{1}{2}}^{-t_z}(1). \tag{9.2}$$

Here $\psi(1)$ and $\phi(1)$ are the spin-spatial part and isospin part of the wave function for the particle 1. $\psi(2, 3, \ldots, 8)$ is the wave function for the system of the particle 2 to 8. The wave function for the ground state of B^{12}, which has $J = T = 1$, is given by

$$U_{1,1}^{M,-1}(B^{12}) = (\tfrac{1}{8})^{1/2} \sum_P \sum_m \epsilon_P(\tfrac{3}{2}\tfrac{1}{2}\; m\; M - m\,|\,1\; M)$$
$$\times\; \psi_{\frac{3}{2},\frac{1}{2}}^{m,-\frac{1}{2}}(2, 3, \ldots, 8)\,\psi_{\frac{1}{2}}^{M-m}(1)\,\phi_{\frac{1}{2}}^{-\frac{1}{2}}(1). \tag{9.3}$$

Here P means the replacement of a member in $(2, 3, \ldots, 8)$ by 1. ϵ_P has a plus sign (or minus sign) if this permutation is even (or odd). For more detail on constructing nuclear wave functions, the reader can refer to textbooks such as [10, 11].

Now, let us find the reduced nuclear matrix elements $[1\; l\; 1]$, $l = 0$ and 2, etc., explicitly. Inserting Eqs. (9.2) and (9.3) into Eq. (9.1), and integrating over the $3 \times (8 - 1)$ dimensions, we have the result

$$([1\; l\; 1], \text{etc.}) = \sum_m (-)^{\frac{1}{2}-m}(\tfrac{3}{2}\tfrac{1}{2}\; m\; M - m\,|\,1\; M) \int \psi_{\frac{1}{2}}^{M-m\dagger}\,\Omega\,\psi_{\frac{3}{2}}^{m}\,d\mathbf{r}, \tag{9.4}$$

with the single particle wave function

$$\psi_j^m = R \sum_\mu (1\tfrac{1}{2}\; m - \mu\; \mu\,|\,jm)\, Y_{1\,m-\mu}(\hat{r})\chi_\mu \tag{9.5}$$

and the muon radial wave function

$$G_{-1} = 2(\alpha Z m'_\mu)^{3/2} \exp(-\alpha Z m'_\mu r). \tag{9.6}$$

Here χ_μ represents the spin wave function with its magnetic quantum number μ, while the radial wave function R is the oscillator wave function for the $1p$ state,

$$R = Nr \exp(-r^2/2b^2) \tag{9.7}$$

with the normalization constant $N^2 = \frac{8}{3}\pi^{-1/2}b^{-5}$. Equation (9.4) represents the matrix element for the single particle transition from the orbit $1p_{3/2}$ to $1p_{1/2}$.

The reduced nuclear matrix elements become more explicit if we insert Eq. (9.5) and the operator Ω given in Table 8.3 into Eq. (9.4). For example, we have

$$[1\;1\;1p] = \sum_m \sum_\mu \sum_{\mu'} \sum_{M'} (-)^{\frac{1}{2}-m}(\tfrac{3}{2}\tfrac{1}{2}\; m\; M - m\,|\,1\; M)$$
$$\times\; (1\;1\; M'\; M - M'\,|\,1\; M)(1\tfrac{1}{2}\; M - m - \mu\; \mu\,|\,\tfrac{1}{2}\; M - m)$$
$$\times\; (1\tfrac{1}{2}\; -m - \mu'\; \mu'\,|\,\tfrac{3}{2} - m)$$
$$\times \int [\chi_\mu\, Y_{1\,M-m-\mu}(\hat{r})R]^\dagger G_{-1}\, j_1(qr)\, Y_{1\,M-M'}(\hat{r})$$
$$\times\; [\mathbf{Y}_{1\,M'}(\boldsymbol{\nabla})\chi_{\mu'} Y_{1\,-m-\mu'}(\hat{r})R]\, r^2\, dr\, d\Omega. \tag{9.8}$$

After summation over all magnetic quantum numbers except M, Eq. (9.8) becomes

$$[1\ 1\ 1\ p] = -\sum_l \pi^{-1}\cdot 2^{-1/2}\cdot 3(2l+1)(l\ 1\ 0\ 0\,|\,1\ 0) \times W(1\ 1\ 1\ 1, l\ 1)W(1\ \tfrac{3}{2}\ 1\ \tfrac{1}{2}, \tfrac{1}{2}\ 1) \times \int RG_{-1}j_1(qr)(D_lR)r^2\,dr, \tag{9.9}$$

with

$$D_2 = (\tfrac{2}{5})^{1/2}\left(\frac{d}{dr}-\frac{1}{r}\right), \qquad D_1 = 0, \qquad D_0 = -\left(\frac{d}{dr}+\frac{2}{r}\right), \tag{9.10}$$

Similarly, we have

$$[0\ 1\ 1\ p] = \sum_l \pi^{-1}\cdot 3^{1/2}(2l+1)(l\ 1\ 0\ 0\,|\,1\ 0)[W(\tfrac{3}{2}\ 1\ \tfrac{1}{2}\ 1, \tfrac{1}{2}\ l)]^2 \times \int RG_{-1}j_1(qr)(D_lR)r^2\,dr, \tag{9.11}$$

and

$$[1\ l\ 1] = -\pi^{-1}\cdot 3[3(2l+1)]^{1/2}(l\ 1\ 0\ 0\,|\,1\ 0)\begin{pmatrix} 1 & l & 1 \\ \tfrac{1}{2} & 1 & \tfrac{3}{2} \\ \tfrac{1}{2} & 1 & \tfrac{1}{2} \end{pmatrix} \times \int R^2G_{-1}j_l(qr)r^2\,dr. \tag{9.12}$$

Integrating over the radial coordinate, we have the final results

$$Z_o[1\ 0\ 1] = -(\tfrac{1}{3})^{1/2}\pi^{-1}[(1-\tfrac{1}{6}\lambda)\exp(-\lambda/4) - \zeta(1-\tfrac{1}{2}\lambda+\tfrac{1}{10}\lambda^2)] = -0.138, \tag{9.13}$$

$$Z_o[1\ 0\ 1\ -] = -(\tfrac{1}{3})^{1/2}\pi^{-1}[(1-\tfrac{1}{6}\lambda)\exp(-\lambda/4) - \zeta(\tfrac{4}{3}-\tfrac{3}{5}\lambda+\tfrac{4}{35}\lambda^2)] = -0.135, \tag{9.14}$$

$$Z_o[1\ 2\ 1] = (\tfrac{1}{6})^{1/2}(2\pi)^{-1}[\tfrac{1}{6}\lambda\exp(-\lambda/4) - \zeta(\tfrac{1}{5}\lambda-\tfrac{2}{35}\lambda^2)] = 0.00482, \tag{9.15}$$

$$Z_o[1\ 2\ 1\ +] = (\tfrac{1}{6})^{1/2}(2\pi)^{-1}[\tfrac{1}{6}\lambda\exp(-\lambda/4) + \zeta(\tfrac{1}{3}-\tfrac{3}{10}\lambda+\tfrac{1}{14}\lambda^2)] = 0.00584, \tag{9.16}$$

$$Z_o[0\ 1\ 1\ p] = (\tfrac{1}{3})^{1/2}\frac{q}{2\pi}[\tfrac{5}{6}(1-\tfrac{1}{10}\lambda)\exp(-\lambda/4) - \zeta(1-\tfrac{2}{5}\lambda+\tfrac{1}{14}\lambda^2)] = 0.00575M, \tag{9.17}$$

$$Z_o[1\ 1\ 1\ p] = (\tfrac{1}{2})^{1/2}\frac{q}{2\pi}[\tfrac{1}{3}\exp(-\lambda/4) - \zeta(\tfrac{1}{3}-\tfrac{1}{10}\lambda+\tfrac{1}{70}\lambda^2)] = 0.00301M, \tag{9.18}$$

with

$$Z_o^{-1} = 2(\alpha Z m'_\mu)^{3/2} \qquad \text{and} \qquad Z = 6, \tag{9.19}$$

$$\lambda = \tfrac{2}{5}\langle r^2\rangle q^2 = q^2b^2, \tag{9.20}$$

and

$$\zeta = \tfrac{8}{3}\,\alpha Z\,\frac{m'_\mu}{q}\left(\frac{\lambda}{\pi}\right)^{1/2}. \tag{9.21}$$

Here $\langle r^2\rangle^{1/2}$ is the root mean square radius and M is the nucleon mass. For numerical evaluation, we have used the following values

$$\langle r^2\rangle^{1/2} = 2.52 \text{ fm} \qquad \text{and} \qquad q = 91.4 \text{ MeV}/c. \tag{9.22}$$

Since we have the numerical results for the reduced nuclear matrix elements, we can now evaluate the partial transition rate, Eqs. (8.90) and (8.91), using Eqs. (9.13)-(9.18) and the coupling constants

$$C_V = 0.972C_V^\beta, \qquad C_A = 0.999C_A^\beta, \qquad \text{and} \qquad C_A^\beta = -1.21C_V^\beta. \tag{9.23}$$

Here the muon coupling constants C_i are assumed to be a little smaller than those for beta decay C_i^β due to the difference in the momentum transfer. The calculated capture rate in this way is about five times larger than the experimental data. This is because the assumed nuclear wave functions, Eqs. (9.2) and (9.3), are not accurate enough. The situation is the same for the calculation of the beta-decay rate in $B^{12} \to C^{12}$. Now, let us assume that the reduction factors which should appear in an accurate calculation are identical for beta decay and muon capture. Then we can derive the theoretical value of the muon capture rate from the calculated one by multiplying by the normalization constant

$$\frac{W_\beta^{\text{exp}}}{W_\beta^{\text{calc}}},$$

where W_β^{exp} and W_β^{calc} are the beta-decay rates found in experiments and by calculation, respectively. That is, the theoretical value W is defined as

$$W = W^{\text{calc}} \times (W_\beta^{\text{exp}}/W_\beta^{\text{calc}}), \tag{9.24}$$

which should be compared with the experimental capture rate.

We have already studied the beta-decay rate W_β in Chapter 2; it is given by

$$W_\beta = \frac{2J_i + 1}{2J_f + 1}\int_1^{E_o} P_\beta F(Z, E)Ep(E_o - E)^2\, dE \tag{9.25}$$

for the transition from the state J_f to the state J_i. Here P_β has the same meaning as Eq. (8.91), except for $q = m'_\mu = 0$. For the allowed transition, we have

$$W_\beta = (2\pi^3)^{-1} f(Z, E_o)[C_V^2(\textstyle\int 1)^2 + C_A^2(\int \sigma)^2],$$

or

$$W_\beta = \frac{8}{\pi}\,\frac{2J_i + 1}{2J_f + 1}\, f(Z, E_o)(C_V^2[0\;0\;0]_\beta^2 + \tfrac{1}{3}C_A^2[1\;0\;1]_\beta^2), \tag{9.26}$$

where $[k\ 0\ k]_\beta^2$ should be equal to

$$\frac{2J_i+1}{2J_f+1}[k\ 0\ k]_\beta^2 = (Z_o^{-2}[k\ 0\ k]^2 \qquad \text{with} \quad q = m'_\mu = 0) \tag{9.27}$$

and $f(Z, E_o)$ is the integrated Fermi function, as before. Finally, the muon capture rate which is normalized with the beta-decay rate is given by

$$W = W_\beta \frac{\pi}{8}\left(\frac{2J_f+1}{2J_i+1}\right)^2 \frac{Pq^2[1-q(m_\mu+AM)^{-1}]}{f(Z,E_o)(C_V^2[0\ 0\ 0]_\beta^2 + \frac{1}{3}C_A^2[1\ 0\ 1]_\beta^2)}, \tag{9.28}$$

where P is as given in Eq. (8.91).

The final result for the muon capture rate in C^{12} is expressed as a function of the induced pseudoscalar coupling constant C_P

$$W = 6.28 + 0.00481[(C_P/C_A) - 21.2]^2 \tag{9.29}$$

in units of 10^3 sec^{-1}. Here we have adopted the experimental data on the beta decay $B^{12} \to C^{12}$,

$$W_\beta = 33.15 \text{ sec}^{-1} \qquad \text{and} \qquad f(Z, E_o) = 5.625 \times 10^5. \tag{9.30}$$

The induced effect due to the assumption of conserved vector current is properly taken into account. Putting

$$C_P/C_A = 8,$$

we have [5]

$$W = 7.12 \times 10^3 \text{ sec}^{-1}, \tag{9.31}$$

which is in fairly good agreement with the experimental data in Table 9.1. We may obtain better agreement by adjusting the coupling constant ratio C_P/C_A.

From the viewpoint of studying nuclear structure, the foregoing analysis is insufficient, since the normalization constant $W_\beta^{\text{exp}}/W_\beta^{\text{calc}}$ is adopted by means of the experimental beta-decay rate. On the other hand, this factor is fully dependent on the detail of the nuclear models, e.g., on the configurations in the nuclear shell model.

9.2 MUON CAPTURE IN C^{12} (NUCLEAR WAVE FUNCTIONS OF B^{12} AND C^{12})

In this section, we try to give a correct muon capture rate in C^{12} without introducing the normalization constant $W_\beta^{\text{exp}}/W_\beta^{\text{calc}}$. This can be done by using general p-shell wave functions with configuration mixing parameters obtained by the dynamic properties of the $A = 12$ system, as described by Cohen and Kurath [12]. As a result, the muon captures take place in either $p_{1/2}$ or $p_{3/2}$

protons, by introducing both $p_{1/2}$ and $p_{3/2}$ neutrons as the final states. As a matter of fact, the Cohen–Kurath wave functions with their original configuration mixing parameters reproduce 88% of the muon capture rate. We can adjust these parameters so as to find an excellent fit to the experimental data on the muon capture rate without producing any appreciable change of the level scheme.

Exact Formula for Muon Capture Rate

To perform this investigation, we use an explicit formula, Eq. (9.32), for the muon capture rate with well-defined nuclear matrix elements including the large and small components of the muon wave functions. This formula can, in principle, be derived from Eqs. (8.47), (8.72)-(8.83), and Table 8.2. Equation (9.32) gives the capture rate for the case of the $0^+ \to 1^+$ transition which corresponds to the muon capture in C^{12} [13].

$$
\begin{aligned}
W = {} & \left[1 - \frac{q}{m_\mu + AM}\right] q^2 \left(\frac{C_A}{C_A^\beta}\right)^2 \left(\frac{C_A^\beta}{C_V^\beta}\right)^2 G^2 \frac{mc^2}{\hbar} \\
& \times \Bigg(\Bigg\{ -2^{1/2}[1\ 0\ 1, 1] + (\tfrac{1}{3})^{1/2} \frac{2C_V}{C_A M} [1\ 1\ 1p, 1] \\
& + \mu \frac{qC_V}{3C_A M} (2^{1/2}[1\ 0\ 1, 2] - [1\ 2\ 1, 2]) - 2^{1/2} \frac{1}{M} [0\ 1\ 1p, 1] \\
& - (\tfrac{1}{2})^{1/2} \frac{q}{3M} ([1\ 0\ 1, 3] + 2^{1/2}[1\ 2\ 1, 3]) \\
& + (\tfrac{1}{2})^{1/2} \frac{qC_P}{3C_A M} ([1\ 0\ 1, 2] + 2^{1/2}[1\ 2\ 1, 2]) \Bigg\}^2 \\
& + \Bigg\{ -2^{1/2}[1\ 2\ 1, 1] - (\tfrac{2}{3})^{1/2} \frac{C_V}{C_A M} [1\ 1\ 1p, 2] \\
& - 2^{1/2} \mu \frac{qC_V}{6C_A M} (2^{1/2}[1\ 0\ 1, 4] - [1\ 2\ 1, 4]) - \frac{2}{M} [0\ 1\ 1p, 2] \\
& - \frac{q}{3M} ([1\ 0\ 1, 5] + 2^{1/2}[1\ 2\ 1, 5]) \\
& + \frac{qC_P}{3C_A M} ([1\ 0\ 1, 4] + 2^{1/2}[1\ 2\ 1, 4]) \Bigg\}^2 \Bigg),
\end{aligned}
\tag{9.32}
$$

with

$$q = (m_\mu - E_o)\left[1 - \frac{1}{2}\,\frac{m_\mu}{m_\mu + AM}\right], \tag{9.33}$$

$$\mu = 1 + \mu_p - \mu_n = 4.706. \tag{9.34}$$

W is given in units of $\sec^{-1}$ by multiplying a factor

$$\frac{mc^2}{\hbar} = 0.7763 \times 10^{21}\ \sec^{-1}. \tag{9.35}$$

The following coupling constants are now adopted.

$$C_V = 0.972 C_V^\beta = 0.972 G(\hbar^3/m^2 c), \tag{9.36}$$

$$G = 2.966(1 \pm 0.17\%) \times 10^{-12}, \tag{9.37}$$

$$C_A = 0.999 C_A^\beta, \tag{9.38}$$

$$C_A^\beta / C_V^\beta = -1.24 \pm 0.01. \tag{9.39}$$

The nuclear matrix elements $[S\ L\ J,\ n]$ or $[S\ L\ J\ p,\ n]$ are defined similarly to $[S\ L\ J]$ or $[S\ L\ J\ p]$ in Chapter 8. Here the last number n in the brackets is just a serial number, and it specifies different matrices whose interaction operators are slightly different from each other. The matrix elements are explicitly given in Table 9.2, with the operators Ω as in the definition (8.95).

Table 9.2 Operators Ω in the Definition (8.95) of Reduced Nuclear Matrix Elements Which Appear in Eq. (9.32)[a]

	Ω
[1 0 1, 1]	$\{j_0 g + \frac{1}{3} j_1 f\}\mathbf{Y}^m_{101}(\hat{r}, \sigma)$
[1 0 1, 2] [1 0 1, 3]	$\left\{j_0 g + j_1 g' \pm \left[j_1 - \frac{2}{\rho} j_0\right] f \mp j_0 f'\right\}\mathbf{Y}^m_{101}(\hat{r}, \sigma)$
[1 0 1, 4] [1 0 1, 5]	$\left\{j_0 g + j_1 g' \pm \left[j_1 - \frac{1}{\rho} j_2\right] f \pm j_2 f'\right\}\mathbf{Y}^m_{101}(\hat{r}, \sigma)$
[0 1 1 p, 1]	$i\{j_1 g + j_0 f\}\mathbf{Y}^m_{011}(\hat{r})\,\sigma\cdot\mathbf{p}$
[0 1 1 p, 2]	$i\{j_1 g - j_2 f\}\mathbf{Y}^m_{011}(\hat{r})\,\sigma\cdot\mathbf{p}$
[1 1 1 p, 1]	$i\{j_1 g - j_0 f\}\mathbf{Y}^m_{111}(\hat{r}, \mathbf{p})$
[1 1 1 p, 2]	$i\{j_1 g + j_2 f\}\mathbf{Y}^m_{111}(\hat{r}, \mathbf{p})$
[1 2 1, 1]	$\{j_2 g + \frac{1}{3} j_1 f\}\mathbf{Y}^m_{121}(\hat{r}, \sigma)$
[1 2 1, 2] [1 2 1, 3]	$\left\{j_2 g - j_1 g' \mp \left[j_1 + \frac{1}{\rho} j_0\right] f \pm j_0 f'\right\}\mathbf{Y}^m_{121}(\hat{r}, \sigma)$
[1 2 1, 4] [1 2 1, 5]	$\left\{j_2 g - j_1 g' \mp \left[j_1 - \frac{4}{\rho} j_2\right] f \mp j_2 f'\right\}\mathbf{Y}^m_{121}(\hat{r}, \sigma)$

[a] The subscripts s, which denote nucleon number, are all dropped for simplicity. The spherical Bessel functions $j_L(qr)$ are abbreviated as j_L. The upper and lower signs in an expression refer to the upper and lower numbers in the brackets of the matrix, respectively. g and f are large and small components of the muon radial wave functions. The prime means the derivative with respect to $\rho = qr$.

Exact Muon Wave Functions and Oscillator Strength Parameter b

The radial wave function of the p-shell nucleons is strongly dependent on the magnitude of the oscillator strength parameter b in Eq. (9.7). This is usually given by the electron scattering data from the extended charge distribution of the nucleus. To be consistent with this nuclear charge distribution, the muon wave functions are no longer of the type of the point charge in Eq. (8.89). In the following, we use the symbols f and g for the small and large components of the radial wave functions F_{-1} and G_{-1}. They are obtained as the solutions to the Dirac equation with a Coulomb potential of the finite size nucleus, for which the shape of the nuclear charge distribution $\rho(r)$ is assumed to be harmonic oscillator type,

$$\rho(r) = \frac{1}{3\pi^{3/2}b^3}\left[1 + \frac{4}{3}\left(\frac{r}{b}\right)^2\right]\exp\left[-\left(\frac{r}{b}\right)^2\right] \tag{9.40}$$

with

$$\langle r^2\rangle^{1/2} = 2.40 \text{ fm (or equivalently } b = 1.63 \text{ fm).} \tag{9.41}$$

This type of radial dependence has been adopted by Crannell in conformity with electron scattering data [14]. It is also confirmed that the nuclear charge distribution in Eq. (9.40) is consistent with the energy of the muonic x ray ($2p_{3/2} \rightarrow 1s_{1/2}$). In fact, Narumi and Konishi made a calculation and obtained 75.28 keV for it [15] in comparison with the experimental value 75.25 ± 0.15 keV [16].

It is interesting to note that the elastic electron scattering and the inelastic scattering due to excitations of the 4.43- and 9.64-MeV states of C^{12} have been studied by Crannell experimentally[14], and all data are in agreement with the rms radius

$$\langle r^2\rangle^{1/2} = 2.40 \pm 0.02 \text{ fm.} \tag{9.42}$$

That is, the data are in favor of the oscillator strength parameter $b = 1.63$ fm. On the other hand, in the analysis [17] of the inelastic electron scattering data due to the excitation of the 15.1-MeV state of C^{12} (which is the isobaric analog state of the ground state of B^{12}), the best fit is obtained with $b = 1.9$ fm. This is, however, based on an analysis with the single particle shell model. It was shown by Hiro-Oka [18] that a good fit is also obtained with $b = 1.64$ fm if the nuclear model given in the next subsection is used.

The calculated values of f and g by the Runge–Kutta method are shown in Fig. 9.1 as functions of qr [13]. Calculations of f and g under the assumption of a Fermi distribution of the nuclear charge, which gives $\langle r^2\rangle^{1/2} = 2.40$ fm, do not show any signficant change.

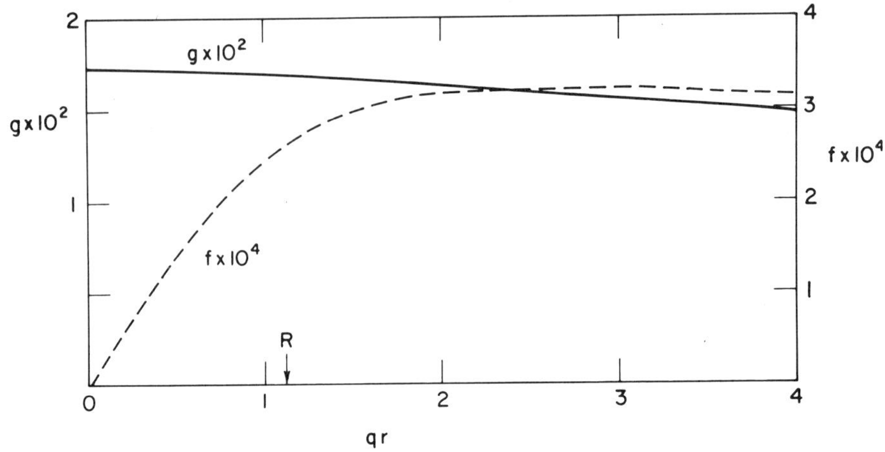

Fig. 9.1 Radial wave functions of the muon in the K orbit of C^{12}. The large and small components g and f are obtained as the solutions of the Dirac equation with a Coulomb potential of the finite-size nucleus for which the shape of the nuclear charge is assumed as in Eq. (9.40). $g \times 10^2$ and $f \times 10^4$ are plotted against qr, where the neutrino energy is $q = 178.59$, in the left-hand and right-hand ordinates, respectively. The arrow R corresponds to the nuclear surface [13]. (The minus signs are dropped in the right-hand ordinate.)

Reduced Nuclear Matrix Elements and Nuclear Wave Functions of C^{12} and B^{12}

For the ground state wave functions of C^{12} and B^{12}, we adopt the most general p-shell configurations studied by Cohen and Kurath [12]. In this framework, there are five independent states with $J = T = 0$ which are made up of eight p-shell nucleons. Similarly there are eight states with $J = T = 1$. We express each state by the symbol

$$|(p_{1/2})^m (p_{3/2})^n, T, J\rangle,$$

which represents the state of J and T constructed with m nucleons in the $p_{1/2}$ shell and n nucleons in the $p_{3/2}$ shell. If there are several states for the same set of quantum numbers m, n, J, and T, they are labeled with α, β, or γ. For example, there are three states with $J = T = 1$ which are made up of $(p_{1/2})^3(p_{3/2})^5$. The wave functions of C^{12} and B^{12} are now expressed by the following five and eight components, respectively.

$$\begin{aligned}\psi(C^{12}, T=0, J=0) = {} & C_1|(p_{1/2})^4(p_{3/2})^4, T=0, J=0\rangle \\ & + C_2|(p_{1/2})^3(p_{3/2})^5, T=0, J=0\rangle \\ & + C_3|(p_{1/2})^2(p_{3/2})^6, \alpha, T=0, J=0\rangle \\ & + C_4|(p_{1/2})^2(p_{3/2})^6, \beta, T=0, J=0\rangle \\ & + C_5|(p_{3/2})^8, T=0, J=0\rangle, \end{aligned} \tag{9.43}$$

$$\begin{aligned}\psi(\mathrm{B}^{12}, T=1, J=1) &= B_1|(p_{1/2})^4(p_{3/2})^4, T=1, J=1\rangle \\ &+ B_2|(p_{1/2})^3(p_{3/2})^5, \alpha, T=1, J=1\rangle \\ &+ B_3|(p_{1/2})^3(p_{3/2})^5, \beta, T=1, J=1\rangle \\ &+ B_4|(p_{1/2})^3(p_{3/2})^5, \gamma, T=1, J=1\rangle \\ &+ B_5|(p_{1/2})^2(p_{3/2})^6, \alpha, T=1, J=1\rangle \\ &+ B_6|(p_{1/2})^2(p_{3/2})^6, \beta, T=1, J=1\rangle \\ &+ B_7|(p_{1/2})^2(p_{3/2})^6, \gamma, T=1, J=1\rangle \\ &+ B_8|(p_{1/2})^1(p_{3/2})^7, T=1, J=1\rangle. \end{aligned} \tag{9.44}$$

Here all the components are ortho-normal and the coefficients C_i ($i = 1, 2, \cdots, 5$) and B_i ($i = 1, 2, \cdots, 8$), are determined so as to fit the nuclear level schemes in the p-shell nuclei, by the method of the Slater determinant. This way of finding the coefficients C_i and B_i has been programmed, e.g., by Soga for computer calculation [13].

Using the wave functions (9.43) and (9.44), we can evaluate the reduced nuclear matrix elements by the standard technique of the nuclear shell model as a linear combination of those for single particle states. For example, we have a relation

$$[S\,L\,J, n] = \sum_i \sum_j \sum_{j_p} \sum_{j_n} B_i C_j D(i, j, j_n, j_p)[S\,L\,J, n]_{j_n, j_p}. \tag{9.45}$$

Here j_n and j_p are the spins of the proton and neutron states, respectively. The factor D is the sum of the products of two fractional parentage coefficients, which are the geometrical factors appearing in the decomposition of the 8-particle states into the 1- and 7-particle states, each of these states being antisymmetrized. Furthermore, $[S\,L\,J, n]_{j_n, j_p}$ are the reduced nuclear matrix elements in the single particle transition from j_p to j_n. They have forms similar to Eq. (9.12) for $j_p = \frac{3}{2} \rightarrow j_n = \frac{1}{2}$.

Results

As we described in the last subsection, the coefficients, B_i and C_i are determined in the analysis of the nuclear level schemes, with small uncertainties which are fixed with the data on the appropriate nuclear quantities, such as nuclear static moments or transition rates. Recently, Kurath obtained the wave function $\psi(\mathrm{B}^{12})$ with the coefficients B_i in Table 9.3 [19], which give good agreement with the experimental value of the magnetic moment of the ground state of B^{12}, 1.003 ± 0.001 nuclear magnetons [20]. In Table 9.3 are also listed the coefficients C_i for the wave function $\psi(\mathrm{C}^{12})$ which were obtained in the analysis of the muon capture rate in [13]. With these nuclear wave functions and the muon wave functions in preceding subsections, the calculated muon capture rate in Eq. (9.32) is plotted against the oscillator strength para-

Table 9.3 Configuration Mixing Parameters C_i and B_i in the Nuclear Wave Functions $\psi(C^{12})$ and $\psi(B^{12})$ (Eqs. (9.43) and (9.44))[a]

i	1	2	3	4	5	6	7	8
C_i	0.3031	0.2429	0.2572	0.6177	0.6335			
B_i	0.0393	0.3284	0.1375	−0.2563	−0.2339	0.1715	−0.0740	−0.8463

[a] Numerical values are given in [13].

meter b in Fig. 9.2 for the fixed value of the induced pseudoscalar coupling constant $C_P/C_A = 8$, or against C_P/C_A in Fig. 9.3 for the fixed value $b = 1.64$ fm.

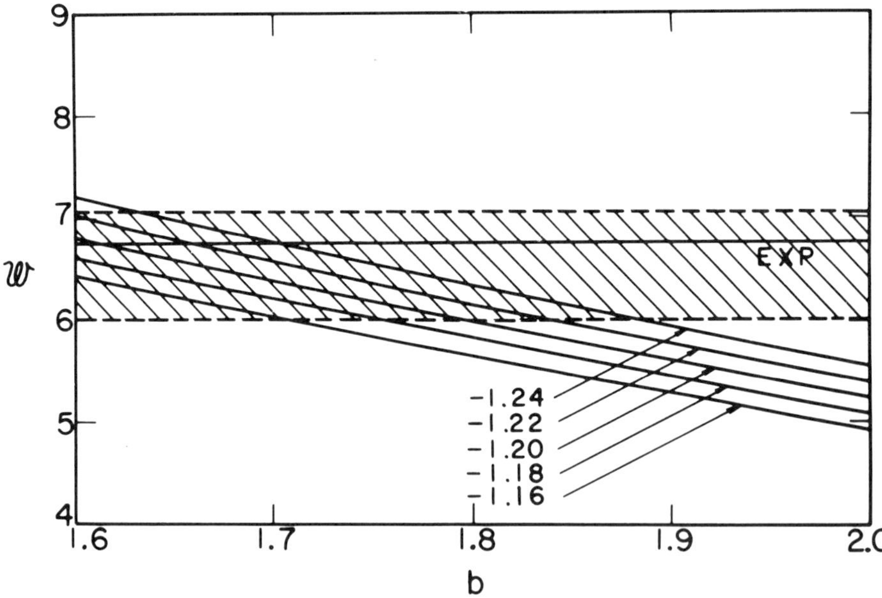

Fig. 9.2 Experimental and theoretical values of the muon capture rate for $C^{12} + \mu^- \to B^{12} + \nu$ plotted against the oscillator strength parameter b for the fixed value of the coupling constant ratio $C_P/C_A = 8$. The five different values adopted for C_A^β/C_V^β are indicated by the arrows. W and b are in units of 10^3 sec^{-1} and fm, respectively [13].

Figure 9.2 indicates that the choice of $b = 1.64$ fm is consistent, while Fig. 9.3 shows that the dependence on C_P/C_A is relatively weak and we have

$$5 \lesssim C_P/C_A \lesssim 30. \tag{9.46}$$

For a particular choice of parameters $b = 1.64$ fm and $C_P/C_A = 8$, we have the muon capture rate in C^{12} and the $ft_{1/2}$ value in the beta decay of B^{12}, as follows.

$$W = 6.67 \times 10^3 \text{ sec}^{-1}, \tag{9.47}$$

and

$$ft_{1/2} = 1.18 \times 10^4 \text{ sec}, \tag{9.48}$$

in an excellent agreement with the experimental data [1, 21]

$$W = \left(6.75 \begin{matrix} +0.30 \\ -0.75 \end{matrix}\right) \times 10^3 \text{ sec}^{-1}, \tag{9.49}$$

and

$$ft_{1/2} = (1.18 \pm 0.007) \times 10^4 \text{ sec}. \tag{9.50}$$

Finally, we conclude that a consistent explanation of the muon capture rate in C^{12}, the beta-decay rate in B^{12}, the magnetic moment of B^{12}, the elastic

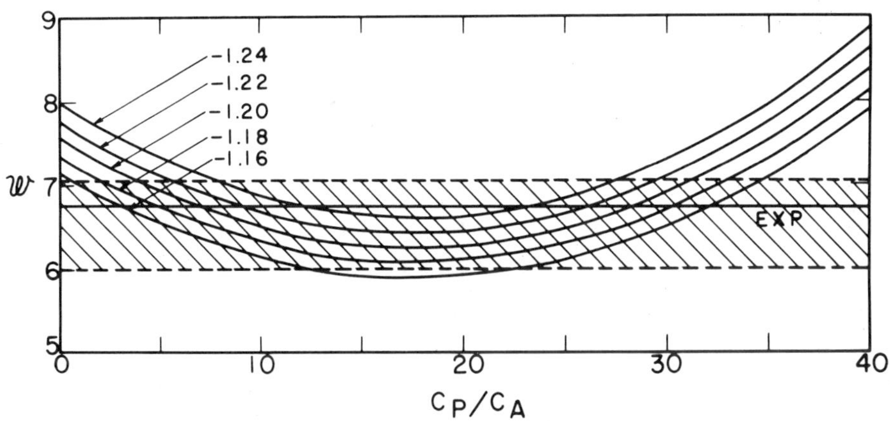

Fig. 9.3 Experimental and theoretical values of the muon capture rate for $C^{12} + \mu^- \rightarrow B^{12} + \nu$ plotted against C_P/C_A for the fixed value of the oscillator strength parameter $b = 1.64$ fm. The five different values for C_A^β/C_V^β are indicated by the arrows. W is in units of 10^3 sec^{-1} [13].

scattering data from C^{12}, and all the inelastic electron scattering data from C^{12} (including those due to excitation of the 15.1-MeV state of C^{12}) is successfully obtained by using the nuclear wave functions of the ground states of C^{12} and B^{12}, which are given by Eqs. (9.43) and (9.44) and Table 9.3.

9.3 MUON CAPTURE IN O^{16}

The other famous examples of muon capture reactions are the partial transitions in O^{16} (see the level scheme in Fig. 9.4). There are four low-

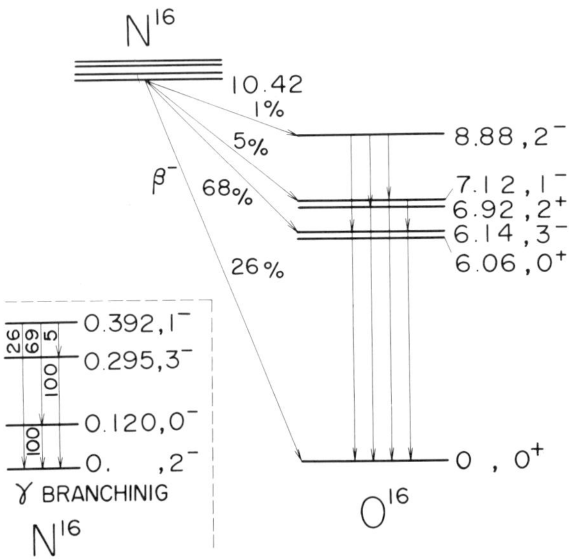

Fig. 9.4 Level scheme of O^{16} and N^{16}.

lying states in N^{16}, all of which are, in principle, excited in the case of muon absorption in the ground state of O^{16},

$$\mu^- + O^{16}(0^+) \rightarrow B^{12}(0^-, 1^-, 2^-, 3^-) + \nu. \tag{9.51}$$

According to the classification in section 8.4, the transitions to the 0^-, 1^-, and 2^- states are the first forbidden, while the transition to the 3^- state is the third forbidden. From the simple estimation in Eq. (8.97), the last is approximately 10^{-3} times slower than the first three. This is not too clear in experiments. (And some of the calculations give a factor of about 10^{-1}–10^{-2}, instead of 10^{-3}; see the end of this section.)

The partial capture rates for these four states can be measured by detecting the deexcitation gamma rays in N^{16}. Such experiments have been performed by Berkeley, Columbia, and Louvain groups, and the data are summarized in Table 9.4.

Table 9.4 Experimental Muon Capture Rates in O^{16} in Units of 10^3 sec^{-1}

Laboratory	$0^+ \to 0^-$	$0^+ \to 1^-$	$0^+ \to 2^-$	$0^+ \to 3^-$
Berkeley[a]	1.6 ± 0.2	1.4 ± 0.2	—	—
Columbia[b]	1.1 ± 0.2	1.88 ± 0.10[e]	6.3 ± 0.7	—
Louvain[c]	$0.850 \begin{matrix} +(0.060+0.085)^f \\ -0.060 \end{matrix}$	$1.850 \begin{matrix} +(0.170+0.185)^f \\ -0.170 \end{matrix}$	7.87 ± 0.76[d]	$\leqslant 0.080$

[a] A. Astbury, L. B. Auerback, D. Cutts, R. J. Estering, D. A. Jenkins, N. H. Lipman, and R. E. Shafer, *Nuovo Cimento* 33, 1020 (1964).

[b] R. C. Cohen, S. Devons, and A. D. Kanaris, *Nucl. Phys.* 57, 255 (1964).

[c] J. P. Deutsch, L. Grenacs, J. Lehmann, P. Lipnik, and P. C. Macq, *Phys. Letters* 29B, 66 (1969).

[d] J. P. Deutsch, L. Grenacs, P. Igo-Kemenes, P. Lipnik, and P. C. Macq, *Nuovo Cimento* 25B, 557 (1967).

[e] Originally 1.73 ± 0.10, with the $1^- \to 0^-$ gamma branching ratio 0.75. It is corrected for the new branching ratio 0.65 (which is adopted in references *a* and *c*), instead of the old value 0.75.

[f] In the quoted errors, the upper limit contains the intrinsic error of the measurement plus the absolute calibration uncertainty.

In the early days, the purpose of the investigation of these transitions was to discover various induced effects in the weak interaction Eqs. (8.25) and (8.26):

(a) The capture rate for $0^+ \to 0^-$ is independent of the nuclear vector current, and is very sensitive to the induced pseudoscalar coupling constant C_P. The hypothesis of the partially conserved axial vector current [22] or the Goldberger-Treiman relation [23] could be tested in this process, which predicts

$$C_P = (7\text{–}8)C_A. \tag{9.52}$$

(b) The capture rates for $0^+ \to 1^-$ and 3^- are independent of C_P, and they depend on the weak magnetism coupling constant g_V in Eq. (8.60). The hypothesis of conserved vector current could be tested in these processes.

(c) The capture rate for $0^+ \to 2^-$ depends on both C_P and g_V. The consistency of the theory of weak interactions could be tested in this reaction and in beta decay of the ground state of N^{16}, which is a unique first forbidden transition.

Since the foregoing questions have been mostly answered in atomic and molecular hydrogen, we shall discuss mainly the nuclear structure problems.

There are a number of theoretical investigations of the reaction (9.51). Shapiro and Blokhintsev first pointed out that the ratio of the capture rates $W(0^+ \to 0^-)$ to $W(0^+ \to 1^-)$ is independent of nuclear model, when the nucleon velocity terms of the interaction are neglected [24]. It was, however, pointed out by Duck [25], and Ericson, Sens, and Rood [26] that the nucleon velocity terms make an essential contribution to the capture rates in the first forbidden transitions. The calculations of the partial muon capture rates have been performed on the basis of the shell model with single particle-hole interactions. Duck and Ericson *et al.* calculated the capture rates using the wave functions of the independent particle model and those of Elliott and Flowers [27] with the effective interaction Hamiltonian of Fujii and Primakoff [2]. They neglected the radial dependence of the muon wave functions in the K shell. Gillet and Jenkins [28] adopted the effective interaction, Eqs. (8.25) and (8.26), the formulation in section 8.4, and the relativistic muon wave functions for the point charge. They adopted nuclear wave functions which were obtained by Gillet and Vinh Mau [29] on the basis of the shell model with a one particle-one hole pair scheme (Tamm-Dancoff model) and in the random phase approximation (RPA model) [30]. (These models appear in the next subsection.) It was shown that the RPA model is in better agreement with experiment than the other one. Rho [31] calculated the capture rates by adopting the Migdal theory [32], and obtained quite good agreement with experiment. Strictly speaking, we could not apply the Migdal theory to light nuclei like O^{16}, since the theory assumed a large level density. In another approach which is free from nuclear model, Kim [33] treated O^{16} and $N^{16}(2^-)$ as elementary particles and determined nuclear form factors from the gamma-decay rate, beta-decay rate, and inelastic electron scattering with excitation of the 2^- state. He obtained an agreement between theory and experiment, as far as this transition is concerned.

In the following, we show the calculation of capture rates in the Tamm-Dancoff model and the RPA model, with a relatively simple model for the O^{16} wave function. We can see a variety of calculated capture rates for different wave functions. Generally speaking, good agreement with experiment is obtained in the cases of $0^+ \to 0^-$ and $0^+ \to 1^-$, while definite disagreement is found in the case of $0^+ \to 2^-$. This is due to inaccuracy in the ground state wave function of O^{16}. Calculations with the modified wave functions and the results are given at the end of this section.

Nuclear Wave Functions and Matrix Elements

In order to calculate the muon capture rates in O^{16}, we have to know the wave functions of O^{16} and N^{16}. First, we discuss the wave functions in the Tamm-Dancoff approximation, where we take into consideration only a subspace of particular configurations from the complete set of states. In our case, we have the shell model with the 1p–1h (one particle-one hole) pair scheme. In this model, the odd-parity state of N^{16} with spin J and its projection M is given in the representation of the second quantization as

$$|JM\rangle = \sum_{j_p j_h} X_{j_p j_h} (j_p j_h m_p m_h \mid JM) \xi^\dagger_{j_p m_p} \xi^\dagger_{j_h m_h} \mid 0 \rangle. \tag{9.53}$$

Here, the summation means the sum over the particle states $2s$ and $1d$ and over the hole states $1p$. The vacuum state $|\,0\,\rangle$ represents the closed shell for the $1s$ and $1p$ orbits. That is the ground state of O^{16}. The $\xi_{jm}^{\dagger}$'s are the quasi-particle creation operators with the spin j and its z component m. They are related to the true particle creation and annihilation operators, $a_{j_p m_p}^{\dagger}$ and $a_{j_h, -m_h}$ as follows.

$$\xi_{j_p m_p}^{\dagger} = a_{j_p m_p}^{\dagger} \qquad \text{and} \qquad \xi_{j_h m_h}^{\dagger} = (-)^{j_h + m_h} a_{j_h, -m_h}, \tag{9.54}$$

where the particle creation and annihilation operators $a^{\dagger}$ and a satisfy the anticommutation relations

$$a_{\alpha}^{\dagger} a_{\beta} + a_{\beta} a_{\alpha}^{\dagger} = \delta_{\alpha\beta}, \qquad a_{\alpha}^{\dagger} a_{\beta}^{\dagger} + a_{\beta}^{\dagger} a_{\alpha}^{\dagger} = a_{\alpha} a_{\beta} + a_{\beta} a_{\alpha} = 0, \tag{9.55}$$

α or β standing for a pair of quantum numbers j and m. The configuration mixing parameters $X_{j_p j_h}$ in the Tamm–Dancoff method are determined by diagonalizing the energy matrix with effective interaction in the subspace of the 1p–1h configurations. The normalization of the state vector (9.53) is given by

$$\sum_{j_p j_h} |X_{j_p j_h}|^2 = 1 \tag{9.56}$$

and the interaction operators Ω_{JM} are expressed by the creation and annihilation operators as follows.

$$\Omega_{JM} = \sum_{\alpha\beta} \langle \alpha\,|\,\Omega_{JM}\,|\,\beta \rangle a_{\alpha}^{\dagger} a_{\beta}. \tag{9.57}$$

The matrix element of the operator Ω_{JM} (with quantum numbers J and M) which induces the transition from the state $|\,0\,\rangle$ to the state $|\,JM\,\rangle$ is written by using Eqs. (9.53)-(9.57)

$$\langle JM\,|\,\Omega_{JM}\,|\,0 \rangle = \sum_{j_p j_h} X_{j_p j_h} \langle j_p \,\|\, \Omega_J \,\|\, j_h \rangle [(2j_p + 1)/(2J + 1)]^{1/2} \tag{9.58}$$

Here $\langle j_p \,\|\, \Omega_J \,\|\, j_h \rangle$ is the reduced matrix element for the single particle states with spin j_p and j_h.

In the Tamm–Dancoff approximation described above, the residual interactions between 1p–1h states are taken into account only for the excited states (the states of N^{16}), and not for the ground state (the ground state of O^{16}). However, if we discuss the entire Hamiltonian of the nucleus, the true ground state as the eigenstate of this Hamiltonian must have a predisposition to produce various kinds of p–h (particle-hole) pairs. The method in which the p–h pairs are equally taken into account for the ground state as well as excited states is called the new Tamm–Dancoff model or the random phase approximation (RPA). The latter name comes from the neglect of the matrix elements of the residual interaction between the 1p–1h and 1p–1p (one particle–one particle) states, etc., except for those between the 1p–1h and 1p–1h states.) Here, we leave the detailed formulation for the RPA in the standard textbooks, e.g., in [34]; see also [30]. The ground state of the RPA model, $|0\rangle_{\rm RPA}$, is no longer the same as the vacuum state $|0\rangle$ of the shell model. It has the following approximate form with the creation operator $A^\dagger$ of the p–h pair

$$|0\rangle_{\rm RPA} \approx c_0|0\rangle + c_2 A^\dagger A^\dagger |0\rangle + c_4 A^\dagger A^\dagger A^\dagger A^\dagger |0\rangle + \cdots. \tag{9.59}$$

Here the coefficients $c_0, c_2, c_4, \ldots$ are percentages of the 0p–0h, 2p–2h, 4p–4h, . . ., states. (More generally, various kinds of 2p–2h states, etc., should be included.) We do not need explicit form of Eq. (9.59) in the following. It should be noted that the coefficients c_i are generally small numbers compared with c_0. Now, the creation operator of the 1p–1h pair is explicitly given by

$$A^\dagger_{j_p j_h} = \sum_{j_p j_h} (j_p j_h m_p m_h | JM) a^\dagger_{j_p m_p} (-)^{j_h + m_h} a_{j_h, -m_h}, \tag{9.60}$$

and its annihilation operator is

$$A_{j_p j_h} = (A^\dagger_{j_p j_h})^\dagger. \tag{9.61}$$

Instead of Eq. (9.53), the excited JM state in the RPA is now given by

$$|JM\rangle_{\rm RPA} = Q^\dagger |0\rangle_{\rm RPA} \tag{9.62}$$

where $Q^\dagger$ is

$$Q^\dagger = \sum_{j_p j_h} (X_{j_p j_h} A^\dagger_{j_p j_h} - (-)^M Y_{j_p j_h} A_{j_p j_h}) \tag{9.63}$$

with

$$\sum_{j_p j_h} (|X_{j_p j_h}|^2 - |Y_{j_p j_h}|^2) = 1. \tag{9.64}$$

$A_{j_p j_h}$ in Eq. (9.61) is defined for M. Therefore, we have to use $A_{j_p j_h}$ for $-M$ in Eq. (9.63). The ground state satifies a relation

$$Q|0\rangle_{\rm RPA} = 0. \tag{9.65}$$

From Eqs. (9.57), (9.62), and (9.63), we have the matrix element in the RPA model

$$\langle JM|\Omega_{JM}|0\rangle_{\rm RPA} = \sum_{j_p j_h} [X_{j_p j_h} \langle j_p \| \Omega_j \| j_h \rangle (2j_p + 1)^{1/2} + (-)^{j_p - j_h} Y_{j_p j_h} \langle j_h \| \Omega_J \| j_p \rangle (2j_h + 1)^{1/2}] (2J+1)^{-1/2}. \tag{9.66}$$

In the equations above the isospin operators are not explicitly written. The configuration mixing parameters X were calculated by Elliott and Flowers [27], and X and Y by Gillet and Vinh Mau [29] with the residual interaction, which is of the central force

$$V(\mathbf{r}_1 - \mathbf{r}_2) = V_o f\left(\frac{|\mathbf{r}_1 - \mathbf{r}_2|}{\mu}\right)(W + BP_\sigma + MP_\sigma P_\tau - HP_\tau) \tag{9.67}$$

with

$$W + B + M + H = 1.$$

Here P_σ and P_τ are spin and isospin exchange operators and W, B, M, and H are the relative strengths of the Wigner, Bartlett, Majorana, and Heisenberg forces. V_o represents the potential depth of the residual interaction; μ is the range of the interaction. The actual form of $f(x)$ is assumed to be

$$f(x) = \exp(-x^2). \tag{9.68}$$

As the potential parameters, Elliott and Flowers took the Rosenfeld mixture, while Gillet and Vinh Mau left the relative strengths of the different forces and potential depth as free parameters and determined their values from a least-squares fit for nine energy levels of O^{16}. These potential parameters are summarized in Table 9.5, and the results for X and Y are given in Table 9.6,

Table 9.5 Residual Interaction Used in Calculating O^{16} Wave Functions†

	V_o (MeV)	W	B	M	H	μ/b
Elliott (EF)[a]	−40	−0.13	0.46	0.93	−0.26	0.90
Gillet (G)[b]	−40	0.35	−0.1	0.35	0.4	1.0

† In this table the potential is defined by $V(r) = f(r/\mu)V_O(W + BP_\sigma + MP_\sigma P_T - HP_T)$. P_σ and P_T and spin and isospin exchange operators; $f(r/\mu)$ is a radial factor; V_O is the potential depth; W, B, M, and H are magnitudes of the different forces; b is the oscillator strength parameter; and μ is the range of the force.

[a] J. P. Elliott and B. H. Flowers, *Proc. Roy Soc.* A242, 57 (1957).

[b] V. Gillet and N. Vinh Mau, *Nucl. Phys.* 54, 321 (1964).

where I_P, I_{EF}, and I_G refer to the wave functions of the pure shell model, Elliott and Flowers, and Gillet and Vinh Mau, respectively. The latter two wave functions are obtained in the Tamm-Dancoff model, while II_G refers to the wave function obtained by Gillet and Vinh Mau in the RPA model.

The radial wave functions f and g for the muon are again the numerical solutions of the Dirac equation with a Coulomb potential of the finite-size nucleus. (This is studied in [35, 36].) The shape of the nuclear charge is assumed to be harmonic oscillator type

$$\rho(r) = \tfrac{1}{4}(\pi b)^{-3/2}[1 + 2(r/b)^2]\exp[-(r/b)^2]. \tag{9.69}$$

Table 9.6 Configuration Mixing Parameters X and Y of the Wave Functions[a] of N^{16}

	X						Y					
Model	$2s_{1/2}$ $1p_{-1/2}$	$1d_{5/2}$ $1p_{-1/2}$	$1d_{3/2}$ $1p_{-1/2}$	$2s_{1/2}$ $1p_{-3/2}$	$1d_{5/2}$ $1p_{-3/2}$	$1d_{3/2}$ $1p_{-3/2}$	$2s_{1/2}$ $1p_{-1/2}$	$1d_{5/2}$ $1p_{-1/2}$	$1d_{3/2}$ $1p_{-1/2}$	$2s_{1/2}$ $1p_{-3/2}$	$1d_{5/2}$ $1p_{-3/2}$	$1d_{3/2}$ $1p_{-3/2}$
0^-												
I_P	1.00											
I_{EF}	1.00					−0.05						
I_G	0.999					0.055						
II_G	0.999					0.053	−0.012					0.012
1^-												
I_P	1.00											
I_{EF}	0.98		0.01	−0.16	−0.08	−0.02						
I_G	0.995		−0.008	0.026	−0.096	−0.026						
II_G	0.996		−0.006	0.026	−0.090	−0.019	0.001		−0.009	−0.012	−0.008	0.008
2^-												
I_P		1.00										
I_{EF}		0.98	−0.10	0.06	0.14	0.09						
I_G		0.983	0.007	0.054	0.174	0.035						
II_G		0.985	0.007	0.051	0.166	0.034		−0.026	−0.001	0.009	0.026	0.015
3^-												
I_P		1.00										
I_{EF}		0.98			−0.18	0.06						
I_G		0.998			−0.062	−0.011						
II_G		0.999			−0.059	−0.010		0.000			−0.004	0.029

[a] I_P, I_{EF}, and I_G denote the wave functions of the *j-j* coupling shell model, of Elliott and Flowers, and Gillet and Vinh Mau in the Tamm–Dancoff model, and II_G denotes the wave function of Gillet and Vinh Mau in the RPA model. X and Y are defined in Eqs. (9.62) and (9.63). $1p_{-1/2}$ should read $(1p_{1/2})^{-1}$, etc.

Investigation of this shape of the nuclear charge distribution has been performed by the Stanford group in the elastic electron scattering experiments [37], and the best fit has been found for the oscillator strength parameter

$$b = 1.75 \text{ fm}. \tag{9.70}$$

This corresponds to the rms radius of O^{16}

$$\langle r^2 \rangle^{1/2} = 2.70 \text{ fm}. \tag{9.71}$$

The explicit formulas for muon capture rates for reactions (9.51) will not be written here, since they can easily be derived from Eqs. (8.47), (8.49), (8.52), and (8.88), and Table 8.2.

Results

Extensive calculations of muon capture rates were performed by Ohtsubo [38, 39], using the nuclear and muon wave functions given in the preceding subsection. Calculated capture rates for the transitions $0^+ \to 0^-$, $0^+ \to 1^-$, $0^+ \to 2^-$, and $0^+ \to 3^-$ are listed in Table 9.7 for four kinds of nuclear model. (In Table 9.7, the capture rates for $0^+ \to 3^-$ are those given by Gillet and Jenkins [28], who adopted a different approximation for the muon wave functions from that of Ohtsubo.) In the Table 9.7, the coupling constants

Table 9.7 Calculated Muon Capture Rates in O^{16} for Different Nuclear Models[a]

Model	$0^+ \to 0^-$	$0^+ \to 1^-$	$0^+ \to 2^-$	$0^+ \to 3^-$
I_P	2.04	4.88	25.7	0.187[b]
I_{EF}	2.50	3.48	15.0	0.163[b]
I_G	1.58	2.74	15.5	0.196[b]
II_G	1.59	2.52	13.8	0.182[b]

[a] Rates are in units of 10^3 sec^{-1} and are from [38, 39]. I_P, I_{EF}, and I_G refer to the simple shell model and the Tamm–Dancoff models of Elliott and Flowers and Gillet and Vinh Mau, respectively. II_G refers to the last authors' RPA model. For the coupling constants, we assume $C_A^\beta/C_V^\beta = -1.18$, C_P/C_A, and the CVC theory here.

[b] From [28].

in the interaction Hamiltonian, Eqs. (8.59) and (8.60), are fixed as follows.

$$C_V = 0.972 C_V^\beta, \qquad C_A = 0.999 C_A^\beta, \qquad C_A^\beta = -1.18 C_V^\beta,$$

$$C_A = 8 C_A, \qquad 2Mg_V = (\mu_p - \mu_n) C_V = 3.7 C_V. \tag{9.72}$$

Now, let us discuss the capture rates for different partial transitions separately.

The muon capture rate for the transition $0^+ \to 0^-$ is strongly dependent on the induced pseudoscalar coupling constant, and is plotted against C_P in Fig. 9.5. Good agreement with the Columbia and Berkeley data is obtained with

$$5 \lesssim C_P/C_A \lesssim 14 \tag{9.73}$$

[35, 39]. The magnitude of C_P is limited in a narrower range if we adopt either the Columbia or Berkeley data only. In fact, the Louvain group, whose

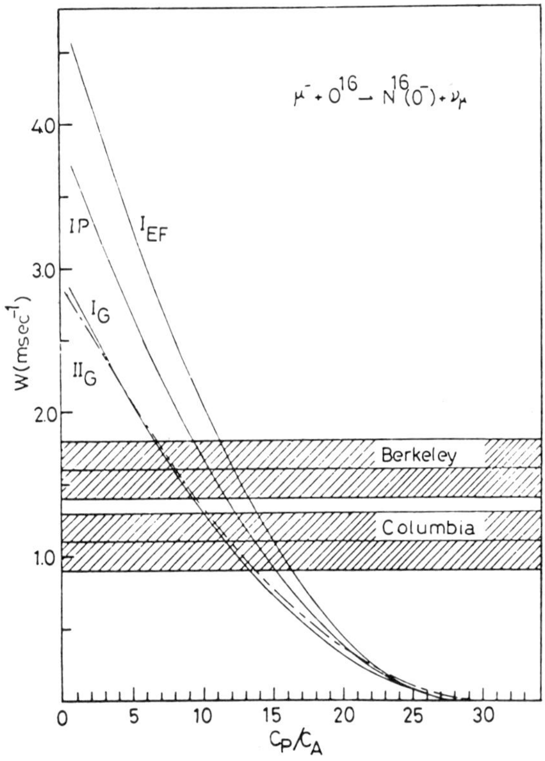

Fig. 9.5 Experimental and theoretical values of the muon capture rate for $\mu^- + O^{16} \to N^{16}(0^-) + \nu$ plotted against C_P/C_A. The other coupling constants are those given in Eq. (9.72). I_p, I_{EF}, I_G, and II_G refer to the four different nuclear models in Table 9.6. W is in units of $10^3\ \text{sec}^{-1}$ [39].

data are in fair agreement with the Columbia data, recently obtained the value [40]

$$C_P = (10.6 \pm 1.0)C_A \tag{9.74}$$

from the observed $W(0^+ \to 0^-)/W(0^+ \to 1^-)$ ratio on the basis of the Rho-Migdal theory [31, 32]. This value for C_P is still subject to discussion, on theoretical grounds as well as regarding the discrepancy observed in the experimental data summarized in Table 9.4. Nevertheless, we can conclude that (1) the capture rate for $0^+ \to 0^-$ does not change significantly with different nuclear models, and (2) the magnitude of C_P derived from the dispersion-theoretic approach [23] is consistent with the capture rate in this particular transition.

In the case of $0^+ \to 1^-$, the calculated capture rates are a little too high. This is, however, remedied by modifying the nuclear wave function of N^{16} slightly. We shall confine ourselves to the wave function II_G. The wave function ψ for the 1^- state is now assumed to be a linear combination of ψ_0 and ψ_n, ψ_0 and ψ_n being the wave functions of the ground state and the nth excited state with the same spin, isospin, and parity. We examine two kinds of wave functions for the 1^- state

$$\text{(a)}\quad \psi = (1 - a^2)^{1/2}\psi_0 - a\psi_1, \qquad \text{(b)}\quad \psi = (1 - a^2)^{1/2}\psi_0 - a\psi_3. \tag{9.75}$$

The wave function ψ_3 is known as the giant resonance level in photoabsorption. In Fig. 9.6, the capture rates are plotted against the mixing parameter

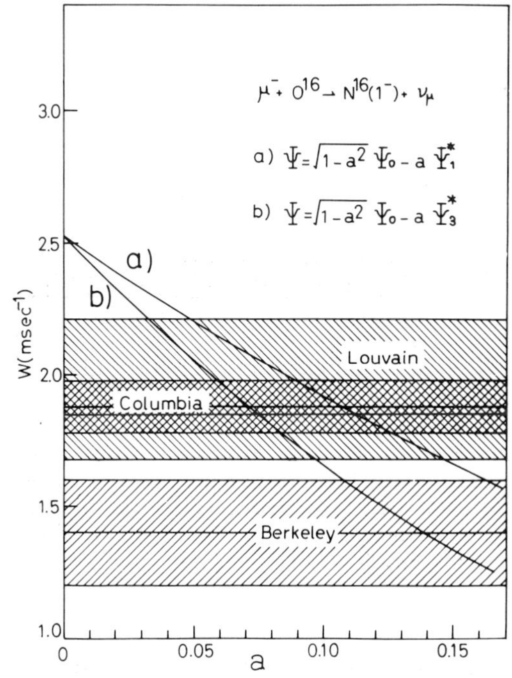

Fig. 9.6 Experimental and theoretical values of the muon capture rate for $\mu^- + O^{16} \to N^{16}(1^-) + \nu$ plotted against the mixing parameter a in Eq. (9.75). ψ_0, ψ_1, and ψ_3 are the wave functions of the lowest, first, and third excited states with spin and parity 1^- of N^{16}, for model II_G. The coupling constants are those given in Eq. (9.72). W is in units of 10^3 $\sec^{-1}$ [39].

for these wave functions. A good fit with the Columbia data is obtained with

$$0.12 \lesssim a \lesssim 0.16 \qquad \text{and} \qquad 0.08 \lesssim a \lesssim 0.10, \tag{9.76}$$

for the wave functions (a) and (b), respectively. For a good fit with the Berkeley data, we need a considerably larger mixing parameter a.

The calculated capture rate for $0^+ \to 2^-$ is plotted against C_P in Fig. 9.7.

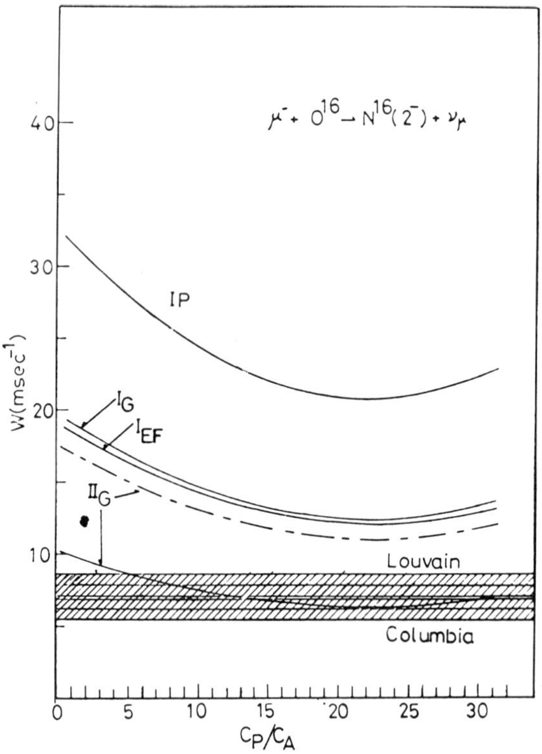

Fig. 9.7 Experimental and theoretical values of the muon capture rate for $\mu^- + O^{16} \to N^{16}(2^-) + \nu$ plotted against C_P/C_A. The other coupling constants are those given in Eq. (9.72). I_p, I_{EF}, I_G, and II_G refer to four different nuclear models in Table 9.6. The solid curve II_G is the capture rate normalized by the experimental beta-decay rate: $W = W^{\text{calc}} W_\beta^{\text{exp}} (W_\beta^{\text{calc}})^{-1}$. W is in units of 10^3 sec^{-1} [39].

Although the nuclear model II_G is better than the others, theoretical values are about twice as large as experimental data. A study of configuration mixing similar to Eq. (9.75) was performed. It was, however, unsuccessful, since the required value of a becomes unreasonably large. In this connection, there is a definite discrepancy between theory and experiment if we adopt any nuclear model listed in Table 9.6. If we normalize the calculated muon capture rates by the beta-decay rate of N^{16} [41],

$$W_\beta^{\exp} = (2.45 \pm 0.70) \times 10^{-2}\ \text{sec}^{-1}, \tag{9.77}$$

in the manner given in Eq. (9.24), we have the following results, which are almost independent of nuclear wave functions,

$$\begin{aligned} W &= (7.80 \pm 2.23) \times 10^3\ \text{sec}^{-1} && \text{for } I_\mathrm{P} \\ &= (8.06 \pm 2.30) \times 10^3\ \text{sec}^{-1} && \text{for } I_\mathrm{EF} \\ &= (7.92 \pm 2.26) \times 10^3\ \text{sec}^{-1} && \text{for } I_\mathrm{G} \\ &= (7.87 \pm 2.25) \times 10^3\ \text{sec}^{-1} && \text{for } II_\mathrm{G}. \end{aligned} \tag{9.78}$$

Strong Ground State Correlation

In conclusion of the foregoing discussion, the RPA model II_G which includes the ground state correlation in O^{16} in part gives a better explanation of the experimental data on three partial transition rates of muon capture in O^{16}, compared with the other three models. Furthermore, the discrepancy in theory and experiment for the transition $0^+ \to 2^-$ suggests that we take into account the strong ground state correlation in O^{16}, rather than that considered in the preceding subsection.

Theoretical work along this line was performed by Walker, who assumed the ground state of O^{16} containing the 2p–2h component and also amended the odd-parity states so as to contain the 3p–3h admixture [42]. The results of his calculation indicate that the configuration mixing model can appreciably lower the partial muon capture rates to the 0^-, 1^-, and 2^- states, in conformity with experimental data. Similar work was also done by Green and Rho [43, 44], which we introduce here briefly. In this work, they adopted the wave function of O^{16} discussed by Brown and Green [45]. The ground state of O^{16} has deformed components. Thus the wave function contains (in addition to the dominant 0p–0h component) 2p–2h, 4p–4h, etc. components. In the calculation by Green and Rho, they assumed an approximate form

$$\begin{aligned} |0^+\rangle &= \alpha\,|\text{0p–0h}\rangle + \beta\,|(d_{5/2}^2)^{T=1,J=0}(1p_{1/2}^{-2})^{1,0}\rangle \\ &\quad + \gamma\,|(2s_{1/2}^2)^{1,0}(1p_{1/2}^{-2})^{1,0}\rangle. \end{aligned} \tag{9.79}$$

Here the state $|\text{0p–0h}\rangle$ means the vacuum state $|0\rangle$ in Eq. (9.53) or (9.59), while the other two states correspond to the states $A^\dagger A^\dagger|0\rangle$ in Eq. (9.59). The reason for neglecting those configurations where particles and holes are separately coupled to (T, J) other than $T = 1, J = 0$ is based on the observation by Zamick [46] that they lie much higher in energy and their coupling to the $|\text{0p–0h}\rangle$ state is much weaker. The original wave function given by Brown and Green has the form 0.874(0p–0h) + 0.469(2p–2h) + 0.130(4p–4h).

Dropping 4p–4h component and considering the two particles in the 2p–2h component to be the combination of $(d_{5/2}^2)^{T=1,J=0}$ and $(2s_{1/2}^2)^{T=1,J=0}$ that results from a 2 x 2 diagonalization with the Kuo-Brown O^{18} matrix elements (which is $0.924(1d_{5/2}^2)^{T=1,J=0} + 0.382(2s_{1/2}^2)^{1,0}$ [47]), Green and Rho obtained the following parameters in Eq. (9.79):

$$\alpha = 0.88, \qquad \beta = 0.44, \qquad \gamma = 0.18. \tag{9.80}$$

In the calculation of the nuclear matrix elements for muon capture, these parameters play a role similar to that of Y in the RPA model. The essential difference is as follows. In the RPA model, we take into account two body residual interactions among only the 1p–1h pairs, so that Y's are generally small. On the other hand, in the strong ground correlation model, as in [45, 47], we take all two-body residual interactions which include, e.g., those between 1p–1h and 1p–1p pairs, so that the parameters β and γ become appreciably large; besides, α differs from unity significantly.

For the ground state and low-lying states of N^{16}, Green and Rho adopted those given by Kuo [48]. These wave functions are described by the 1p–1h combinations in Table 9.8. They have been calculated with the matrix

Table 9.8 Wave Functions of N^{16} by Kuo [48][a]

State	$2s_{1/2}$ $1p_{-1/2}$	$1d_{5/2}$ $1p_{-1/2}$	$1d_{3/2}$ $1p_{-1/2}$	$2s_{1/2}$ $1p_{-3/2}$	$1d_{5/2}$ $1p_{-3/2}$	$1d_{3/2}$ $1p_{-3/2}$
0^-	0.997					0.076
1^-	0.975		0.073	−0.064	−0.198	0.009
2^-		0.962	−0.097	0.048	0.226	0.106
3^-		0.942			−0.329	−0.073

[a] The two-body residual interactions are the Hamada-Johnston potentials. The effect of 2p–2h states $|\, 1d\ 1f\ 1p^{-2}\rangle$ are indirectly included. In the table, $1p_{-1/2}$ should read $(1p_{1/2})^{-1}$, etc.

elements for two-body residual interactions [47] from the Hamada-Johnston potential. The effects of 2p–2h states of the form $|\, 1d\ 1f\ 1p^{-2}\rangle$ are being indirectly included in the N^{16} wave functions. Such states do not appear explicitly in the N^{16} wave functions. Their effect is simply to change the linear combination of 1p–1h states involved. This change, although small, causes about 20% reduction in the capture rates. The effects of the 3p–3h states are completely neglected. If we denote by W_o the capture rate in the absence of 2p–2h correlations in the O^{16} ground state, then the corresponding rates are simply given by

$$\begin{aligned} W(0^+ \to 0^-) &= W_o(0^+ \to 0^-)[\alpha - (12)^{-1/2}\gamma]^2, \\ W(0^+ \to 1^-) &= W_o(0^+ \to 1^-)[\alpha + (12)^{-1/2}\gamma]^2, \\ W(0^+ \to 2^-) &= W_o(0^+ \to 2^-)[\alpha + \tfrac{1}{6}\beta]^2, \\ W(0^+ \to 3^-) &= W_o(0^+ \to 3^-)[\alpha - \tfrac{1}{6}\beta]^2, \end{aligned} \tag{9.81}$$

when these correlations are present. Green and Rho finally gave the numerical values of the capture rates

$$W(0^+ \to 0^-) = 1.01 \times 10^3 \text{ sec}^{-1}, \qquad W(0^+ \to 1^-) = 1.45 \times 10^3 \text{ sec}^{-1},$$
$$W(0^+ \to 2^-) = 6.87 \times 10^3 \text{ sec}^{-1}, \qquad W(0^+ \to 3^-) = 0.12 \times 10^3 \text{ sec}^{-1} \tag{9.82}$$

for $C_P/C_A = 8$ and $b = 1.75$ fm. The results are in good agreement with the experimental data listed in Table 9.4. There is still a small amount of reduction in the nuclear matrix elements, which appear in the axial vector hadron current, due to the renormalization caused by the vacuum polarization type. This is also discussed in [44].

The weak interaction matrix element (8.25) and (8.26) can be transformed to the nonrelativistic form for nucleons so as to include $(p/2M)^2$ terms, where p stands for the nucleon momentum inside nucleons. This is the generalization of Eq. (8.59). Ohtsubo calculated the effect of the $(p/2M)^2$ terms on the muon capture rates in O^{16}, and he found that it is not significant [36, 38, 39].

REFERENCES

1 E. J. Maier, R. M. Edelstein, and R. T. Siegel, *Phys. Rev.* **133**, B663 (1964).
2 A. Fujii and H. Primakoff, *Nuovo Cimento* **12**, 327 (1959).
3 H. Primakoff, *Rev. Mod. Phys.* **31**, 319 (1959).
4 L. Wolfenstein, *Nuovo Cimento* **13**, 319 (1959).
5 M. Morita and A. Fujii, *Phys. Rev.* **118**, 606 (1960).
6 G. F. Flamand and K. W. Ford, *Phys. Rev.* **116**, 1591 (1959).
7 L. L. Foldy and J. D. Walecka, *Phys. Rev.* **140**, B1339 (1965).
8 F. Gudden, *Phys. Letters* **10**, 313 (1964); J. Goldemberg, W. C. Barber, F. H. Lewis, Jr., and J. D. Walecka, *Phys. Rev.* **134**, B1022 (1964).
9 C. W. Kim and H. Primakoff, *Phys. Rev.* **140**, B566 (1965).
10 E. Feenberg, *Shell Theory of the Nucleus*, Princeton Univ. Press, Princeton, N. J., 1955.
11 A. de-Shalit and I. Talmi, *Nuclear Sehll Theory*, Academic Press, New York, 1963.
12 S. Cohen and D. Kurath, *Nucl. Phys.* **73**, 1 (1965).
13 M. Hiro-Oka, T. Konishi, R. Morita, H. Narumi, M. Soga, and M. Morita, *Progr. Theoret. Phys.* **40**, 808 (1968).
14 H. Crannell, *Phys. Rev.* **148**, 1107 (1966).
15 T. Konishi and H. Narumi, private communication.
16 G. Backenstoss, S. Charalambus, H. Daniel, H. Koch, G. Poelz, H. Schmitt, and L. Tauscher, *Phys. Letters* **25B**, 547 (1967).
17 F. Gudden, *Phys. Letters* **10**, 313 (1964); B. Dudelzak and R. E. Taylor, *J. Phys. Radium* **22**, 544 (1961); H. Schmid and W. Scholz, *Z. Phys.* **175**, 430 (1963); J. Goldemberg and R. H. Pratt, *Rev. Mod. Phys.* **38**, 311 (1966).
18 M. Hiro-Oka, *Progr. Theoret. Phys.* **43**, 689 (1970).
19 D. Kurath, private communication.

20 K. Sugimoto, K. Nakai, K. Matuda, and T. Minamisono, *J. Phys. Soc. Japan Suppl.* **24**, 669 (1968).

21 T. R. Fisher, *Phys. Rev.* **130**, 2388 (1963); R. W. Kavanagh, *Phys. Rev.* **133**, B1504 (1964).

22 Y. Nambu, *Phys. Rev. Letters* **4**, 380 (1960); J. Bernstein, S. Fubini, M. Gell-Mann, and W. Thirring, *Nuovo Cimento* **17**, 757 (1960).

23 M. L. Goldberger and S. B. Treiman, *Phys. Rev.* **111**, 355 (1958).

24 I. S. Shapiro and L. D. Blokhintsev, *Zh. Eksp. Teoret Fiz.* **39**, 1112 (1960).

25 I. Duck, *Nucl. Phys.* **35**, 27 (1962).

26 T. E. O. Ericson, J. C. Sens, and H. P. C. Rood, *Nuovo Cimento* **31**, 51 (1964).

27 J. P. Elliott and B. H. Flowers, *Proc. Roy. Soc. London* **A242**, 57 (1957).

28 V. Gillet and D. A. Jenkins, *Phys. Rev.* **140**, B32 (1965).

29 V. Gillet and N. Vinh Mau, *Nucl. Phys.* **54**, 321 (1964).

30 K. Sawada, *Phys. Rev.* **106**, 372 (1957).

31 M. Rho, *Phys. Rev.* **161**, 955 (1967); *Phys. Rev. Letters* **18**, 671 (1967).

32 A. B. Migdal, *Nucl. Phys.* **57**, 29 (1966).

33 C. W. Kim, *Phys. Rev.* **146**, 691 (1966).

34 G. E. Brown, *Unified Theory of Nuclear Models and Forces*, Wiley (Interscience), New York, 1967.

35 M. Morita, M. Hiro-Oka, and H. Narumi, *Proceedings of 1966 International Nucl. Physics Conference* (C. D. Goodman *et al.*, eds.), p. 482. Academic Press, New York, 1967.

36 H. Ohtsubo, Ph.D. thesis, Tokyo Institute of Technology, 1967.

37 H. F. Ehrenberg, R. Hofstadter, U. Meyer-Berkhout, D. G. Ravenhall, and S. E. Sobotika, *Phys. Rev.* **113**, 666 (1959).

38 H. Ohtsubo, *Phys. Letters* **22**, 480 (1965).

39 A. Fujii, M. Morita, and H. Ohtsubo, Suppl. *Progr. Theoret. Phys.* Extra Number 303 (1968).

40 J. P. Deutsch, L. Grenacs, J. Lehmann, P. Lipnik, and P. C. Macq, *Phys. Letters* **29B**, 66 (1969).

41 F. Ajzenberg-Selove and J. Lauritsen, *Nucl. Phys.* **11**, 214 (1959).

42 G. E. Walker, *Phys. Rev.* **174**, 1290 (1968).

43 A. M. Green and M. Rho, *Nucl. Phys.* **A130**, 112 (1969).

44 A. M. Green, T. K. Dahlblom, A. Kallio, and M. Rho, *Phys. Letters* **31B**, 189 (1970).

45 G. E. Brown and A. M. Green, *Nucl. Phys.* **75**, 401 (1966).

46 L. Zamick, *Phys. Letters* **19**, 580 (1965).

47 T. T. S. Kuo and G. E. Brown, *Nucl. Phys.* **85**, 40 (1966).

48 T. T. S. Kuo, referred to in [43].

Chapter Ten

RECENT DEVELOPMENTS IN THE STUDY OF BETA DECAY AND RELATED PROCESSES

In this chapter we shall discuss a few problems among recent findings that came to the author's attention after completion of the main part of this book. These include experiments on gamma decay or electromagnetic excitation of the state which is isobaric analog to the parent nucleus of beta decay, recent techniques for testing second-class currents, and experimental evidence of the presence of parity-nonconserving nuclear forces.

10.1 $ft_{1/2}$ VALUES STUDIED FROM GAMMA DECAYS OF ISOBARIC ANALOG STATES AND (e, e′p) REACTIONS

As is shown in section 4.3, the strength of the weak magnetism in beta decay and that of the $M1$ gamma transition from the isobaric analog state (IAS) are proportional to each other. Explicitly, the matrix element for an $M1$ transition from the IAS is proportional to

$$\langle \mathrm{f} | M1 | \mathrm{IAS} \rangle = \langle \mathrm{f} | \sum_{\mathrm{p}} \mathrm{l}^{(k)} + (1+\mu_{\mathrm{p}}) \sum_{\mathrm{p}} \sigma^{(k)} + \mu_{\mathrm{n}} \sum_{\mathrm{n}} \sigma^{(k)} | \mathrm{IAS} \rangle. \quad (10.1)$$

Here, $|\mathrm{IAS}\rangle$ and $|\mathrm{f}\rangle$ represent the IAS and the final state wave functions. p or n refers to the summation over the number of protons or neutrons. Using the isospin formalism (see Appendix A.8), we rewrite Eq. (10.1) as

$$\langle \mathrm{f} | \sum_{k=1}^{A} \{ \tfrac{1}{2} \mathrm{l}^{(k)} (1+\tau_z^{(k)}) + \tfrac{1}{2}(1+\mu_{\mathrm{p}}) \sigma^{(k)} (1+\tau_z^{(k)}) + \tfrac{1}{2}\mu_{\mathrm{n}} \sigma^{(k)} (1-\tau_z^{(k)}) \} | \mathrm{IAS} \rangle. \quad (10.2)$$

Now, k runs over all nucleons. In our case, the transition occurs between the states with different T values. Therefore, the matrix element effective for the $M1$ transition is given by

$$\langle \mathrm{f} | \sum_{k=1}^{A} \tfrac{1}{2} \tau_z^{(k)} \{ \mathrm{l}^{(k)} + (1+\mu_{\mathrm{p}}-\mu_{\mathrm{n}}) \sigma^{(k)} \} | \mathrm{IAS} \rangle \quad (10.3)$$

with $1 + \mu_p - \mu_n = 4.7$. Since $|\,\mathrm{IAS}\,\rangle = T_+\,|\,\mathrm{i}\rangle(2T)^{-1/2}$, the $M1$ matrix element of the transition from IAS is expressed as

$$\text{Expression (10.3)} \approx \langle\,\mathrm{f}\,|\left[\sum_{k=1}^{A}\tfrac{1}{2}\tau_z^{(k)}\{\mathbf{l}^{(k)} + 4.7\sigma^{(k)}\},\,T_+\right]|\,\mathrm{i}\,\rangle\,(2T)^{-1/2}$$

$$= \langle\,\mathrm{f}\,|\sum_{k=1}^{A}\tau_+^{(k)}\{\mathbf{l}^{(k)} + 4.7\sigma^{(k)}\}|\,\mathrm{i}\,\rangle\,(2T)^{-1/2}. \tag{10.4}$$

Here T is the isospin of the initial state $|\,\mathrm{i}\rangle$, and we have used the relation

$$T_-\,|\,\mathrm{f}\rangle = (\langle\,\mathrm{f}\,|\,T_+)^\dagger = 0. \tag{10.5}$$

This relation holds when the isospin is a good quantum number in the final state and $T_{fz} = -T_f$, T_f and T_{fz} being the isospin of the final state and its z component. Assuming the same orbital momenta for the neutron and proton, we obtain a negligible value for the $l^{(k)}$ term as compared with that of $\sigma^{(k)}$ term. In this approximation, we have

$$\langle\,\mathrm{f}\,|\,M1\,|\,\mathrm{IAS}\,\rangle \approx \frac{4.7}{(2T)^{1/2}}\langle\,\mathrm{f}\,|\sum_{k=1}^{A}\tau_+^{(k)}\sigma^{(k)}\,|\,\mathrm{i}\rangle. \tag{10.6}$$

That is, the $M1$ matrix is proportional to the Gamow–Teller matrix. This rule is violated when a decaying neutron has the same j as a proton has.

In this connection, Sakai presented the idea of using the $M1$ gamma transitions from the compound IAS resonances in (p, γ) reactions, and of comparing them with the Gamow–Teller transitions of the corresponding parent nucleus [1]. The experiment was performed by Sakai and his collaborators in the reaction $\mathrm{Mn}^{55}(\mathrm{p}, \gamma)\mathrm{Fe}^{56}$ [2]. As is shown in Fig. 10.1 this reaction excites three states which lie about 11–12 MeV above the ground state. They are isobaric analog states of the low-lying levels of Mn^{56}. Sakai *et al.* measured the comparative lifetime for each $M1$ transition. From this, the magnitude of the $M1$ matrix and the $\log ft_{1/2}$ are obtained except for a constant. It is adjusted so as to fit the $\log ft_{1/2}$ obtained from the beta-decay data at a given state. This is indicated as the standard in Table 10.1. There, the results are given only for the 1349-keV and 1439-keV resonances. In general, agreement of $\log ft_{1/2}$ values measured from beta decay and gamma transitions is good. This fact is in favor of the existence of IAS in medium nuclei, and of a close relationship between the electromagnetic and weak interactions. (Note, however, that the approximation (10.6) is adopted.)

Now, let us discuss similar problems in the first forbidden beta decays. In 1963, Fujita proposed measurement of the $E1$ transition from IAS induced

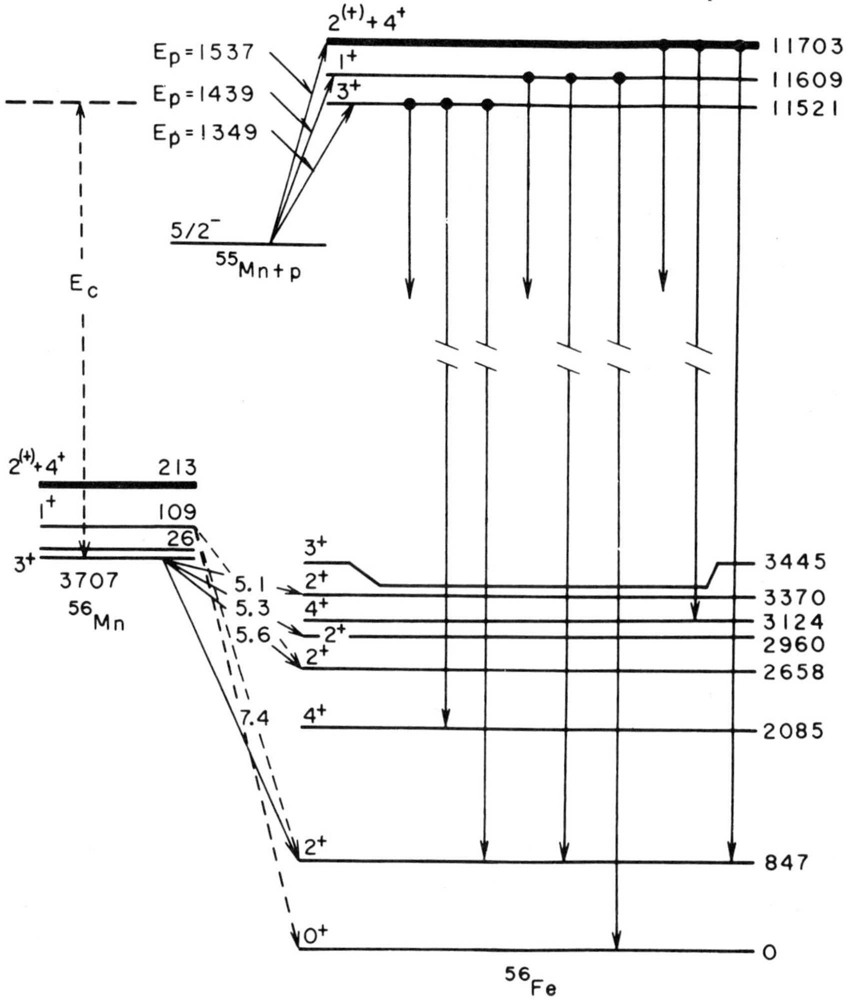

Fig. 10.1 Gamow–Teller of Mn^{56} and the related $M1$ transitions from the isobaric analog states in the Mn^{55} (p, γ)Fe^{56} reaction. The excitation energies are in units of kiloelectron volts (keV). E_c is the Coulomb displacement energy. Figure is provided by Sakai (private communication).

by the (p, n) reaction to determine the first forbidden matrix elements of beta decay [3]. Experiments following this idea were recently carried out by Ejiri and his collaborators [4]. They measured the $E1$ gamma rays from the $2f_{7/2}$ IAS to the $2d_{5/2}$ ground state in Pr^{141} ($N = 82$). This gamma transition corresponds to the first forbidden beta decay, $Ce^{141} \rightarrow Pr^{141}$ (see Fig. 10.2). The $2f_{7/2}$ resonance analog to the ground state of Ce^{141} was excited by the

Table 10.1 Energies of Main Gamma Rays in $Mn^{55}(p, \gamma)Fe^{56}$ Reactions, and Comparison of $(\log ft_{1/2})_\gamma$ Obtained from These Reactions with $\log ft_{1/2}$ from Beta Decays [31]

Final State		1349-keV 3^+ Resonance		1439-keV 1^+ Resonance	
Energy (KeV)	Spin Parity	$(\log ft_{1/2})_\gamma$	$\log ft_{1/2}$	$(\log ft_{1/2})_\gamma$	$\log ft_{1/2}$
0	0^+			5.0 ± 0.02	5.0 (standard)
846.6	2^+	6.44 ± 0.16	7.4	5.23 ± 0.03	5.7
2084.6	4^+	6.11 ± 0.16			
2657.5	2^+	5.75 ± 0.11	5.6	> 6.52	6.2
2959.9	2^+	5.39 ± 0.06	5.4	5.43 ± 0.06	
3122.2	4^+	5.62 ± 0.15			
3370.1	2^+	5.10 ± 0.05	5.1 (standard)	5.48 ± 0.10	
3444.5	3^+	5.89 ± 0.23			
4048.5	3^+				
4100.0	2^+				
4297.9	4^+	5.05 ± 0.09	Energetically impossible		

proton capture reaction on Ce^{140} at $E_p = 9.75$ MeV. The transition probability is

$$\Gamma_\gamma = 13.0 \pm 1.4 \text{ eV}. \tag{10.7}$$

This value is about 25 times smaller than the theoretical value based on the *j-j* coupling single particle calculation.

In the first forbidden beta decay with spin change unity, there are four matrices contributing and the transition rate becomes complicated in its expression (see section 6.3). Formulas become simpler if we omit the matrix $\int B_{ij}$ and the energy dependence of the spectral shape factor from the other matrices. (We call this the ξ approximation [5, 6]. We do not introduce this approximation in section 6.3, since it sometimes introduces a large amount of theoretical error.) In this approximation, the $ft_{1/2}$ value (2.165) for the first forbidden transition is given by

$$ft_{1/2} = 6146/|M|^2 \text{ sec}, \tag{10.8}$$

with

$$M = i(\alpha Z/2R) \int \mathbf{r}[1 - \Lambda - (C_A/C_V)\Lambda_1], \tag{10.9}$$

$$\Lambda_1 = \frac{i \int \sigma \times \mathbf{r}}{\int \mathbf{r}}, \tag{10.10}$$

$$\Lambda = -\frac{i\int \alpha}{(\alpha Z/2R)\int \mathrm{r}}. \qquad (10.11)$$

Here $\alpha Z/2R$ is half of the Coulomb energy at the surface of the daughter nucleus. The expression for Λ is given by Eq. (4.58) in the CVC theory. In our case $\Lambda = 2.4$. We can derive the matrix M in Eq. (10.8) from the beta-

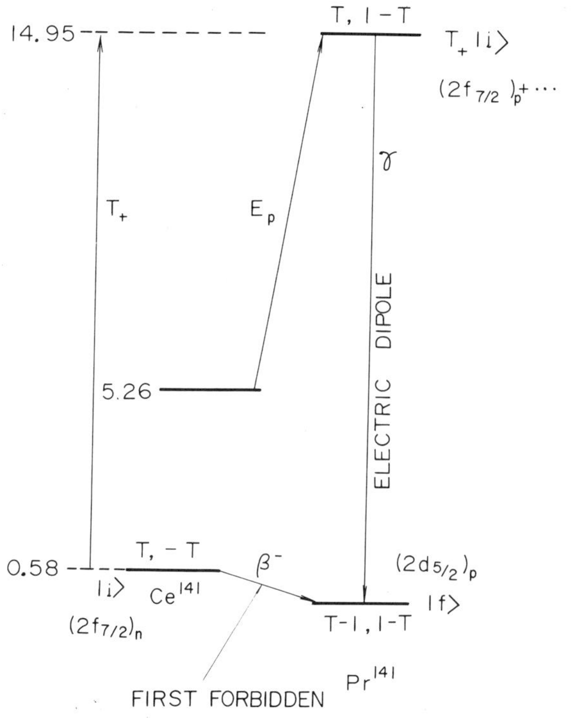

Fig. 10.2 Schematic diagram of the isobaric analog state and the electric dipole gamma and analogous first forbidden beta transitions. The initial state $|i\rangle$, analog state $T_+|i\rangle$, and final state $|f\rangle$ have configurations $(2f_{7/2})_n$, $(2f_{7/2})_p + (2f_{7/2})_n\,[\Sigma 1p\text{-}1h]$, and $(2d_{5/2})_p$, respectively, plus the Ce^{140} core.

decay data, and $\int \mathrm{r}$ from Eq. (10.7) of the gamma-decay data. Substituting these three quantities into Eq. (10.9), Ejiri and his co-worker found

$$\Lambda_1 = 0.8 \pm 0.1, \qquad (10.12)$$

in accord with the value $\Lambda_1 = 1$ evaluated by using the j–j coupling shell model

$$\Lambda_1 = j_i(j_i + 1) - l_i(l_i + 1) - j_f(j_f + 1) + l_f(l_f + 1). \qquad (10.13)$$

Here, the subscripts i and f stand for the initial and final nucleons, respectively. This means that the hindrance factor for $\int \sigma \times r$ is about the same as for $\int r$. In other words, the CVC and Ahrens-Feenberg approximation [7], which was adopted in the derivation of Eq. (4.58), are consistent with this experiment as long as we use the shell model value of Λ_1.

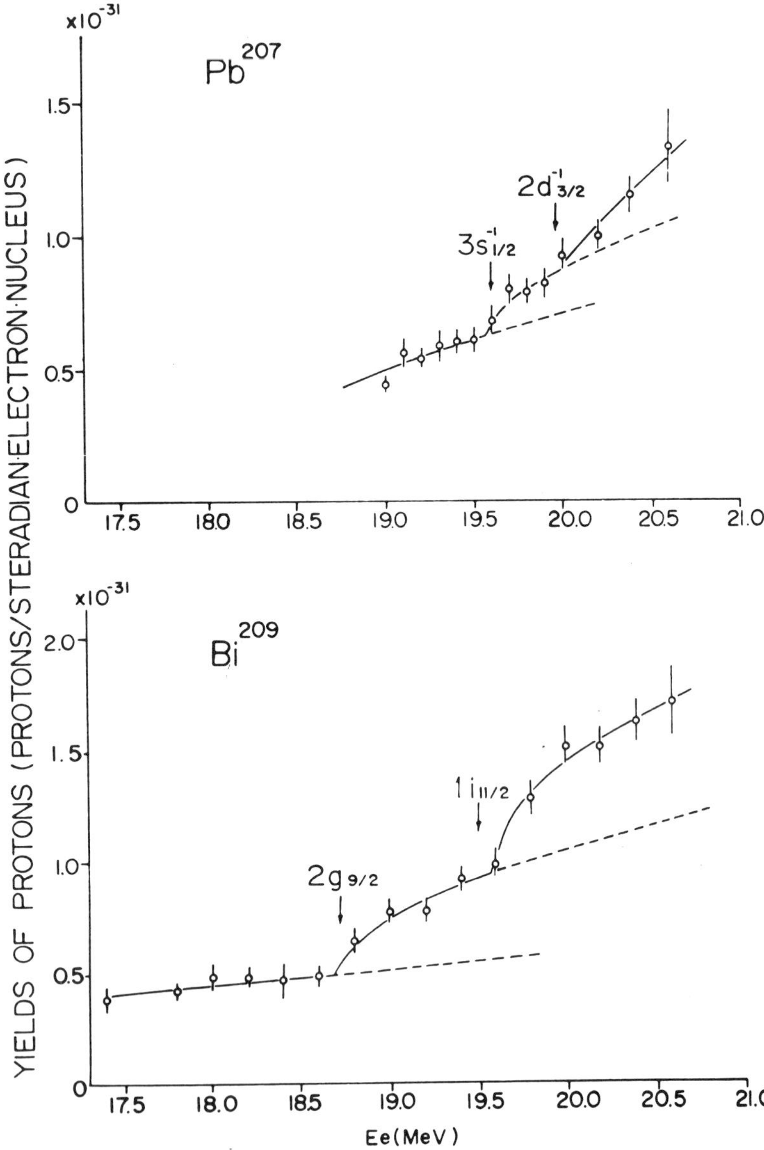

Fig. 10.3 Yield curves of the Pb^{207} (e, e′p) and Bi^{209} (e, e′p) reactions at $\theta = 125.3°$. The positions of IAS are shown by arrows with the single particle shell states [38].

The radiative widths of the IAS of the ground states of Ba^{139}, Ce^{141}, Pm^{148}, Pb^{207}, and Bi^{209} were also determined from the yield curve of the (e, e′p) reaction by Shoda and his collaborators [8, 9]. The experiments were performed by using the Tohoku Linear Accelerator. The radiation width can be calculated from the increment of the proton yields over the continuous resonance yield curve (dashed line in Fig. 10.3). Figure 10.3 shows yield curves of the Pb^{207} (e, e′p) and Bi^{209} (e, e′p) reactions. From these experimental results, Shoda *et al.* obtained the following values of the matrix element $\int \mathbf{r}$.

For $Tl^{207}(3s_{1/2}) \to Pb^{207}(3p_{1/2})$,

$$\begin{aligned} \left|\int \mathbf{r}\right| &= 1.5 \pm 0.2 \qquad \text{exp.[8]}, \\ &= 2.7 \qquad \text{theory[10]}. \end{aligned} \tag{10.14}$$

For $Pb^{209}(2g_{9/2}) \to Bi^{209}(1h_{9/2})$,

$$\begin{aligned} \left|\int \mathbf{r}\right| &= 2.0 \pm 0.2 \qquad \text{exp.[8]}, \\ &= 9.4 \times 10^{-2} \qquad \text{theory[10]}. \end{aligned} \tag{10.15}$$

Numerical values are in units of femtometers (fm) here. Theoretical values are those given by Damgaard and Winther [10]. Agreement of theory and experiment is not necessarily good.

In a paper [9], Shoda *et al.* analyzed experimental data with the precise formulas of beta decay [5]. In the case of $Ce^{141} \to Pr^{141}$, they obtained

$$\left|\frac{M}{(\alpha Z/2R)\int \mathbf{r}}\right| = 0.50 \pm 0.04. \tag{10.16}$$

It is in good agreement with the theoretical value 0.57 based on the *j–j* coupling single particle shell model and with the other experimental value 0.48 ± 0.07 given by Ejiri *et al.* [4]. In the other cases, $Ba^{139}(1f_{7/2}) \to La^{139}(1g_{7/2})$ and $Pm^{148}(1^-) \to Sm^{148}(0^+)$, agreement of theory and experiment is not obtained.

To conclude this section, we emphasize the importance of the IAS and the possibility of studying beta decay from experiments on electromagnetic processes, or vice versa.

10.2 TESTS FOR SECOND-CLASS CURRENTS IN NUCLEAR BETA DECAY AND PRESENT STATUS OF $ft_{1/2}$-VALUE EXPERIMENTS IN MIRROR GAMOW–TELLER TRANSITIONS

In section 5.10, we studied that a possible existence of second-class currents which correspond to the *h* terms in Eq. (5.161) implies the *G*-parity noninvariance of the weak currents. If this is the case, a mirror pair of allowed beta decays $\alpha \to \beta + e^- + \bar{\nu}$ and $\tilde{\alpha} \to \tilde{\beta} + e^+ + \nu$ has an asymmetry of $ft_{1/2}$ values. (Here $\tilde{\alpha}$ is the mirror of α and $\tilde{\beta}$ the mirror of β. For example, $\tilde{p} = n$ and $\tilde{\Sigma}^+ = \Sigma^-$.) If we could ignore the electromagnetic violation of isospin conservation, $\tilde{\alpha}$ and α would have the same masses (as would $\tilde{\beta}$ and β). In the absence of the second-class currents, the net rates for the mirror processes would be identical. Of course, we should take into account electromagnetic effects, especially mass shifts (and therefore the difference of E_o) and the Coulomb interaction difference between mirror decays. After correcting for these effects, we have the $ft_{1/2}$ values for the mirror decays that should be equal in the absence of second-class currents. In the presence of second-class currents, we have the asymmetry δ of the $ft_{1/2}$ values

$$\delta = \frac{(ft_{1/2})^+}{(ft_{1/2})^-} - 1 = -\frac{4h_A}{3C_A}\,(E_o^+ + E_o^-) \tag{10.17}$$

in mirror Gamow–Teller decays. Here the superscript plus refers to positron decay and the minus to negatron decay. The induced tensor coupling constant h_A is defined in the second line of Eq. (5.161). Several experimental data on δ measured by Wilkinson *et al.* [11] were shown in Eq. (5.170) numerically. They also found

$$h_A/C_A \approx -2 \times 10^{-3}$$

by plotting experimental values of δ against $(E_o^+ + E_o^-)$, as is seen in Eq. (10.17). In this section, we introduce new proposals for testing second-class currents in nuclear beta decay by Holstein and Treiman [12], and the present status of $ft_{1/2}$-value experiments in mirror Gamow–Teller transitions [13]. In connection with these problems, we also introduce a preliminary result on the asymmetries of the beta-ray angular distribution in polarized B^{12} and N^{12} [14].

Let us first introduce the recently proposed experiments by Holstein and Treiman [12] for testing second-class currents in nuclear beta decays. As was discussed in section 5.10, the $ft_{1/2}$-value asymmetry δ may be non-vanishing for different reasons, since we have, e.g.,

$$\frac{(ft_{1/2})^+}{(ft_{1/2})^-} = \left(\frac{(\int \sigma)^-}{(\int \sigma)^+}\right)^2 \left[1 - \frac{4h_A}{3C_A}\,(E_o^+ + E_o^-)\right]. \tag{10.18}$$

The asymmetry δ in Eq. (10.17) is derived under the assumption that $(\int \sigma)^+/(\int \sigma)^- = 1$. Without such an assumption, δ may not be zero even if the second-class currents are absent. In order to avoid such ambiguity, Holstein and Treiman proposed to perform precision measurements of spectral functions $C_i(E)$ (which will be given later) in allowed beta decays in which the parent and daughter nuclei belong to a common isomultiplet. The energy dependence of $C_i(E)$ changes with the magnitude of the coupling constant h_A. Comparison of mirror decays is not required.

In order to understand the details of the proposals by Holstein and Treiman, we study the formulas for the beta-ray spectrum and beta-ray angular distribution in oriented nuclei [12]. The interaction Hamiltonian in Eqs. (5.159) and (5.161) is adopted together with the conserved vector current hypothesis. As a consequence, we have no induced scalar ($h_V = 0$) and a relation

$$g_V/f_V = 3.7/M$$

for the neutron beta decays. Furthermore, Holstein and Treiman assumed a relation

$$h_A/f_A \approx g_V/f_V$$

for the induced tensor term, although they have no physical ideas which serve to relate the induced tensor coupling constant h_A to other coupling constants. In the case of a transition between members of an isomultiplet $(T, T_{iz}) \to (T, T_{fz})$, with spin j, the formulas are dependent on four coupling constants

$$a = C_V \int 1, \qquad b = \mu A\,[(j+1)/j]^{1/2} \int 1,$$
$$c = C_A \int \sigma, \qquad d = \mu' A\,[(j+1)/j]^{1/2} \int 1. \tag{10.19}$$

Here the Fermi matrix element is given by Eq. (7.6)

$$\int 1 = [T(T+1) - T_{iz}T_{fz}]^{1/2}. \tag{10.20}$$

In Eq. (10.19), A is the mass number and μ is the isovector part of the total magnetic moment measured in units of nuclear magnetons. That is,

$$\mu = \frac{\mu_i - \mu_f}{T_{iz} - T_{fz}} \tag{10.21}$$

where μ_i and μ_f are the magnetic moments of the initial and final state nuclei. Four coupling constants a, b, c, and d in Eq. (10.19) are the main vector, weak magnetism, main axial vector, and induced tensor term, respectively. The numerical coefficient μ' is hopefully assumed to be of the order of μ, although there is no reason to do so [12]. The formulas are given as follows.

Beta-Ray Spectrum

$$P(E)\,dE = \frac{1}{2\pi^3} F(\pm Z, E) pE(E_o - E)^2 C(E)\,dE \tag{10.22}$$

with the spectral shape factor

$$C(E) = a^2 + c^2 - \frac{2E_o}{3AM}(c^2 \pm cb \pm cd) + \frac{2E}{3AM}(3a^2 + 5c^2 \pm 2cb)$$
$$- \frac{1}{3AME}(2c^2 \pm 2cb \pm cd). \tag{10.23}$$

Here the upper sign refers to the electron and the lower sign to the positron.

Beta-Ray Angular Distribution in Oriented Nuclei

$$W(\theta) = C(E) + C_1(E)\frac{\langle m \rangle}{j}\frac{p}{E}\cos\theta + C_2(E)\left[1 - \frac{3\langle m^2 \rangle}{j(j+1)}\right]$$
$$\times \left[\cos^2\theta - \tfrac{1}{3}\right]\frac{p^2}{E^2}. \tag{10.24}$$

Here the angle θ is the direction of the momentum of the beta particle with respect to the nuclear orientation axis. $\langle m \rangle$ and $\langle m^2 \rangle$ are the average values of the magnetic quantum number m and its square m^2, respectively. The spectral functions $C_i(E)$ are

$$C_1(E) = \left(\frac{j}{j+1}\right)^{1/2}\left[2ac - \frac{2E_o}{3AM}(ac \pm ab \pm ad) + \frac{2E}{3AM}(7ac \pm ab \pm ad)\right]$$
$$- \left(\frac{1}{j+1}\right)^{1/2}\left[\pm c^2 - \frac{2E_o}{3AM}(\pm c^2 + cb + cd)\right.$$
$$\left. + \frac{E}{3AM}(\pm 11c^2 + 5cb - cd)\right]. \tag{10.25}$$

$$C_2(E) = \frac{E}{2AM}(c^2 \pm cb \mp cd). \tag{10.26}$$

An idea for testing second-class current is as follows. In our case of beta decays, the coupling constants a and c are known. Furthermore, the coupling constant b can be determined by measuring the energy dependence of the spectral shape factor $C(E)$ that contains bE/M, but not dE/M. Finally, since a, b, and c are known, we can determine the magnitude of d by measuring the energy dependence of spectral functions $C_1(E)$ and $C_2(E)$ of the beta-ray angular distribution in oriented nuclei. A nonzero value of d corresponds to the existence of second class-currents, and hence to the G-parity non-conservation in beta decays. In this connection, Holstein and Treiman [12] calculated formulas for the beta-ray spectrum, electron–neutrino angular correlation, beta-ray angular distribution in oriented nuclei, and beta–gamma angular correlation as functions of four coupling constants a, b, c, and d.

Along the same lines, Nakamura, Sato, and Igarashi [15] gave also formulas for the beta-ray spectrum and electron–neutrino angular correlation in oriented nuclei. The latter includes, of course, the beta-ray angular distribution in oriented nuclei. They emphasized importance of performing experiments in the pure Gamow–Teller transitions. According to their assumption, the beta decay due to the induced tensor term is mediated by the B meson with mass $M_B = 1220$ MeV in the derivative coupling scheme. Comparing the vector interaction which is mediated by the ρ meson in their theory, they obtained a relation

$$\left|\frac{h_A}{C_A}\right| \sim \frac{m}{M_B}\left(\frac{M_\rho}{M_B}\right)^2,$$

m and $M_\rho = 760$ MeV being the electron and ρ meson mass, respectively. If this is the case, $|h_A/C_A| \sim 10^{-3}$ or 10^{-4}.

The next problem which we have to introduce here is the recent measurement of the excitation spectra of Be^8 following the beta decays of Li^8 and B^8 by Wilkinson and Alburger [13]. It permits the comparison of mirror Gamow–Teller transitions over a wide range of $E_o^+ + E_o^-$, largely free from the uncertainties associated with possible T_z dependence of nuclear overlaps. (That is, the ratio of the Gamow–Teller matrix elements may not be unity. This effect arises because the last proton in the positron-emitting nucleus is less tightly bound than the associated last neutron in the electron-emitting mirror. The wave function of this last proton is, in the light nuclei, despite the Coulomb barrier, more spread out radially than that of the last neutron, and therefore their overlaps with the nucleus into which they transform on beta decay differ. Consequently, $(\int \sigma)^+ < (\int \sigma)^-$ and $(ft_{1/2})^+ > (ft_{1/2})^-$ even in the absence of second-class currents.)

Now let us consider the case of a common residual even-even nucleus of $T_z = 0$. Neutrons and protons are both tightly bound in the final $T_z = 0$ state, but in the initial $T_z = \pm 1$ states the last nucleons are weakly bound, the more weakly bound ones in the proton-rich nucleus. This induces the different overlaps with $T_z = 0$. If it were possible to gradually change the mass for $T_z = 0$ leaving unchanged the masses for $T_z = \pm 1$, so long as the nucleus in a $T_z = 0$ state remained strongly bound, their functions would change little and their overlaps with $T_z = \pm 1$ would change little. But $E_o^+ + E_o^-$ should change. If the nonzero value of δ were due to the second-class currents, it would change proportionally to $E_o^+ + E_o^-$. On the other hand, if δ were due to the difference of the Gamow–Teller matrix elements, it would change little as we changed the mass of $T_z = 0$. Actually we can do this only in the case of Be^8. The final states of the Li^8 and B^8 beta decays are the first excited state of Be^8. This state is tremendously broad and extends for many MeV in excitation. Wilkinson and Alburger effectively changed the mass of the $T_z = 0$ nucleus by measuring the excitation in Be^8 to which the beta transitions lead. The spectra of Be^8 excitation by positron (electron) transitions are called $N^{\pm}(E_x)$ without reference to normalization. E_o^+ and E_o^- are given if E_x is known. Since Be^8 is unstable against breakup into two alpha particles, the excitation E_x is found by measuring the kinetic energy of the alpha particles. Wilkinson and Alburger's experiment [13] consists of an accurate measurement of the alpha-particle spectra $N^{\pm}(E_x)$ following the beta decays of Li^8 and B^8. If we have full mirror symmetry, we have

$$\left[\frac{N^-(E_x)/f(E_o^-)}{N^+(E_x)/f(E_o^+)}\right] = \text{const.} \times \frac{(ft_{1/2})^+}{(ft_{1/2})^-} = \text{const.} \tag{10.27}$$

This holds well even for $\delta \neq 0$, if the failure of mirror symmetry is due to the difference in the Gamow–Teller matrix elements. If $\delta \neq 0$ is due to the presence of the second-class current ($h_A \neq 0$), we have

$$\left[\frac{N^-(E_x)/f(E_o^-)}{N^+(E_x)/f(E_o^+)}\right] = \text{const.} \times \left[1 - \frac{4h_A}{3C_A}(E_o^+ + E_o^-)\right]. \tag{10.28}$$

Wilkinson and Alburger plotted the quantity of the left-hand side in Eq. (10.27) or (10.28) against $(E_o^+ + E_o^-)$ (see Fig. 10.4). They found no evidence

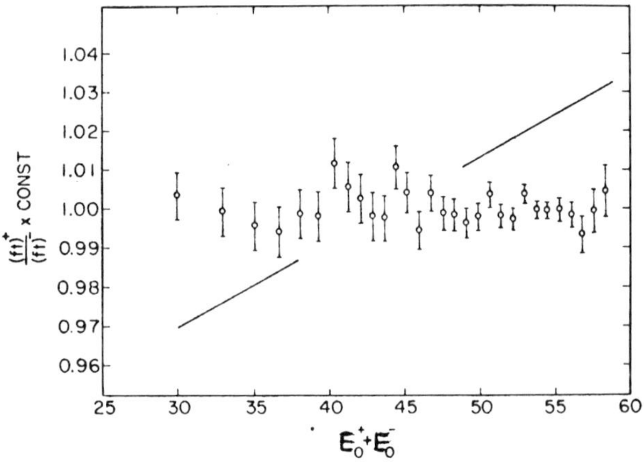

Fig. 10.4 $[N^-(E_x)/f(E_o^-)]/[N^+(E_x)/f(E_o^+)] = \text{const.} \times (ft_{1/2})^+/(ft_{1/2})^-$ as a function of $E_o^+ + E_o^-$. The solid lines show the slope due to a second-class current $|h_A/C_A| = 1.6 \times 10^{-3}$ such as is obtained by the absolute lifetimes [13].

for second-class currents, and obtained a value $|h_A/C_A| < 6 \times 10^{-4}$ at the 99% confidence level from the chi-square fitting of the points of Fig. 10.4 by slopes. They also remeasured the lifetime of Li^8 and B^8 via the alpha particles and obtained the $ft_{1/2}$-value asymmetry δ to be 0.107 ± 0.011 (that corresponds to $h_A/C_A = -1.6 \times 10^{-3}$ if δ is due to the h_A term).

Wilkinson [16] also computed the nuclear overlaps for the known $\delta \neq 0$ nuclei, using appropriate nuclear models. Theoretical values of δ due to this origin are about five times smaller than experimental data in most of the cases. He concluded that the observed $ft_{1/2}$-value asymmetries must be due either to a fundamental weak interaction effect or to a nuclear structure effect of some type not yet quantitatively discussed.

Finally we wish to mention a preliminary result of the asymmetry of the beta-ray angular distribution

$$A(E) = \frac{W(0) - W(\pi)}{W(0) + W(\pi)}$$

in polarized B^{12} and N^{12}. Sugimoto *et al.* [14] measured the energy dependence of $A(E)$ where

$$A(E) = \mp \frac{\langle m \rangle}{j} [1 + kE].$$

Preliminary data include

$$k \sim 1 \times 10^{-3}/\text{MeV} \quad \text{for } B^{12} \qquad \text{and} \qquad 2 \times 10^{-3}/\text{MeV} \quad \text{for } N^{12}.$$

Here the absolute value of k may change after obtaining greater precision. The weak magnetism (the b term in Eq. (10.19)) gives $k \sim \pm 1 \times 10^{-3}/\text{MeV}$ where the upper sign refers to B^{12} and the lower sign to N^{12}. In the presence of the induced tensor term, the magnitude of k may change. But k has still the same absolute values and different signs for different charges of electrons. This is again a theoretically unknown effect in beta decay, if the sign of k is plus for N^{12}.

10.3 EXPERIMENTAL EVIDENCE FOR PARITY-NONCONSERVING NUCLEAR FORCES

As is shown in Chapter 4, the current-current theory of weak interactions which is proposed by Feynman and Gell-Mann [17] and generalized by Cabibbo [18] indicates that the weak interactions involve the four-nucleon coupling (see Eq. (4.78)). Since this coupling represents a weak nuclear force, nuclear levels may not be states of pure parity, and there exist small admixtures of wave functions of parity opposite to that of the strong interaction such that

$$\psi = \psi_{\text{normal}} + F\psi_{\text{abnormal}}. \tag{10.29}$$

Here ψ_{normal} has the regular parity and ψ_{abnormal} has the opposite parity with a fractional amplitude F. Parity-violating nuclear potentials suitable for nuclear model calculations were derived from the current–current theory of weak interactions by Blin-Stoyle [19] and Michel [20]. The magnitude of F estimated by the Michel potential is usually of the order of 10^{-7}. Thus, the parity-nonconserving component is expected to produce very small experimental effects. Since Henley [21] and Hamilton [22] have written comprehensive reviews of the theoretical background for parity-nonconserving nuclear forces and the experimental status at that time, we will not repeat them. Here we simply describe types of possible experiments and recent experimental evidence relating to parity-nonconserving nuclear forces.

We can discuss possible parity experiments for nuclear forces after the analogy of those for beta decay. We again measure pseudoscalar quantities under the space reflection, as we did in the parity experiments of beta decay (see Table 10.2). We detect certain asymmetries in nuclear gamma transitions.

Table 10.2 Possible Parity Experiments for Nuclear Forces in Comparison with Those for Beta Decay[a]

Beta Decay	Nuclear Forces	Possible Effect	Order of F
Longitudinal polarization of beta ray	Circular polarization of gamma ray	Nonzero value of circular polarization	F
Beta-ray angular distribution in polarized nuclei	Gamma-ray angular distribution in polarized nuclei	Asymmetry	F
Beta–gamma circular polarization correlation	Beta–gamma direction correlation	Asymmetry	F
	Gamma–gamma directional correlation	Asymmetry	F^2
	Intensites of satellites in Mössbauer-type experiment	Asymmetry in the intensities	F

[a] In the third column, asymmetry means the presence of the cosine-odd terms in the angular distribution or angular correlation.

Because of the parity mixing in nuclear states (see Eq. (10.29)), there is a mixing of, e.g., the electric dipole and magnetic dipole transitions. As a consequence, the gamma ray from polarized nuclei has a certain degree of circular polarization, called the gamma-ray helicity. The possible experiment is, therefore, the detection of the circular polarization of the gamma ray as that of the longitudinal polarization of beta ray. The experimentally observable effect is about $0.01F$ where a factor of 0.01 is inherent in the circular polarization measurements. The asymmetry of the gamma–ray angular distribution in polarized nuclei is a direct experiment on parity nonconservation as in beta decay. The third experiment is the beta–gamma directional correlation in unoriented nuclei. This is possible because the intermediate state is polarized as a consequence of parity nonconservation in beta decay. The subsequently emitted gamma rays have an angular distribution with respect to the polarization axis of the intermediate state. Therefore, this experiment is reduced to the second one. The fourth and fifth experiments are measurements of the asymmetry of the gamma–gamma directional correlation in unoriented nuclei and the asymmetry of the intensities of satellites in Mössbauer-type experiments. These can be understood if we note that the gamma ray has a helicity of the order of F.

In order to estimate possible experimental effects, we need the formula for the gamma-ray angular distribution from polarized nuclei. This is derived from Eqs. (6.53) and (6.130), together with a multiplication factor

$$(-\tau)^{\delta+\delta'+L+L'}$$

where $\delta = 1$ for electric and $\delta = 0$ for magnetic radiation. The factor above is not necessarily zero because of the parity maxing in nuclear states. The result is

$$\mathscr{W}(\theta) = \sum_{\delta L} \sum_{\delta' L'} \sum_{n} \hat{f}_n(j) F_n(LL'j_1 j)(-\tau)^{\delta+\delta'+L+L'+n} \times \langle j_1 \| \delta L \| j \rangle \langle j_1 \| \delta' L' \| j \rangle P_n(\cos\theta), \tag{10.30}$$

where δL should read EL for electric and ML for magnetic radiation. The angle θ is the direction of emission of the gamma ray with respect to the initial nuclear spin **j**. After summing over the circular polarization variable $\tau = \pm 1$ in Eq. (10.30), we have the gamma-ray distribution function $W(\theta)$. It contains the terms with

$$\delta + \delta' + L + L' = \text{even} \qquad \text{and} \qquad n = \text{even}. \tag{10.31}$$

They are square terms of transition amplitudes from the parity-conserving or parity-nonconserving state. $W(\theta)$ also contains the terms with

$$\delta + \delta' + L + L' = \text{odd} \qquad \text{and} \qquad n = \text{odd}. \tag{10.32}$$

They are interferences between transition amplitudes from the parity-conserving and parity-nonconserving states.

The intensity of the gamma ray from unoriented nuclei is proportional to

$$I(\tau) = \sum_{\delta\delta'} \sum_{L} \langle j_1 \| \delta L \| j \rangle \langle j_1 \| \delta' L \| j \rangle (-\tau)^{\delta+\delta'}, \tag{10.33}$$

which is the $n = 0$ term of Eq. (10.30) obtained by averaging over the angle θ. (Note that there are no interferences between different L values.) For a single L, the degree of the circular polarization is, therefore, given by

$$P_\gamma = -\frac{2\epsilon}{1+\epsilon^2} \qquad \text{with} \qquad \epsilon = \frac{\langle j_1 \| EL \| j \rangle}{\langle j_1 \| ML \| j \rangle}.$$

Since we have already studied several circular polarization experiments in section 4.4, we focus our attention on the measurements of forward-backward asymmetries $\mathscr{A}$ of gamma rays from polarized nuclei performed by the Los Alamos group by using a He^3-He^4 dilution refrigerator. Here we define $\mathscr{A}$ as

$$\mathscr{A} = 2\frac{W(0) - W(\pi)}{W(0) + W(\pi)} = 2\frac{N}{D} \tag{10.34}$$

with

$$N = [W \text{ with condition } (10.32)], \qquad D = [W \text{ with condition } (10.31)],$$

and

$$W = \sum_{\delta L}\sum_{\delta' L'}\sum_{n} \hat{f}_n(j)F_n(LL'j_1j)\langle j_1 \| \delta L \| j\rangle\langle j_1 \| \delta' L' \| j\rangle.$$

In the case where the preceding radiations from polarized nuclei are unobserved, the angular distribution function of gamma rays is also derived from Eqs. (6.54)-(6.56), etc. Since this is a simple modification of the factor $\hat{f}_n(j)$, we omit the details. For the same reason, the formulas for the other three possible parity experiments are also omitted here.

Now let us come back to the recent experiments performed by Krane, Olsen, Sites, and Steyert [23]. They measured a very large parity-nonconserving forward-backward asymmetry $\mathscr{A}$ in the angular distribution of the 501-keV gamma rays from low-temperature polarized Hf^{180m}. The level scheme of Hf^{180} is shown in Fig. 10.5. A parity-nonconserving nucleon–nucleon interaction would permit admixture of the 1142- and 1085-keV levels. Then, the 501-keV transition would be expected to have an irregular $E2$ transitions admixed with its regular $M2$ and $E3$ transitions, as is seen in the numerator N of Eq. (10.34). This is an advantage of the asymmetry measurement. In fact, the observable effect is $(1 + \xi)$ times larger than that of the circular polarization measurement where we have the interference of $E2$ and $M2$ only. ($\xi = \langle j_1 \| E3 \| j\rangle / \langle j_1 \| M2 \| j\rangle$. In our case, ξ is $+5.5 \pm 0.1$ by the directional correlation techniques [24] in conformity with the experimental value $+5.3 \pm 0.3$ obtained by an anisotropy measurement of gamma rays from polarized nuclei.) The magnitude of the observed effect in this particular transition of Hf^{180m} is enhanced by two nuclear structure factors: (1) The regular transition is sixfold K-forbidden and thus strongly hindered, allowing small irregular admixtures to compete favorably. (2) The proximity of the 8^- and 8^+ levels (in Fig. 10.5) allows substantial mixing by the weak interactions. The observed asymmetry [23] is

$$\mathscr{A} = -(1.66 \pm 0.18)\%$$

at a polarization of 72%, which is seen from the anisotropies of pure $E2$ transitions within the $K = 0$ rotational band. From the asymmetry above, the magnitude of the irregular $E2/M2$ mixing ratio ϵ is deduced to be

$$\epsilon = 0.038 \pm 0.004.$$

An ambiguity of the sign determination is deleted by choosing the sign in conformity with the previous circular polarization measurements [25, 26].

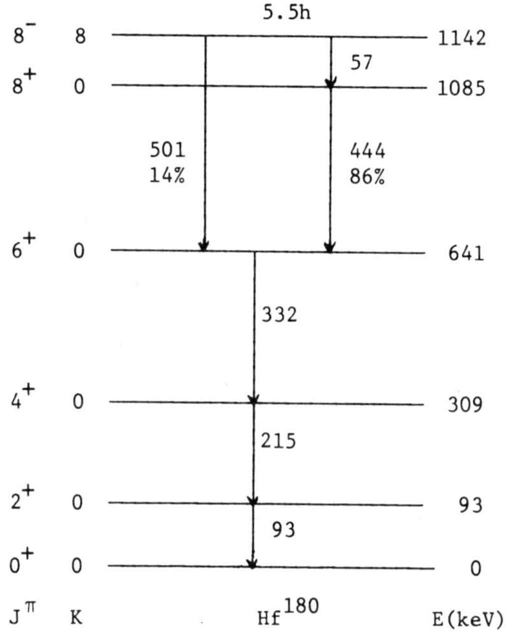

Fig. 10.5 Decay scheme of Hf^{180m}.

As a matter of fact, we expect the degree of circular polarization from unoriented Hf^{180m} to be

$$P_\gamma = \frac{-2\epsilon}{1+\xi^2} = -(2.1 \pm 0.2) \times 10^{-3}$$

from the value of ϵ given above, while experimental values of P_γ are

$$P_\gamma = -(2.8 \pm 0.45) \times 10^{-3} \quad [25],$$
$$= -(2.3 \pm 0.6) \times 10^{-3} \quad [26].$$

The nuclear structure effects were also estimated by Krane *et al.* [23] for the amplitude F in Eq. (10.29). Assuming the parity impurities relevant to the present problem to be due in large part to an admixture of the 8^+ $K = 0$ level into the 1142-keV 8^- level with $K = 8$, the wave function of the

1142-keV level (i.e., the explicit form of Eq. (10.29)) can be described by first-order perturbation theory as

$$|1142\rangle = |8^-\rangle + F\,|8^+\rangle \tag{10.35}$$

where $|8^-\rangle$ and $|8^+\rangle$ are wave functions of the pure parity and F is given by

$$F = \frac{\langle 8^+ | \mathrm{PV} | 8^- \rangle}{E(8^-) - E(8^+)}.$$

Here the denominator is equal to 57 keV and PV is the parity-violating nuclear potential. The magnitude of F is the parity-mixing amplitude, which is of the order of 10^{-7} as Michel [20] estimated, over the hindrance factor due to K forbiddenness. The latter was estimated by Lawson and Segel [27] from consideration of the lifetimes of K-forbidden gamma transitions. In general, for each degree of K forbiddenness a nuclear transition is inhibited by a factor of between 10 and 100. This means that the hindrance of the amplitude is 10^3–10^6 for the sixfold forbidden transition. Consequently, the factor F in Eq. (10.35) is estimated [27] as

$$10^{-10} \gtrsim F \gtrsim 10^{-13}. \tag{10.36}$$

On the other hand, the matrix element of the irregular $E2$ operator is approximated as

$$\langle 641 \| E2 \| 1142 \rangle \approx F\langle 6^+ \| E2 \| 8^+ \rangle \approx F\langle 641 \| E2 \| 1085 \rangle,$$

which can be estimated by the relation between the transition probabilities within a rotational band. The $M2$ matrix is also found experimentally from the lifetime associated with the $M2$ component of the 501-keV radiation. Finally, Krane *et al.* [23] obtained

$$|\epsilon| = F\frac{\langle 6^+ \| E2 \| 8^+ \rangle}{\langle 6^+ \| M2 \| 8^- \rangle} \approx (2 \pm 0.2) \times 10^9\, F$$

and the experimental value of $|\epsilon|$ gives

$$F = (1.9 \pm 0.3) \times 10^{-11}.$$

This is in good agreement with the estimation (10.36) by Lawson and Segel [27]. In conclusion, the existence of the parity-nonconserving nuclear forces is experimentally proved by a very large asymmetry of the angular distribution of the 501-keV gamma ray from polarized Hf^{180m}. The same statement holds by circular polarization measurements of gamma rays from unoriented Hf^{180m} [25, 26].

Historically, an experimental test for possible parity mixing in the 8^- level of Hf^{180} was attempted by Scharff-Goldhaber and McKeown [28] by measuring the L conversion coefficients of the 57-keV transition. It indicated anomalous values which would be explained by the presence of a 10%

irregular $M1$ admixture into the regular $E1$ transition. However, an alternative interpretation in terms of nuclear penetration effects is likewise compatible with the observed conversion coefficients [28–30]. The latter explanation was substantiated by measurement of the circular polarization [29] of the 57-keV gamma ray. The result (6 ± 6)% was not consistent with the result of 56% derived from the 10% $M1$ admixture, and would be consistent with the irregular mixing ratio $\epsilon = M1/E1$ less than 0.06 in magnitude. In this connection, an experiment on the asymmetry $\mathscr{A}$ of the 57-keV gamma ray from polarized Hf^{180m} was performed [31], and the result was $\mathscr{A} = -(19.4 \pm 9.2) \times 10^{-4}$, indicating the lack of any large parity-violating effect for this very strongly hindered $E1$ transition. Krane *et al.* [32] also measured the asymmetries $\mathscr{A}$ of various gamma rays in the decays from low-temperature polarized Gd^{159}, Tb^{161}, and Ta^{182}. The effects of parity-nonconserving forces in these cases were found to be vanishingly small, except for the 1189-keV gamma ray from Ta^{182}, for which they obtained $\mathscr{A} = -(2.8 \pm 1.7) \times 10^{-4}$.

REFERENCES

1 M. Sakai, Institute for Nuclear Study, Univ. of Tokyo Rept. INSJ-102 (1966).
2 M. Sakai, R. Bertini, and C. Gehringer, *Nucl. Phys.* **A157**, 113 (1970).
3 J. I. Fujita, *Brookhaven Natl. Lab. Rept.* BNL 887C-39 (1963), p. 340; *Phys. Letters* **24B**, 123 (1967).
4 H. Ejiri, P. Richard, S. Ferguson, R. Heffner, and D. Perry, *Phys. Rev. Letters* **21**, 373 (1968); *Nucl. Phys.* **A128**, 388 (1969).
5 M. Morita and R. Morita, *Phys. Rev.* **109**, 2048 (1958).
6 T. Kotani, *Phys. Rev.* **114**, 795 (1959).
7 T. Ahrens and E. Feenberg, *Phys. Rev.* **86**, 64 (1952); M. Yamada, *Progr. Theoret. Phys.* **9**, 268 (1953).
8 K. Shoda, A. Suzuki, M. Sugawara, T. Saito, H. Miyase, and S. Oikawa, *Phys. Rev.* **C3**, 1999 (1972).
9 K. Shoda, A. Suzuki, M. Sugawara, T. Saito, H. Miyase, and S. Oikawa, *Phys. Rev.* **C3**, 2006 (1972).
10 J. Damgaard and A. Winther, *Nucl. Phys.* **54**, 615 (1964).
11 D. H. Wilkinson, *Phys. Letters* **31B**, 447 (1970); D. H. Wilkinson and D. E. Alburger, *Phys. Rev. Letters* **24**, 1134 (1970); D. E. Alburger and D. H. Wilkinson, *Phys. Letters* **32B**, 190 (1970); D. H. Wilkinson, D. E. Alburger, D. R. Goosman, K. W. Jones, E. K. Warburton, G. T. Gravey, and R. H. Williams, *Nucl. Phys.* **A166**, 661 (1971).
12 B. R. Holstein and S. B. Treiman, *Phys. Rev.* **C3**, 1921 (1971).
13 D. H. Wilkinson and D. E. Alburger, *Phys. Rev. Letters*, **26**, 1127 (1971).
14 K. Sugimoto, T. Minamisono, J. Göring, and I. Tanihata, private communication.
15 M. Igarashi, *Progr. Theoret. Phys.* **48**, 1237 (1972); S. Nakamura, S. Sato, and M. Igarashi, *Progr. Theoret. Phys.* **48**, 1899 (1972).
16 D. H. Wilkinson, *Phys. Rev. Letters* **27**, 1018 (1971).
17 R. P. Feynman and M. Gell-Mann, *Phys. Rev.* **109**, 193 (1958); M. Gell-Mann, *Rev. Mod. Phys.* **31**, 834 (1959).
18 N. Cabibbo, *Phys. Rev. Letters* **10**, 531 (1963).

19 R. J. Blin-Stoyle, *Phys. Rev.* **118,** 1605 (1960).
20 F. C. Michel, *Phys. Rev.* **133,** B329 (1964).
21 E. M. Henley, *Ann. Rev. Nucl. Sci.* **19,** 367 (1969).
22 W. D. Hamilton, *Progr. Nucl. Phys.* **10,** 1 (1969).
23 K. S. Krane, C. E. Olsen, J. R. Sites, and W. A. Steyert, *Phys. Rev. Letters* **26,** 1579 (1971); *Phys. Rev.* **C4,** 1906 (1971).
24 E. Bodenstedt, H. J. Körner, E. Gerdau, J. Radeloff, C. Günther, and G. Strubs, *Z. Phys.* **165,** 57 (1961).
25 B. Jenschke and P. Bock, *Phys. Letters* **31B,** 65 (1970).
26 E. D. Lipson, F. Boehm, and J. C. Vanderleeden, *Phys. Letters* **35B,** 307 (1971).
27 R. D. Lawson and R. E. Segel, *Phys. Rev. Letters* **16,** 1006 (1966).
28 G. Scharff-Goldhaber and M. McKeown, *Phys. Rev.* **158,** 1105 (1967).
29 H. Paul, M. McKeown, and G. Scharff-Goldhaber, *Phys. Rev.* **158,** 1112 (1967).
30 R. Hager and E. Seltzer, *Phys. Letters* **20,** 180 (1966).
31 K. S. Krane, C. E. Olsen, and W. A. Steyert, *Phys. Rev.* **C5,** 1663 (1972).
32 K. S. Krane, C. E. Olsen, J. R. Sites, and W. A. Steyert, *Phys. Rev.* **C4,** 1942 (1971); K. S. Krane, J. R. Sites, and W. A. Steyert, *Phys. Rev.* **C5,** 1104 (1972).

APPENDIX

A.1 DEFINITION OF ANGULAR MOMENTUM IN QUANTUM MECHANICS

In the classical theory the angular momentum of a particle about a point O is defined as

$$\mathbf{L} = \mathbf{r} \times \mathbf{p},$$

where $\mathbf{r}$ is the position vector of the particle with respect to O and $\mathbf{p}$ is its linear momentum. In this case, the three components L_x, L_y, L_z can take any finite values. In quantum mechanics, the momentum is replaced by the differential operator $(\hbar/i)\boldsymbol{\nabla}$. As a consequence, the angular momentum operators $\mathbf{L}$ satisfy the commutation relations

$$[L_x, L_y] = i\hbar L_z, \quad \text{and cyclic.} \tag{A.1}$$

It is seen that the operator $\mathbf{L}$ is the conjugate of the angular variable in the three-dimensional rotation in the following way. Under the rotation of the coordinate system through an infinitesimal angle $\Delta\theta$ around the z axis, the coordinates $\mathbf{r}$ are transformed into $\mathbf{r}'$,

$$x' = x + y\,\Delta\theta, \qquad y' = -x\,\Delta\theta + y, \qquad z' = z.$$

The wave function of the physical system $\psi(\mathbf{r})$ is also transformed into

$$\psi(\mathbf{r}) \to R\psi(\mathbf{r}) = \psi(\mathbf{r}')$$

where

$$\begin{aligned}
\psi(\mathbf{r}') &= \psi(x + y\,\Delta\theta, -x\,\Delta\theta + y, z) \\
&= \psi(x, y, z) + \Delta\theta\left(y\frac{\partial}{\partial x} - x\frac{\partial}{\partial y}\right)\psi(x, y, z) \\
&= \left(1 - i\frac{\Delta\theta}{\hbar}L_z\right)\psi(\mathbf{r}).
\end{aligned}$$

That is,

$$R = \left(1 - i\frac{\Delta\theta}{\hbar} L_z\right). \tag{A.2}$$

Therefore, L_z is the operator of the infinitesimal rotation, which is conjugate with the angular variable. For the rotation of the finite angle θ around the z axis, we perform n times the infinitesimal rotation $\Delta\theta (= \theta/n)$. Then,

$$R = \lim_{n\to\infty} \left(1 - i\frac{\Delta\theta}{\hbar} L_z\right)^n = \lim_{n\to\infty} \left(1 - i\frac{\theta}{n\hbar} L_z\right)^n$$

$$= \exp\left(-i\frac{\theta}{\hbar} L_z\right). \tag{A.3}$$

R is the operator which rotates the coordinate system through angle θ around the z axis. For the general case of the rotation axis $\mathbf{n}$ and the rotation angle θ, we have

$$R = \exp\left(-i\frac{\theta}{\hbar}\mathbf{n}\cdot\mathbf{L}\right). \tag{A.3$'$}$$

Now, starting from the commutation relations (A.1), we shall see that this gives a greater content to our theory than just assuming that all properties of $\mathbf{L}$ are expressed in terms of differential operators. In particular, this definition permits the existence of the spin. We now write the angular momentum operators as $\mathbf{J}$ and the commutation relations (A.1) as

$$[J_x, J_y] = i\hbar J_z, \qquad \text{and cyclic.} \tag{A.4}$$

From these relations, we derive

$$[\mathbf{J}^2, J_i] = 0, \qquad i = x, y, z. \tag{A.5}$$

Thus, it is possible to diagonalize the square and any one component of the angular momentum. We choose a representation where $\mathbf{J}^2$ and J_z are diagonal. In this representation, J_x and J_y are nondiagonal.

The eigenfunctions ψ_{jm} are labeled j and m, which are related to the eigenvalues for $\mathbf{J}^2$ and J_z as follows,

$$\mathbf{J}^2\psi_{jm} = j(j+1)\hbar^2\psi_{jm}, \qquad J_z\psi_{jm} = m\hbar\psi_{jm}. \tag{A.6}$$

Here j can be

$$j = 0, \tfrac{1}{2}, 1, \tfrac{3}{2}, 2, \cdots.$$

For a particular j, $2j + 1$ values of m are allowed,

$$m = j, j-1, \cdots, -j+1, -j. \tag{A.7}$$

The derivation of these properties can be found in the literature, e.g., in [1].

If necessary, the eigenfunctions are labeled with additional quantum numbers γ which are the eigenvalues of those operators Γ which together with $\mathbf{J}^2$ and J_z form a complete commuting set for the system.

A.2 ADDITION OF ANGULAR MOMENTA

In quantum mechanical calculations, we very often have to consider the vector addition $\mathbf{J} = \mathbf{J}_1 + \mathbf{J}_2$ of two angular momenta $\mathbf{J}_1$ and $\mathbf{J}_2$ that commute with each other, $J_{1i}J_{2j} - J_{2j}J_{1i} = 0, i, j = x, y, z$. $\mathbf{J}_1$ and $\mathbf{J}_2$ separately satisfy the commutation relations $J_{1x}J_{1y} - J_{1y}J_{1x} = i\hbar J_{1z}$, etc. These angular momenta could refer to independent particles, or to the noninteracting spin and orbital angular momenta of the same particle. Then the state of the system can be represented in terms of the scheme

$$\phi(j_1, j_2, m_1, m_2, \gamma)$$

in which j_1 and m_1 are quantum numbers labeling the eigenfunctions of $\mathbf{J}_1^2$ and J_{1z}; j_2 and m_2 eigenfunctions of $\mathbf{J}_2^2, J_{2z}$; and γ eigenfunctions of the set of observables which commute with $\mathbf{J}_1$ and $\mathbf{J}_2$.

It is possible to find a second representation in which $\mathbf{J}_1^2, \mathbf{J}_2^2, \mathbf{J}^2$, and J_z are diagonal and are labeled j_1, j_2, j, and m. The eigenfunctions of this set are denoted by

$$\psi(j_1, j_2, j, m).$$

The number of independent states is $(2j_1 + 1)(2j_2 + 1)$, corresponding to the number of ways of independently assigning m_1 and m_2 to the ϕ's,

$$m_1 = j_1, j_1 - 1, \cdots, -j_1, \qquad m_2 = j_2, j_2 - 1, \cdots, -j_2. \tag{A.8}$$

Now, we will see how to characterize ψ with the $(2j_1 + 1)(2j_2 + 1)$ values of j and m. Since $j_z = j_{1z} + j_{2z}$, it is apparent that possible values of m are $m_1 + m_2$. The largest value of m is therefore $j_1 + j_2$, and this occurs once, when $m_1 = j_1$ and $m_2 = j_2$. This shows that the largest value of j is $j_1 + j_2$, and there is only one such state. The next largest value of m is $j_1 + j_2 - 1$, and this occurs twice: when $m_1 = j_1$ and $m_2 = j_2 - 1$, and when $m_1 = j_1 - 1$ and $m_2 = j_2$. One of the linearly independent combinations of these two states must be associated with the new state for which $j = j_1 + j_2$, since for this j value there must be m values ranging from $j_1 + j_2$ to $-j_1 - j_2$ by integral steps. The other combination cannot be associated with this or a larger j. The second combination is associated with $j = j_1 + j_2 - 1$. By repeating this procedure, we can see that the possible values of j are

$$j = j_1 + j_2, \; j_1 + j_2 - 1, \cdots, |j_1 - j_2|. \tag{A.9}$$

The number of states ψ assigned with j and m is

$$\sum_{j=|j_1-j_2|}^{j_1+j_2} (2j+1) = (2j_1+1)(2j_2+1). \tag{A.10}$$

A.3 CLEBSCH-GORDAN COEFFICIENTS

The two sets of eigenfunctions are written briefly as

$$\psi(j_1, j_2, j, m, \gamma) = \psi_{jm} \qquad \text{and} \qquad \phi(j_1, m_1, j_2, m_2, \gamma) = \phi_{m_1 m_2}. \tag{A.11}$$

The simplest example of ϕ is

$$\phi_{m_1 m_2} = \psi_{j_1 m_1} \psi_{j_2 m_2} \tag{A.12}$$

where the $\psi_{j_i m_i}$'s are eigenfunctions of the two particles, while the $\phi_{m_1 m_2}$'s are those of the resultant system. ψ_{jm} 's are connected with the $\phi_{m_1 m_2}$'s by a unitary transformation

$$\psi_{jm} = \sum_{m_1, m_2} (j_1 j_2 m_1 m_2 | jm) \phi_{m_1 m_2}. \tag{A.13}$$

The transformation coefficients $(j_1 j_2 m_1 m_2 | jm)$ are called the Clebsch-Gordan coefficients (or Wigner's 3-j coefficients). In the literature, these coefficients are proved to be real, and are expressed by many different notations.

$$\begin{aligned}(j_1 j_2 m_1 m_2 | jm) &= (j_1 j_2 m_1 m_2 | j_1 j_2 jm)\\ &= C(j_1 j_2 j, m_1 m_2 m)\\ &= (-)^{j_2 - j_1 - m}(2j+1)^{1/2}\begin{pmatrix} j_1 & j_2 & j \\ m_1 & m_2 & -m\end{pmatrix}\\ &= s^{j_1 j_2}_{j m_1 m_2} = C^{jm}_{j_1 m_1\, j_2 m_2}, \quad \text{etc.}\end{aligned} \tag{A.14}$$

Since the two sets of eigenfunctions are orthonormalized, the Clebsch-Gordan coefficients are also orthonormalized.

$$\phi_{m_1 m_2} = \sum_{j,m} (j_1 j_2 m_1 m_2 | jm) \psi_{jm}, \tag{A.15}$$

$$\sum_{j,m} (j_1 j_2 m_1 m_2 | jm)(j_1 j_2 m_1' m_2' | jm) = \delta_{m_1 m_1'} \delta_{m_2 m_2'}, \tag{A.16}$$

$$\sum_{m_1, m_2} (j_1 j_2 m_1 m_2 | jm)(j_1 j_2 m_1 m_2 | j'm') = \delta_{jj'} \delta_{mm'}. \tag{A.17}$$

In Eq. (A.16), Σ_m is redundant because of the conservation of the magnetic quantum number. We must ensure that the right-hand side is not equal to $(2j+1)\delta_{m_1 m_1'}\delta_{m_2 m_2'}$, even if we take Σ_m. If we understand the Clebsch-Gordan coefficients as elements of a real orthogonal matrix O of order $(2j_1+1)(2j_2+1)$,

$$OO^T = O^T O = E, \tag{A.18}$$

the summations over j, m in Eq. (A.16) and over m_1, m_2 in Eq. (A.17) are generally required for matrix calculation. A special case of $j_1 = j_2 = \frac{1}{2}$ is illustrated below.

$$(\tfrac{1}{2}\,\tfrac{1}{2}\,m_1 m_2 \,|\, j\,m) = O_{m_1,m_2,jm} = \begin{array}{cc|cccc} & j & 1 & 1 & 0 & 1 \\ & m & 1 & 0 & 0 & -1 \\ m_1 & m_2 & & & & \\ \hline \frac{1}{2} & \frac{1}{2} & 1 & 0 & 0 & 0 \\ -\frac{1}{2} & \frac{1}{2} & 0 & \frac{1}{\sqrt{2}} & \frac{-1}{\sqrt{2}} & 0 \\ \frac{1}{2} & -\frac{1}{2} & 0 & \frac{1}{\sqrt{2}} & \frac{1}{\sqrt{2}} & 0 \\ -\frac{1}{2} & -\frac{1}{2} & 0 & 0 & 0 & 1 \end{array} . \tag{A.19}$$

For permutation of three angular momenta, the Clebsch-Gordan coefficients transform as

$$\begin{aligned} (j_1 j_2 m_1 m_2 \,|\, jm) &= (-)^{j_1+j_2-j}(j_1 j_2 -m_1 -m_2 \,|\, j -m) \\ &= (-)^{j_1+j_2-j}(j_2 j_1 m_2 m_1 \,|\, jm) \\ &= (-)^{j_2+m_2}\left(\frac{2j+1}{2j_1+1}\right)^{1/2}(j j_2 -m\; m_2 \,|\, j_1 -m_1). \end{aligned} \tag{A.20}$$

Therefore, the resultant eigenfunction obtained from the individual eigenfunctions of two particles depends on how the two angular momenta are added into the resultant one. That is, there are two kinds of resultant eigenfunctions, which differ by a phase factor.

$$\psi_{jm} = \sum_{m_1,m_2} (j_1 j_2 m_1 m_2 \,|\, j\,m)\psi_{j_1 m_1}\psi_{j_2 m_2},$$

$$\phi_{jm} = \sum_{m_1,m_2} (j_2 j_1\; m_2\; m_1 \,|\, j\,m)\psi_{j_2 m_2}\psi_{j_1 m_1},$$

$$\phi_{jm} = (-)^{j_1+j_2-j}\psi_{jm}. \tag{A.21}$$

The following restrictions on the variables in the Clebsch-Gordan coefficient come from the definition itself.

(1) $-j_i \leqslant m_i \leqslant j_i$ for $i = 1, 2$ and no subscript.
(2) j_1, j_2, j satisfy the triangular inequalities

$$|j_1 - j_2| \leqslant j \leqslant j_1 + j_2$$

which we abbreviate as (j_1, j_2, j); see Fig. A.1.

(3) The magnetic quantum number is conserved:

$$m_1 + m_2 = m.$$

Otherwise the coefficients are zero.

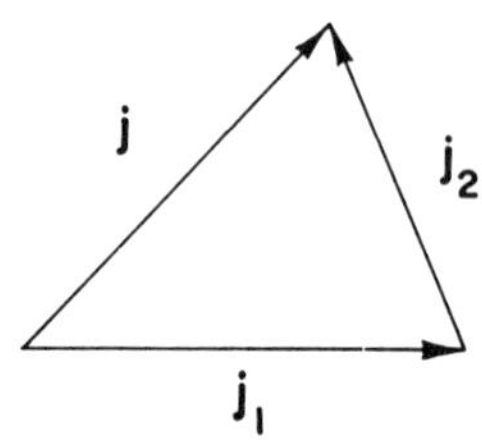

Fig. A.1 Triangle of three angular momenta.

As an example, the product of two spherical harmonics is expanded in a series of spherical harmonics with the Clebsch-Gordan coefficients.

$$Y_{L_1M_1}(\theta, \phi) Y_{L_2M_2}(\theta, \phi) = \sum_L \left(\frac{(2L_1 + 1)(2L_2 + 1)}{4\pi(2L + 1)} \right)^{1/2}$$

$$\times (L_1\, L_2\, M_1\, M_2 \,|\, L\, M)(L_1\, L_2\, 0\, 0 \,|\, L\, 0) Y_{LM}(\theta, \phi). \qquad \text{(A.22)}$$

From this, we can derive the integration of the product of three spherical harmonics over the angular variables.

$$\int Y^*_{LM}(\theta, \phi) Y_{L_1M_1}(\theta, \phi) Y_{L_1M_2}(\theta, \phi) \sin\theta \, d\theta d\phi$$

$$= \left(\frac{(2L_1 + 1)(2L_2 + 1)}{4\pi(2L + 1)} \right)^{1/2} (L_1\, L_2\, M_1\, M_2 \,|\, L\, M)(L_1\, L_2\, 0\, 0 \,|\, L\, 0). \qquad \text{(A.23)}$$

The derivation of the algebraic form of the Clebsch–Gordan coefficients is given by Wigner [2] and Racah [3]. We omit the details, but give the final result shown in the book by Condon and Shortley [4, p. 75],

$$(j_1\, j_2\, m_1\, m_2 \,|\, j\, m) = \delta_{m,\,m_1+m_2}$$
$$\times \left(\frac{(j+j_1-j_2)!(j-j_1+j_2)!(j_1+j_2-j)!(j+m)!(j-m)!(2j+1)}{(j+j_1+j_2+1)!(j_1-m_1)!(j_1+m_1)!(j_2-m_2)!(j_2+m_2)!} \right)^{1/2}$$
$$\times \sum_{\kappa} \frac{(-)^{\kappa+j_2+m_2}(j+j_2+m_1-\kappa)!(j_1-m_1+\kappa)!}{(j-j_1+j_2-\kappa)!(j+m-\kappa)!\kappa!(\kappa+j_1-j_2-m)!}. \tag{A.24}$$

There are several different forms for the summation of the dummy index. Formula (A.24) can be greatly simplified for fixed values of j_2. Such simple formulas are also given for $j_2 = \frac{1}{2}, 1, \frac{3}{2}, 2$ in [4, pp. 73–78] and for $j_2 = \frac{5}{2}, 3, \frac{7}{2}$ in [5]. The general formula, Eq. (A.24), can be easily programmed for computers, and the subroutine of the Clebsch–Gordan coefficients is available in most laboratories.

A.4 THE WIGNER–ECKART THEOREM

The Wigner–Eckart theorem states that the dependence of the matrix element on the magnetic quantum numbers is entirely represented by the Clebsch–Gordan coefficient

$$(j_2\, m_2 \,|\, T_{lm} \,|\, j_1\, m_1) = (j_1\, l\, m_1\, m \,|\, j_2\, m_2)\langle j_2 \| T_l \| j_1 \rangle, \tag{A.25}$$

where $\langle j_2 \| T_l \| j_1 \rangle$ is called the reduced matrix element of the set of tensor operators T_{lm}. The selection rules for angular momenta and their magnetic quantum numbers for the matrix elements are also entirely given by the Clebsch–Gordan coefficient. The theorem was formulated by Wigner and Eckart [2, 6]. The magnetic quantum number independence of $\langle j_2 \| T_l \| j_1 \rangle$ is proved by evaluating the matrix element of the commutator of $J_\pm$ and T_{lm}

$$[J_\pm, T_{lm}].$$

The detailed proof can be seen in [7].

The reduced matrix element is sometimes differently defined with an extra constant factor which is independent of the m's. For example, it is defined in [3] as

$$\langle j_2 \| T_l \| j_1 \rangle_{\text{Racah}} = (2j_2+1)^{1/2}\langle j_2 \| T_l \| j_1 \rangle. \tag{A.26}$$

A.5 RACAH COEFFICIENTS

Let us now consider the coupling of three angular momenta $\mathbf{j}_1, \mathbf{j}_2, \mathbf{j}_3$ to give a resultant $\mathbf{j}$. There are at least two ways to carry out this coupling:

$$\text{(I)}\quad \mathbf{j}_1+\mathbf{j}_2=\mathbf{j}_{12},\quad \mathbf{j}_{12}+\mathbf{j}_3=\mathbf{j};\qquad \text{(II)}\quad \mathbf{j}_2+\mathbf{j}_3=\mathbf{j}_{23},\quad \mathbf{j}_1+\mathbf{j}_{23}=\mathbf{j}. \tag{A.27}$$

The angular momenta $\mathbf{j}, \mathbf{j}_1, \mathbf{j}_2, \mathbf{j}_3, \mathbf{j}_{12}, \mathbf{j}_{23}$ make a tetrahedron: see Fig. A.2.

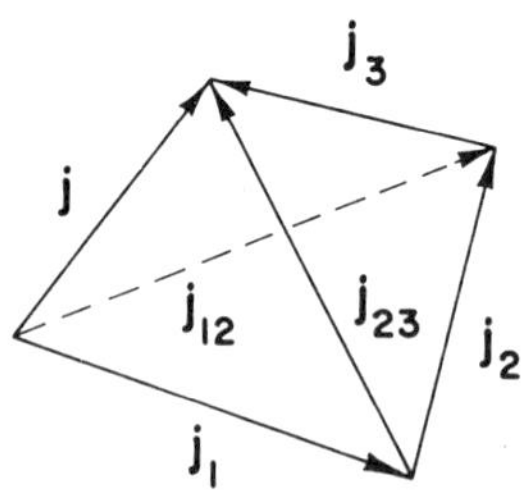

Fig. A.2 Tetrahedron of six angular momenta.

The Racah coefficients are components of the transformation matrix between the two sets of eigenfunctions (I) and (II). That is,

$$\underbrace{\psi((j_1j_2)j_{12}, j_3, j, m)}_{\text{(I)}} = \sum_{j_{23}} (\underbrace{j_1, (j_2j_3)j_{23}, j}_{\text{(II)}} \mid \underbrace{(j_1j_2)j_{12}, j_3, j}_{\text{(I)}}) \times \underbrace{\psi(j_1, (j_2j_3)j_{23}, j, m)}_{\text{(II)}}. \tag{A.28}$$

Here the m independence of the transformation coefficients can be proved by multiplying both sides by the angular momentum operator $J_+ = J_x + iJ_y$. We must be careful of the sequence of the coupling of the angular momenta, since this determines the phase of the wave function, which is one of the most important factors in the calculation of matrix elements. We take the phase factors as defined in the sequence above. Two sets of eigenfunctions are given as follows.

$$\begin{aligned}\psi(\text{I}) &= \sum_{m_{12}, m_3} \psi_{j_{12}m_{12}} \psi_{j_3m_3} (j_{12}\, j_3\, m_{12}\, m_3 \mid j\, m) \\ &= \sum_{m_1, m_2, m_3, m_{12}} \psi_{j_1m_1}\psi_{j_2m_2}\psi_{j_3m_3}(j_1\, j_2\, m_1\, m_2 \mid j_{12}\, m_{12}) \\ &\quad \times (j_{12}\, j_3\, m_{12}\, m_3 \mid j\, m),\end{aligned} \tag{A.29}$$

$$\psi(\mathrm{II}) = \sum_{m_1, m_{23}} \psi_{j_1 m_1} \psi_{j_{23} m_{23}} (j_1\, j_{23}\, m_1\, m_{23} | j\, m)$$

$$= \sum_{m_1, m_2, m_3, m_{23}} \psi_{j_1 m_1} \psi_{j_2 m_2} \psi_{j_3 m_3} (j_2\, j_3\, m_2\, m_3 | j_{23}\, m_{23})$$

$$\times (j_1\, j_{23}\, m_1\, m_{23} | j\, m). \tag{A.30}$$

For simplicity, we use

$$\psi(\mathrm{I}) = \sum_{j_{23}} \psi(\mathrm{II})(\mathrm{II} | \mathrm{I}), \tag{A.31}$$

$$(\mathrm{II} | \mathrm{I}) = (j_1, (j_2 j_3) j_{23}, j \,|\, (j_1 j_2) j_{12}, j_3, j)$$

$$= [(2j_{12} + 1)(2j_{23} + 1)]^{1/2} W(j_1\, j_2\, j\, j_3, j_{12}\, j_{23}). \tag{A.32}$$

The transformation coefficient (II | I) is called the normalized Racah coefficient, while the Racah coefficient is defined by the second line of Eq. (A.32). As is shown in the following calculations, the Racah coefficient is given by the product of four 3-j coefficients summed over the magnetic quantum numbers. Now substituting the explicit forms of the eigenfunctions (A.29) and (A.30) into (A.31) and using the orthonormality of the eigenfunctions for individual particles, we have

$$(j_1\, j_2\, m_1\, m_2 | j_{12}\, m_{12})(j_{12}\, j_3\, m_{12}\, m_3 | j\, m)$$

$$= \sum_{j_{23}} (\mathrm{II} | \mathrm{I})(j_2\, j_3\, m_2\, m_3 | j_{23}\, m_{23})(j_1\, j_{23}\, m_1\, m_{23} | j\, m). \tag{A.33}$$

All summations over magnetic quantum numbers are dropped here, because we have fixed values for m_1, m_2, m_3, and m. Equation (A.33) may also be written as

$$(j_1\, j_2\, m_1\, m_2 | j_{12}\, m_1 + m_2)(j_{12}\, j_3\, m_1 + m_2\, m_3 | j\, m)$$

$$= \sum_{j_{23}} (\mathrm{II} | \mathrm{I})(j_2\, j_3\, m_2\, m_3 | j_{23}\, m_2 + m_3)(j_1\, j_{23}\, m_1\, m_2 + m_3 | j\, m). \tag{A.34}$$

Next, multiply both sides of Eq. (A.34) by $(j_2\, j_3\, m_2\, m_3 | j_{23}\, m_2 + m_3)$ and sum over m_2, m_3, keeping $m_2 + m_3$ fixed.

$$\sum_{m_2, m_3} (j_1\, j_2\, m_1\, m_2 | j_{12}\, m_1 + m_2)(j_{12}\, j_3\, m_1 + m_2\, m_3 | j\, m)$$

$$\times (j_2\, j_3\, m_2\, m_3 | j_{23}\, m_2 + m_3)$$

$$= (\mathrm{II} | \mathrm{I})(j_1\, j_{23}\, m_1\, m_2 + m_3 | j\, m), \qquad m_2 + m_3 = \text{fixed}. \tag{A.35}$$

Here we have used the orthonormality of the Clebsch-Gordan coefficient, Eq. (A.17). We again multiply both sides of Eq. (A.35) by $(j_1 j_{23} m_1 m_2 + m_3 | j\, m)$ and sum over m_1 and $m_2 + m_3$, keeping m fixed:

$$\sum_{m_1, m_2, m_3} (j_1\, j_2\, m_1\, m_2 | j_{12}\, m_1 + m_2)(j_2\, j_3\, m_2\, m_3 | j_{23}\, m_2 + m_3)$$
$$\times (j_{12}\, j_3\, m_1 + m_2\, m_3 | j\, m)(j_1\, j_{23}\, m_1\, m_2 + m_3 | j\, m)$$
$$= (\mathrm{II} | \mathrm{I}) \equiv [(2j_{12} + 1)(2j_{23} + 1)]^{1/2} W(j_1\, j_2\, j\, j_3, j_{12}\, j_{23}). \qquad \text{(A.36)}$$

Here we applied Eq. (A.17) once more. The Racah coefficients are now explicitly given by Eq. (A.36). They are real. The last three identities (A.34)–(A.36) are frequently used in nuclear spectroscopic calculations. The following properties of the Racah coefficients are easily seen from the definition (A.36), together with the properties of the Clebsch-Gordan coefficients; see also the tetrahedron composed by the six angular momenta, **a**, **b**, **c**, **d**, **e**, **f** in Fig. A.3.

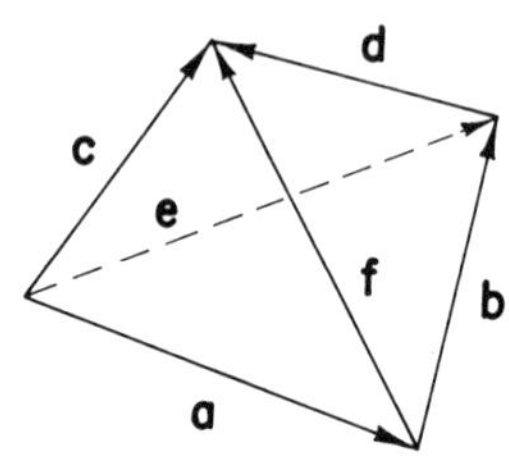

Fig. A.3 Tetrahedron of six angular momenta.

A. The six angular momenta satisfy the triangular inequalities

$$(a, b, e), \quad (b, d, f), \quad (c, d, e), \quad \text{and} \quad (a, c, f). \qquad \text{(A.37)}$$

B. The Racah coefficients are symmetric (expect for sign) for 24 possible arrangements of the six angular momenta:

$$W(abcd, ef) = W(badc, ef) = W(cdab, ef) = W(acbd, fe)$$
$$= (-)^{e+f-a-d} W(ebcf, ad) = (-)^{e+f-b-c} W(aefd, bc). \qquad \text{(A.38)}$$

C. The normalized Racah coefficients $[(2e + 1)(2f + 1)]^{1/2} W(abcd, ef)$ are elements of a real orthogonal matrix element whose row and column are labeled e and f, respectively:

$$\sum_f (2e + 1)(2f + 1) W(abcd, ef) W(abcd, e'f) = \delta_{ee'}. \qquad \text{(A.39)}$$

Racah [3] was able to perform the sum indicated on the left-hand side of Eq. (A.36). His result is

$$W(abcd, ef) = \Delta(abe)\Delta(cde)\Delta(acf)\Delta(bdf)w(abcd, ef), \tag{A.40}$$

with

$$w(abcd, ef) = \sum_z \frac{(-)^{z+a+b+c+d}(z+1)!}{(z-a-b-e)!(z-c-d-e)!(z-a-c-f)!(z-b-d-f)!}$$
$$\times \frac{1}{(a+b+c+d-z)!(a+d+e+f-z)!(b+c+e+f-z)!}.$$

Here, the "triangle" coefficient $\Delta(abc)$ is defined by

$$\Delta(abc) = \left[\frac{(a+b-c)!(a-b+c)!(-a+b+c)!}{(a+b+c+1)!}\right]^{1/2}. \tag{A.41}$$

The most extensive tables for numerical values of the Racah coefficients are given, in fractional form, by Ishizu *et al.* [8]. The subroutine of the Racah coefficient for computer calculation is also available at almost all laboratories.

A.6 NUCLEAR ORIENTATION

Suppose we have a nuclear state of spin j. This state consists of $2j + 1$ magnetic substates, each of which precesses around the z axis of the quantization so that the projection of the spin vector is m. Therefore, the angle which is allowed for this precession is quantized. We define the degree of the orientation for each magnetic substate by

$$\text{degree of orientation} = m/j. \tag{A.42}$$

It is $\cos\theta$, where θ is the angle between the z axis and the spin vector with its magnitude j. (This convection is sometimes adopted in the literature, although the spin vector with the absolute value $[j(j+1)]^{1/2}$ precesses around the z-axis of the quantization, and its projection is m. See, e.g., Fig. 43 in [9].) The average value of the orientation for the nuclear state j is, therefore, given by

$$\cos\theta = \frac{\langle m \rangle}{j} = \frac{1}{2j+1}\sum_{m=-j}^{m=j}\frac{m}{j} = 0. \tag{A.43}$$

Here we assume that the substates are equally populated. This is the case if the energies of the substates are exactly the same. These substates are degenerate, as is shown in Fig. A.4a. They may split if there are some electromagnetic interactions between the nuclear moments and the crystal or external fields, Fig. A.4b. The relative population of each substate is proportional to $\exp(-E_m/kT)$ in thermal equilibrium. Here E_m is the relative energy of the magnetic substates, k is the Boltzmann constant, and T is the temperature. E_m is of the order of 10^{-6} to 10^{-7} eV; on the other hand, kT is about 10^{-4} eV at $T = 1°$ Kelvin (°K). Therefore, at room temperature, the factor $\exp(-E_m/kT)$ is practically unity for all magnetic substates, and they are equally populated. However, when the temperature goes down to the order of 0.01 or 0.001°K, $\exp(-E_m/kT)$ is strongly m dependent. Most of the nuclei are in the magnetic substate (or substates) which has the lowest energy, and very few are in the other substates. As a consequence, $\langle \cos\theta \rangle \neq 0$ and

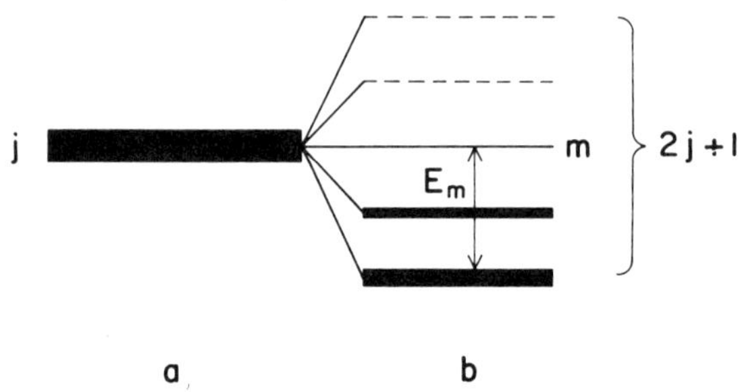

Fig. A.4 Zeeman splitting of nuclear magnetic substates.

the atomic nuclei are oriented. There are two types of nuclear orientation, which are characterized by a_m, the relative population of the magnetic substate. If $a_m = a_{-m}$, and $a_m \neq a_{m'}$ for $|m| \neq |m'|$, the nuclei are *aligned.* If $a_m \neq a_{m'}$ for $m \neq m'$, the nuclei are *polarized.* On the other hand, if a_m is constant for all m, the nuclei are not oriented.

There are several methods for obtaining oriented nuclei. If the nuclear magnetic moment couples with the magnetic field H at the position of nucleus, the energy of the magnetic substate is given by

$$E_m = -\frac{\mu m H}{j}. \tag{A.44}$$

An appreciable degree of the nuclear polarization is obtained if the magnetic field is about 10^5–10^6 gauss, and the temperature is 0.01°–0.001°K. If the nuclear quadrupole moment Q couples with the electric field gradient, the energy of the magnetic substate is given by

$$E_m = 3eQ\frac{\partial^2 V}{\partial z^2}\frac{1}{4j(2j-1)}\left[m^2 - \frac{j(j+1)}{3}\right], \tag{A.45}$$

where V is the electric potential at the nucleus. In this case, E_m is same for m and $-m$. Therefore, nuclear alignment is achieved. A temperature of about 0.001°K is also required.

Besides these methods at low temperatures, we have nonstationary methods of nuclear polarization, if the relaxation time of the nuclear spin is relatively long. These are of the optical pumping techniques. Suppose we have two states, ground and excited, as shown in Fig. A.5. Their spins are 1 and 2, respectively. In the normal condition, the magnetic substates are equally populated, and those of the excited state are empty. Now let us feed a dipole photon of helicity +1 to the ground state. Then, the nucleus will be excited by absorbing the photon. The magnetic quantum number increases by one unit in this process. Furthermore, the excited state emits a dipole photon, and decays back to the ground state. In this decay process, $m = 0, \pm 1$ for the dipole transition. Consequently, the populations of the magnetic

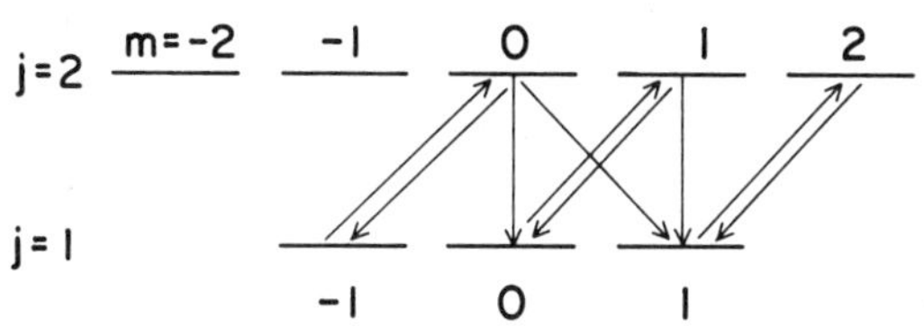

Fig. A.5 Nonstationary method of nuclear orientation.

substates in the final state are not equal, but $a_1 > a_0 > a_{-1}$, as can be seen in Fig. A.5. A low temperature is not required in this case, but it is helpful, since the relaxation time becomes longer at low temperatures. The excited state is also polarized, because the substates of $m = -1, -2$ are not excited.

Nuclear orientation can also be achieved by using the beam of accelerated nucleons and nuclei. For example, we know the direction $\mathbf{p}_{in}$ of the incoming beam, and we can measure the direction $\mathbf{p}_{out}$ of the scattered particles. The orbital angular momentum of these particles is oriented in $\mathbf{p}_{in} \times \mathbf{p}_{out}$, so that a certain amount of the angular momentum is transferred into the target (or residual) nuclei quantum mechanically. It is known that the states with high spin and high energy are excited and are strongly polarized, from observation of the angular distribution of the emitted gamma rays or conversion electron.

The successive nuclear transitions of alpha, beta, and gamma decays are also major sources of oriented nuclei. Cascade gamma transitions are a concrete example. For simplicity, the spin sequence is chosen to be $0 \to 1 \to 0$. The initial state is necessarily unoriented since the spin is zero. The gamma ray emitted from this initial state has only two magnetic states, i.e., the left and right circularly polarized states. This means that the gamma ray carries a magnetic quantum number of $m = -1$ or 1, with respect to the direction of the emission as the z axis of the quantization. As a consequence, the magnetic substates with $m = \pm 1$ of the intermediate state with $j = 1$ are created in this gamma transition, but the substate with $m = 0$ is not, and the intermediate state is aligned. This can be detected by measuring the angular distribution of the second gamma ray, which is emitted from the intermediate state of the spin one, with respect to this particular axis of the nuclear alignment, and consequently with respect to the direction of the emission of the first gamma ray. This will be studied in section A.7 in detail.

In the Mössbauer-type experiments [10], each subtransition, corresponding to the magnetic quantum numbers $m_i \rightarrow m_f$, can be observed as a well-separated absorption line if the hyperfine splittings are sufficiently large in the excited and ground states of nuclei and are appropriately adjusted in the emitter and absorber. (We call this subtransition or absorption line the Mössbauer satellite.) In other words, by observing the Mössbauer satellites, we can select each magnetic substate, which itself is polarized, while the initial states as a whole are not polarized. The degree of polarization of a magnetic substate j_i, m_i is unambiguous and is m_i/j_i. We can consider this as a method for obtaining nuclear polarization, too [11].

A.7 A NAIVE WAY OF UNDERSTANDING ANGULAR DISTRIBUTIONS AND ANGULAR CORRELATIONS

Nuclear radiations from unoriented nuclei have an isotropic angular distribution since no special direction in space is available. However, if the nuclei are oriented, then, in general, radiations have an anisotropic angular distribution with respect to the spin orientation of the initial nuclear state. In this section, we first explain a possible anisotropic angular distribution of gamma rays in oriented nuclei. Next, we apply this consideration to understanding an angular correlation of two gamma rays emitted in cascade transitions. Finally, we replace the first gamma ray by a beta ray to explain beta-gamma angular correlation.

Gamma-Ray Angular Distribution in Oriented Nuclei

Let us assume a decay scheme $1 \xrightarrow{\gamma} 0$, where the excited state of spin one emits a gamma ray which is necessarily dipole, and the ground state has spin zero (see Fig. A.6). We also assume that the initial state is completely polarized ($m = 1$) and the right-handed circularly polarized gamma ray (which has helicity +1) is emitted. The conservation of magnetic quantum numbers requires that one unit of the magnetic quantum number of the initial nuclear state should be carried away by the gamma ray. Since the helicity is +1 in our particular case, the linear momentum of the photon should be parallel

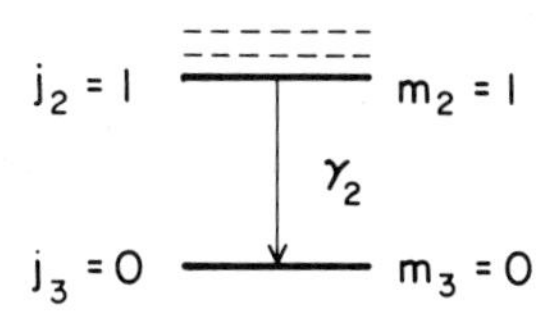

Fig. A.6 Decay scheme. The initial state is completely polarized, and a right-handed circularly polarized gamma ray is emitted.

to its spin and consequently parallel to the initial nuclear spin. This means that the right-handed circularly polarized gamma ray is emitted preferentially in the direction of the spin orientation of the initial nuclear state (see Fig. A.7). In fact, a detailed calculation gives the angular distribution of this

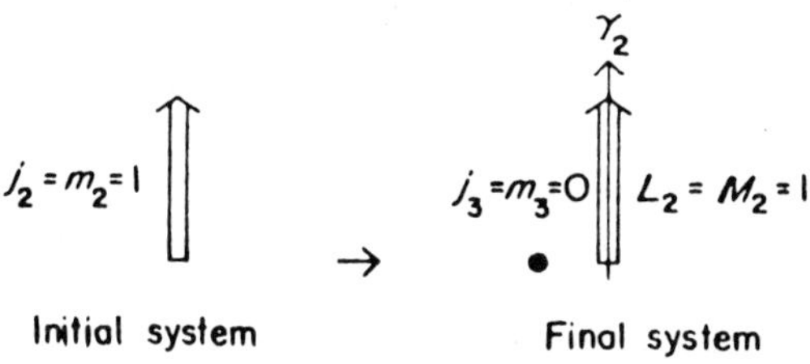

Fig. A.7 Conservation of magnetic quantum numbers. The magnetic quantum number of the initial nuclear state is completely transformed into the emitted gamma ray, so that the right-handed circularly polarized gamma ray is emitted preferentially in the direction of the spin orientation of the initial state.

gamma ray to be

$$1 + 2\cos\theta + \cos^2\theta.$$

Here, θ is the emission angle of the gamma ray with respect to the initial nuclear spin. The left-handed circularly polarized gamma ray, for which the linear momentum and spin are antiparallel, is emitted preferentially antiparallel to the spin orientation of the initial nuclear state. The sign of $\cos\theta$ in the formula above is reversed.

Gamma–Gamma Angular Correlation in Unoriented Nuclei

As a special example, we consider the decay scheme

$$0 \xrightarrow{\gamma_1} 1 \xrightarrow{\gamma_2} 0$$

where two gamma rays are both dipole (see Fig. A.8). Since the initial state has spin zero, this state is necessarily unoriented. We choose the axis of the quantization to be the direction of the emission of the first gamma ray.

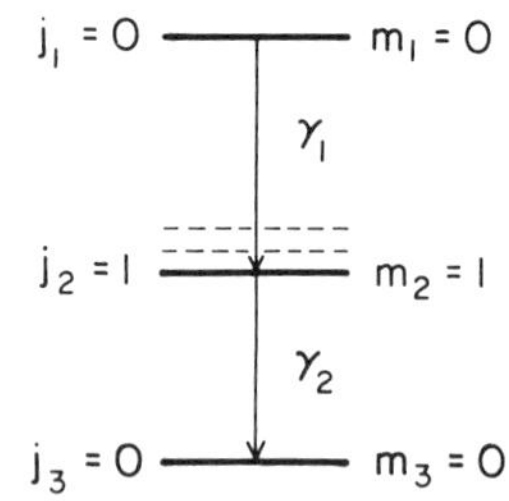

Fig. A.8 Decay scheme. By the emission of the left-handed circularly polarized gamma ray, a perfectly polarized intermediate state is excited. This state has the same property as that of the initial state in Fig. A.6. The energy splitting among the intermediate substates is not necessary.

When the initial state emits a left-handed circularly polarized gamma ray which has the magnetic quantum number $m = -1$, the intermediate state has $m = 1$ because of the conservation of magnetic quantum numbers. That is, the intermediate state of spin one is completely polarized, as in the case of the preceding subsection. The angular distribution of the second gamma ray is anisotropic with respect to the direction of the polarization of the intermediate state, which is parallel to the direction of the emission of the first gamma ray (see Fig. A.9). Consequently, two gamma rays are emitted preferentially parallel (antiparallel) if they have different (the same) circular polarizations.

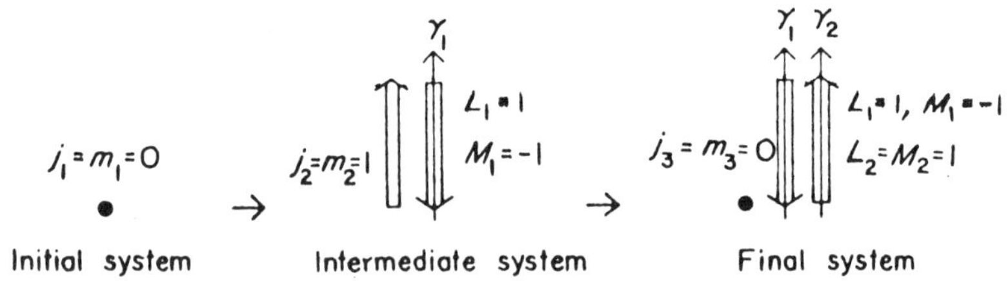

Fig. A.9 Conservation of magnetic quantum numbers. The spin-zero initial state splits into a completely polarized intermediate state and a completely polarized gamma ray. The intermediate state splits again into a spin-zero final state and a completely polarized gamma ray, in the same manner as in Fig. A.7. Finally, two gamma rays are preferentially emitted in the same direction if the polarizations of these two gamma rays are different.

This is summarized in Table A.1.

Table A.1 Angular Correlation of Two Radiations in the Decay Scheme $0 \rightarrow 1 \rightarrow 0$[a]

	γ_1-γ_2	Preferentially Emitted	β-γ
Polarization correlation	Left-Right	Parallel	β^--Right
	Left-Left	Antiparallel	β^--Left
	Right-Right	Antiparallel	β^+-Right
	Right-Left	Parallel	β^+-Left
Directional correlation	Average over the four cases above	Parallel and antiparallel, but not perpendicular	Average over Left and Right

[a] Left (Right) means the left- (right-) handed circularly polarized gamma ray.

Beta-Gamma Angular Correlation

Let us assume the decay scheme $0 \xrightarrow{\beta} 1 \xrightarrow{\gamma} 0$ (see Fig. A.10). Here the first gamma ray in the preceding subsection is replaced by the beta particle,

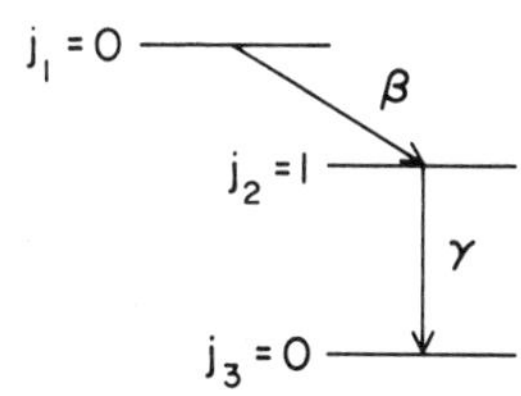

Fig. A.10 Decay scheme.

which is known to be left- (right-) handed for negative (positive) charge. Strictly speaking, the beta particle and either a neutrino or an antineutrino are emitted simultaneously in beta decay. These two leptons make the system spin zero or spin one (the Fermi or Gamow–Teller part). In our particular decay scheme, the dipole gamma ray in the preceding subsection should be replaced by the lepton system of spin one. The negative beta particle and the left-handed circularly polarized gamma ray have the same sign of polarization. They are emitted preferentially antiparallel. As in the case of the gamma–gamma angular correlation, the conservation of magnetic quantum numbers can be satisfied if the neutrino spin is properly taken into account. For the other cases of polarization, see the third and fourth columns in Table A.1.

Directional Correlation

In the directional correlation experiments, we do not observe the circular polarization of gamma ray. We therefore average the angular correlation function over the different signs of circular polarization. As a result, all terms of odd power of $\cos\theta$ disappear in the directional correlations.

A.8 ISOSPIN FORMALISM

In section 2.1, we introduced an operator τ_+ which changes the neutron state to the proton state and annihilates the proton state in the initial nuclear wave function, but which does not change the spatial and spin states of the nucleons. For this operator and related quantities, we introduce a little detail from the isospin formalism. Neglecting the electromagnetic interaction, we can assume that the nuclear forces are charge independent in a good approximation, and the proton and neutron are two states of a particle which we call the nucleon. The two states are represented by two component spinors as

$$\phi_p = \begin{pmatrix} 1 \\ 0 \end{pmatrix} \quad \text{and} \quad \phi_n = \begin{pmatrix} 0 \\ 1 \end{pmatrix}. \tag{A.46}$$

The nucleon wave function ψ_N is generally written as

$$\psi_N = \psi_p \phi_p + \psi_n \phi_n = \psi_p \begin{pmatrix} 1 \\ 0 \end{pmatrix} + \psi_n \begin{pmatrix} 0 \\ 1 \end{pmatrix} = \begin{pmatrix} \psi_p \\ \psi_n \end{pmatrix}, \tag{A.47}$$

ψ_p and ψ_n being the spatial spin wave functions for the proton and neutron, respectively. The spinors ϕ_p and ϕ_n in Eq. (A.46) are similar to the ordinary spinors in the theory of angular momentum. In this theory the ordinary spinors are the eigenfunctions of the fundamental representation of the su_2 group which is homomorphic to the rotation group in the ordinary three-dimensional space. The new spinors (A.46), which are now called isospinors, are again the eigenfunctions of the fundamental representation of the su_2 group which is homomorphic to the rotation group in the three-dimensional charge space (or isospace). This is a fictitious space, since only its z component has a physical meaning which relates to the electric charge. In place of the angular momentum j and its z component m, we have t and t_z in the isospin formalism. The proton and neutron are, therefore, two members of an isodoublet $t = \frac{1}{2}$, and they have $t_z = +\frac{1}{2}$ and $-\frac{1}{2}$, respectively. The isospin operators for nucleons are defined by

$$\mathbf{t} = \tfrac{1}{2}\boldsymbol{\tau} \tag{A.48}$$

where $\boldsymbol{\tau}$ are the Pauli matrices

$$\tau_x = \begin{pmatrix} 0 & 1 \\ 1 & 0 \end{pmatrix}, \quad \tau_y = \begin{pmatrix} 0 & -i \\ i & 0 \end{pmatrix}, \quad \tau_z = \begin{pmatrix} 1 & 0 \\ 0 & -1 \end{pmatrix}. \tag{A.49}$$

The up and down operators for the charge are now defined by

$$\tau_+ = \tfrac{1}{2}(\tau_x + i\tau_y) = (t_x + it_y) = \begin{pmatrix} 0 & 1 \\ 0 & 0 \end{pmatrix}$$

and

$$\tau_- = \tfrac{1}{2}(\tau_x - i\tau_y) = (t_x - it_y) = \begin{pmatrix} 0 & 0 \\ 1 & 0 \end{pmatrix}. \tag{A.50}$$

It is then clear that

$$t_z \begin{pmatrix} \psi_p \\ 0 \end{pmatrix} = \tfrac{1}{2} \begin{pmatrix} \psi_p \\ 0 \end{pmatrix}, \qquad t_z \begin{pmatrix} 0 \\ \psi_n \end{pmatrix} = -\tfrac{1}{2} \begin{pmatrix} 0 \\ \psi_n \end{pmatrix},$$

$$\tau_+ \begin{pmatrix} \psi_p \\ 0 \end{pmatrix} = 0, \qquad \tau_+ \begin{pmatrix} 0 \\ \psi_n \end{pmatrix} = \begin{pmatrix} \psi_n \\ 0 \end{pmatrix},$$

$$\tau_- \begin{pmatrix} \psi_p \\ 0 \end{pmatrix} = \begin{pmatrix} 0 \\ \psi_p \end{pmatrix}, \qquad \tau_- \begin{pmatrix} 0 \\ \psi_n \end{pmatrix} = 0. \tag{A.51}$$

Therefore, τ_+ has a right property for β^- decay, and τ_- for β^+ decay. For a system of nucleons we have the total isospin denoted by **T** which is the vectorial sum of the isospins of the constituents

$$\mathbf{T} = \sum_{k=1}^{A} \mathbf{t}^{(k)}. \tag{A.52}$$

Here the superscript refers to the kth nucleon. Each $\mathbf{t}^{(k)}$ and **T** obeys the commutation relations

$$[t_x^{(k)}, t_y^{(k)}] = it_z^{(k)}, \qquad [T_x, T_y] = iT_z, \qquad \text{and cyclic.} \tag{A.53}$$

The isospin function which has the total isopin T and its z component T_z is denoted by ϕ_{TT_z}. The operators T_z, $\mathbf{T}^2$, T_+, and T_- have the properties

$$\begin{aligned}
T_z\phi_{TT_z} &= T_z\phi_{TT_z}, \\
\mathbf{T}^2\phi_{TT_z} &= (T_x^2 + T_y^2 + T_z^2)\phi_{TT_z} = T(T+1)\phi_{TT_z}, \\
T_+\phi_{TT_z} &= (T_x + iT_y)\phi_{TT_z} = [T(T+1) - T_z(T_z+1)]^{1/2}\phi_{TT_z+1}, \\
T_-\phi_{TT_z} &= (T_x - iT_y)\phi_{TT_z} = [T(T+1) - T_z(T_z-1)]^{1/2}\phi_{TT_z-1}.
\end{aligned} \tag{A.54}$$

Here the eigenvalue of the operator T_z is denoted by the same symbol T_z (or $T_z \pm 1$ for the isospin functions at the ends of the third and fourth equations). These relations are exactly the same as those of the ordinary angular momentum.

A.9 THE DIRAC WAVE EQUATION

In this section we study the Dirac wave equation to describe the motion of a spin $\frac{1}{2}$ particle that has a speed approaching that of light.

At the time when Schrödinger developed his nonrelativistic wave equation, he also proposed an extension of it that meets the requirements of special relativity. This equation follows quite naturally from the transition in classical dynamics for the nonrelativistic relation

$$E = p^2/2m \tag{A.55}$$

between the energy and momentum of a free particle, to the corresponding relativistic relation

$$E^2 = c^2p^2 + m^2c^4, \tag{A.56}$$

where E denotes the total energy of the particle which includes the rest mass. The well-known quantum mechanical operators for E and $\mathbf{p}$ are

$$E = i\hbar \frac{\partial}{\partial t} \qquad \text{and} \qquad p_i = \frac{\hbar}{i}\frac{\partial}{\partial x_i}, \qquad i = 1, 2, 3. \tag{A.57}$$

The first equation gives the nonrelativistic Schrödinger equation

$$-\frac{\hbar^2}{2m}\nabla^2\psi = i\hbar\frac{\partial\psi}{\partial t}, \tag{A.58}$$

and the second equation gives the Klein–Gordan equation

$$\left[\nabla^2 - \left(\frac{mc}{\hbar}\right)^2\right]\psi = \frac{\partial^2\psi}{c^2\,\partial t^2} \qquad \text{or} \qquad \left[\left(\frac{\partial}{\partial x_\mu}\right)^2 - \kappa^2\right]\psi = 0 \tag{A.59}$$

with $\kappa = mc/\hbar$ and $\mu = 1, 2, 3, 4$ where $x_4 = ict$. Dirac [12] approached the problem of finding a relativistic wave equation by starting from the Hamiltonian form

$$i\hbar\frac{\partial}{\partial t}\psi(\mathbf{r}, t) = \mathscr{H}\psi(\mathbf{r}, t). \tag{A.60}$$

In this Appendix we study the free-particle equation. Dirac chose the Hamiltonian in such a way as to make it linear in the space derivatives. The simplest Hamiltonian that is linear in the momentum and mass term is

$$H = -c\boldsymbol{\alpha}\cdot\mathbf{p} - \beta mc^2. \tag{A.61}$$

Substituting Eq. (A.61) into Eq. (A.60), we have

$$(E + c\boldsymbol{\alpha}\cdot\mathbf{p} + \beta mc^2)\psi = 0 \tag{A.62}$$

or

$$\left(i\hbar\frac{\partial}{\partial t} - i\hbar c\boldsymbol{\alpha}\cdot\boldsymbol{\nabla} + \beta mc^2\right)\psi = 0. \tag{A.62$'$}$$

The coefficients α_1, α_2, α_3 and β do not depend on space or time variables and their derivatives. These can be the matrices when ψ has n components. To learn more about α and β we require that any solution of Eq. (A.62) should be a solution of Schrödinger's relativistic equation (A.59).

For this purpose, we multiply Eq. (A.62) on the left by

$$\left(i\hbar\frac{\partial}{\partial t} + i\hbar c\boldsymbol{\alpha}\cdot\boldsymbol{\nabla} - \beta mc^2\right), \tag{A.63}$$

and we have

$$\left[-\hbar^2\frac{\partial^2}{\partial t^2} + \hbar^2c^2\sum_i\left(\alpha_i^2\frac{\partial^2}{\partial x_i^2}\right) - \beta^2m^2c^4 + \hbar^2c^2\sum_{i\neq j}(\alpha_i\alpha_j + \alpha_j\alpha_i)\frac{\partial^2}{\partial x_i\,\partial x_j} + i\hbar mc^3\sum_i(\alpha_i\beta + \beta\alpha_i)\frac{\partial}{\partial x_i}\right]\psi = 0 \tag{A.64}$$

This equation will agree with the Klein-Gordon equation (A.59) if $\boldsymbol{\alpha}$ and β satisfy the relations

$$\begin{aligned} &\alpha_i^2 = \beta^2 = 1, \qquad i = 1, 2, 3, \\ &\alpha_1\alpha_2 + \alpha_2\alpha_1 = \alpha_2\alpha_3 + \alpha_3\alpha_2 = \alpha_3\alpha_1 + \alpha_1\alpha_3 = 0, \\ &\alpha_i\beta + \beta\alpha_i = 0, \qquad i = 1, 2, 3. \end{aligned} \tag{A.65}$$

That is, α_i and β anticommute in pairs, and their squares are unity. They are hermitian, since the Hamiltonian $\mathscr{H}$ in Eq. (A.61) is hermitian. It is convenient to find an explicit representation of these four matrices. We can diagonalize one of them, but the others are not diagonal, since they do not commute with each other.

For simplicity, we find the representation for which the rank is as low as possible. We arbitrarily choose β to be diagonal. Its eigenvalues are +1 and -1, since $\beta^2 = 1$. We arrange the diagonal elements of β so that the +1 eigenvalues are grouped together in a matrix of rank n, and all -1 eigenvalues are grouped in a matrix of rank m. It can be represented symbolically as

$$\beta = \begin{pmatrix} 1 & 0 \\ 0 & -1 \end{pmatrix}, \tag{A.66}$$

which is an abbreviation for

$$\left(\begin{array}{cccc|cccc} 1 & & & & & & & \\ & 1 & & & & 0 & & \\ & & 1 & & & & & \\ & & & \ddots & & & & \\ \hline & & & & -1 & & & \\ & 0 & & & & -1 & & \\ & & & & & & \ddots & \end{array}\right) \begin{array}{l} \left.\vphantom{\begin{array}{c}1\\1\\1\\1\end{array}}\right\} n \\ \left.\vphantom{\begin{array}{c}1\\1\\1\end{array}}\right\} m \end{array} . \tag{A.67}$$

α_i should have the form

$$\alpha_i = \begin{pmatrix} 0 & A_i \\ B_i & 0 \end{pmatrix}, \tag{A.68}$$

where A_i is the $n \times m$ and B_i is the $m \times n$ matrix. From $\alpha_i^2 = 1$ we have

$$A_iB_i = 1 \quad \text{and} \quad B_iA_i = 1, \tag{A.69}$$

where the right-hand side of the first equation is the $n \times n$ unit matrix and of the second the $m \times m$ one. It is easily proved that there are no solutions for A_i and B_i in the cases of $n = 1$, $m = 2$ or $m = 2$, $n = 1$. In the case $n = m = 1$, there are the Pauli spin matrices

$$\sigma_x = \begin{pmatrix} 0 & 1 \\ 1 & 0 \end{pmatrix}, \qquad \sigma_y = \begin{pmatrix} 0 & -i \\ i & 0 \end{pmatrix}, \qquad \sigma_z = \begin{pmatrix} 1 & 0 \\ 0 & -1 \end{pmatrix}, \tag{A.70}$$

with

$$\sigma_x \sigma_y = -\sigma_y \sigma_x = i\sigma_z, \qquad \text{and cyclic.}$$

The fourth matrix, which is independent of the three above, is the unit matrix. Therefore, this set of four matrices cannot satisfy all the relations in Eqs. (A.65).

The next lowest case is $n = m = 2$. If we assume $A_i = B_i$, we have $A_i^2 = 1$ and $A_iA_j + A_jA_i = 0$. These relations can be satisfied by the matrices σ_i. Finally we have an explicit representation for β and $\boldsymbol{\alpha}$,

$$\beta = \begin{pmatrix} 1 & 0 & 0 & 0 \\ 0 & 1 & 0 & 0 \\ 0 & 0 & -1 & 0 \\ 0 & 0 & 0 & -1 \end{pmatrix}, \qquad \alpha_1 = \begin{pmatrix} 0 & 0 & 0 & 1 \\ 0 & 0 & 1 & 0 \\ 0 & 1 & 0 & 0 \\ 1 & 0 & 0 & 0 \end{pmatrix},$$

$$\alpha_2 = \begin{pmatrix} 0 & 0 & 0 & -i \\ 0 & 0 & i & 0 \\ 0 & -i & 0 & 0 \\ i & 0 & 0 & 0 \end{pmatrix}, \qquad \alpha_3 = \begin{pmatrix} 0 & 0 & 1 & 0 \\ 0 & 0 & 0 & -1 \\ 1 & 0 & 0 & 0 \\ 0 & -1 & 0 & 0 \end{pmatrix}. \tag{A.71}$$

Or briefly,

$$\beta = \begin{pmatrix} 1 & 0 \\ 0 & -1 \end{pmatrix}, \qquad \alpha = \begin{pmatrix} 0 & \sigma \\ \sigma & 0 \end{pmatrix}. \tag{A.72}$$

Here 1, 0, σ are 2 x 2 matrices.

The nonrelativistic wave function of the spin $\frac{1}{2}$ particle has two components; it is expressed in the form called the spinor

$$\chi = \begin{pmatrix} a \\ b \end{pmatrix} = a\chi_{\frac{1}{2}} + b\chi_{-\frac{1}{2}}, \qquad |a|^2 + |b|^2 = 1, \tag{A.73}$$

with

$$\chi_{\frac{1}{2}} = \begin{pmatrix} 1 \\ 0 \end{pmatrix} \quad \text{and} \quad \chi_{-\frac{1}{2}} = \begin{pmatrix} 0 \\ 1 \end{pmatrix}.$$

This was already introduced in Eq. (2.34). The Pauli spin matrices are represented by Eq. (A.70), and the two spinors are the eigenfunctions of σ_z:

$$\sigma_z \chi_{\pm\frac{1}{2}} = \pm \chi_{\pm\frac{1}{2}}. \tag{A.74}$$

The relativistic wave equation with the martices α and β in Eq. (A.72) has four-component spinors as the solutions. (The subscript is reserved for the particle name, and the spinor index is given by the superscript.)

$$\psi(\mathbf{r}, t) = \begin{pmatrix} \psi^1 \\ \psi^2 \\ \psi^3 \\ \psi^4 \end{pmatrix} \exp\left[\frac{i(\mathbf{p}\cdot\mathbf{r} - Et)}{\hbar}\right]. \tag{A.75}$$

The wave functions ψ can be obtained if we introduce the bispinors

$$\psi(\mathbf{r}, t) = \begin{pmatrix} v \\ u \end{pmatrix} \exp\left[\frac{i(\mathbf{p}\cdot\mathbf{r} - Et)}{\hbar}\right] \tag{A.76}$$

where u and v are two-component spinors of the form Eq. (A.73). Inserting Eqs. (A.72) and (A.76) into Eq. (A.62), we have

$$-E\begin{pmatrix} v \\ u \end{pmatrix} = \left[c\begin{pmatrix} 0 & \sigma\cdot\mathbf{p} \\ \sigma\cdot\mathbf{p} & 0 \end{pmatrix} + mc^2\begin{pmatrix} 1 & 0 \\ 0 & -1 \end{pmatrix}\right]\begin{pmatrix} v \\ u \end{pmatrix}. \tag{A.77}$$

That is, the equation above is a set of the following coupled equations.

$$v(E + mc^2) = -c\sigma\cdot\mathbf{p}u, \qquad u(E - mc^2) = -c\sigma\cdot\mathbf{p}v. \tag{A.78}$$

These equations are homogeneous in the u and v and have solutions only if the determinant of the coefficients is zero, $E^2 - m^2c^4 - c^2p^2 = 0$, so that the relation between E and $\mathbf{p}$ is in agreement with Eq. (A.56).

If the energy E is chosen as

$$E = E_+ = + (c^2 p^2 + m^2 c^4)^{1/2}, \tag{A.79}$$

then we have

$$v = \frac{-c\boldsymbol{\sigma} \cdot \mathbf{p}}{E_+ + mc^2} u \tag{A.80}$$

from the first equation of (A.78). Here,

$$u = \left(\frac{E_+ + mc^2}{2E_+} \right)^{1/2} \chi \tag{A.81}$$

with χ as in Eq. (A.73). This results in the normalization of ψ,

$$\psi^\dagger \psi = 1. \tag{A.82}$$

This solution is for the positive energy, which is

$$E_+ \to mc^2 \tag{A.83}$$

in the nonrelativistic limit $\mathbf{p} \to 0$. The solution for spin "up" can be given explicitly if we put $a = 1$, $b = 0$, and $\chi = \binom{1}{0}$, and the one for spin "down" if we put $a = 0$, $b = 1$, and $\chi = \binom{0}{1}$.

The solutions for the negative energy

$$E = E_- = -(p^2 c^2 + m^2 c^4)^{1/2} \tag{A.84}$$

are also given by the second equation of (A.78), where

$$u = \frac{c\boldsymbol{\sigma} \cdot \mathbf{p}}{-E_- + mc^2} v. \tag{A.85}$$

Finally, the explicit forms of the solutions of the Dirac equation are given in Table A.2 in the form

$$\psi = \left(\frac{|E| + mc^2}{2|E|} \right)^{1/2} \begin{pmatrix} u^1 \\ u^2 \\ u^3 \\ u^4 \end{pmatrix}. \tag{A.86}$$

The product $\psi\psi^\dagger$ makes a 4 x 4 matrix, and its components are $\psi^\mu \psi^{\nu *}$. It has a property

$$\sum_{\sigma, E} \psi\psi^\dagger = I. \tag{A.87}$$

Here I is the 4 x 4 unit matrix. The sum runs over the two spin states as well as over the positive and negative energy states with the same absolute value of the energy. The relation (A.87) is called the closure property of the wave functions and it is easily confirmed with the wave functions listed in Table I.

Table A.2 Solutions (A.86) of the Dirac Wave Equation (A.62) for Free Particles

	$E_+ = (p^2c^2 + m^2c^4)^{1/2}$		$E_- = -(p^2c^2 + m^2c^4)^{1/2}$	
Spin	Up	Down	Up	Down
u^1	$\dfrac{-cp_z}{E_+ + mc^2}$	$\dfrac{-c(p_x - ip_y)}{E_+ + mc^2}$	1	0
u^2	$\dfrac{-c(p_x + ip_y)}{E_+ + mc^2}$	$\dfrac{cp_z}{E_+ + mc^2}$	0	1
u^3	1	0	$\dfrac{cp_z}{-E_- + mc^2}$	$\dfrac{c(p_x - ip_y)}{-E_- + mc^2}$
u^4	0	1	$\dfrac{c(p_x + ip_y)}{-E_- + mc^2}$	$\dfrac{-cp_z}{-E_- + mc^2}$

A.10 THE DIRAC HOLE THEORY

As we can see in the discussions of section A.9, the particle is treated as capable of having either a positive or a negative energy $E = \pm(p^2c^2 + m^2c^4)^{1/2}$ in free space. Dirac considered [12] that the negative energy states must be treated as "filled up" in representing the vacuum, and only deviations from this condition, by either the occupation of positive energy states, or evacuation of negative energy states, are to be held detectable. An occupied positive energy state is detected as a normal particle, whereas each vacancy (or "hole") of the negative states is detected as an antiparticle. Therefore, a hole of the electron with energy $-E$ and momentum $-\mathbf{p}$ behaves as a positron with energy E and momentum $\mathbf{p}$. In 1949 Feynman presented a theory in which he interpreted the positron as an electron moving backward in the time coordinate.

A.11 THE PROJECTION OPERATOR

It is sometimes necessary to evaluate the matrix

$$\sum_{\sigma} \psi\psi^{\dagger} \tag{A.88}$$

for positive or negative energy states. For this purpose, we introduce the matrix called the projection operator. The projection operator for the positive energy state is characterized by the relations

$$D_+\psi_+ = \psi_+ \quad \text{and} \quad D_+\psi_- = 0, \tag{A.89}$$

where ψ_+ and ψ_- refer to the wave functions for positive and negative energy states, respectively. Similarly, the projection operator for the negative energy state is characterized by

$$D_- \psi_+ = 0, \qquad D_- \psi_- = \psi_- . \tag{A.90}$$

These operators obviously satisfy the relation

$$D_\pm^2 = D_\pm \qquad \text{since} \quad D_\pm^2 \psi_\pm = D_\pm \psi_\pm . \tag{A.91}$$

Therefore, the quantity in Eq. (A.88) becomes the projection operator

$$\sum_\sigma \psi_\pm \psi_\pm^\dagger = D_\pm \sum_\sigma \psi_\pm \psi_\pm^\dagger = D_\pm \sum_{\sigma, E} \psi \psi^\dagger = D_\pm . \tag{A.92}$$

Explicit forms of the projection operators are obtained, in our representation, as follows. From Eqs. (A.61) and (A.62),

$$\mathscr{H} \psi = E \psi \qquad \text{with} \quad \mathscr{H} = -(c\boldsymbol{\alpha} \cdot \mathbf{p} + \beta mc^2). \tag{A.93}$$

If we put $W = |E|$, then the $\psi_\pm$ satisfy the equation

$$\mathscr{H} \psi_\pm = \pm W \psi_\pm \qquad \text{or} \quad \pm \frac{\mathscr{H}}{W} \psi_\pm = \psi_\pm . \tag{A.94}$$

Furthermore, if we construct $D_\pm$ by

$$\begin{aligned} D_\pm &= \tfrac{1}{2}[1 \pm (\mathscr{H}/W)] \\ &= \tfrac{1}{2}\left(1 \mp \frac{c\boldsymbol{\alpha} \cdot \mathbf{p} + \beta mc^2}{W}\right), \end{aligned} \tag{A.95}$$

they satisfy the required properties (A.89) and (A.90). The expression (A.95) is also obtained by the direct calculation of Eq. (A.92) with the solutions in Table A.2. D_- is also rewritten as

$$D_- = \tfrac{1}{2}\left(1 - \frac{c\boldsymbol{\alpha} \cdot \mathbf{p} - \beta mc^2}{W}\right) \text{ for the positron,} \tag{A.96}$$

if the positron has the energy W and the momentum $\mathbf{p}$.

A.12 RELATIVISTIC COVARIANT FORM FOR THE DIRAC EQUATION

The invariance of the Dirac equation under the Lorentz transformation can be understood more clearly if the space and time coordinates are treated with equal footing. Let us introduce x_μ as

$$x_1 = x, \qquad x_2 = y, \qquad x_3 = z, \qquad x_4 = ict. \tag{A.97}$$

This leads to

$$-i\hbar \frac{\partial}{\partial t} = \hbar c \frac{\partial}{\partial x_4}. \tag{A.98}$$

Multiplying $\beta/\hbar c$ from the left in Eq. (A.62′), we have

$$\left(-\beta \frac{\partial}{\partial x_4} - i\beta\boldsymbol{\alpha} \frac{\partial}{\partial \mathbf{x}} + \frac{mc}{\hbar}\right)\psi = 0. \tag{A.99}$$

Introducing a set of the gamma matrices γ_μ,

$$\gamma_k = -i\beta\alpha_k, \qquad k = 1, 2, 3, \qquad \text{and} \qquad \gamma_4 = -\beta, \tag{A.100}$$

the equation becomes

$$\left(\gamma_\mu \frac{\partial}{\partial x_\mu} + \frac{mc}{\hbar}\right)\psi = 0, \tag{A.101}$$

where the repeated Greek letter μ should be summed over 1, 2, 3, 4.

By the definition (A.100), γ_μ are all hermitian and satisfy the relations

$$\gamma_\mu \gamma_\nu + \gamma_\nu \gamma_\mu = 2\delta_{\mu\nu}. \tag{A.102}$$

Matrices which satisfy these relations are called Dirac gamma matrices. Another matrix is also introduced by

$$\gamma_5 = \gamma_1 \gamma_2 \gamma_3 \gamma_4 = -i\alpha_1 \alpha_2 \alpha_3. \tag{A.103}$$

Therefore, γ_5 satisfies

$$\gamma_5^2 = 1 \qquad \text{and} \qquad \gamma_5 \gamma_\mu = -\gamma_\mu \gamma_5. \tag{A.104}$$

Consequently, the equation (A.102) holds for

$$\gamma_\mu \gamma_\nu + \gamma_\nu \gamma_\mu = 2\delta_{\mu\nu}, \qquad \mu, \nu = 1, 2, 3, 4, 5. \tag{A.105}$$

The explicit forms of γ_μ in our representation are

$$\gamma_k = \begin{pmatrix} 0 & -i\sigma_k \\ i\sigma_k & 0 \end{pmatrix}, \quad \gamma_4 = \begin{pmatrix} -1 & 0 \\ 0 & 1 \end{pmatrix}, \quad \gamma_5 = \begin{pmatrix} 0 & 1 \\ 1 & 0 \end{pmatrix}. \tag{A.106}$$

Pauli's Fundamental Theorem

The following theorem is useful for proving the Lorentz covariance of the Dirac equation.

If two sets of matrices γ_μ and γ'_μ ($\mu = 1, 2, 3, 4$) obey the commutation rules

$$\gamma_\mu \gamma_\nu + \gamma_\nu \gamma_\mu = 2\delta_{\mu\nu}, \qquad \gamma'_\mu \gamma'_\nu + \gamma'_\nu \gamma'_\mu = 2\delta_{\mu\nu}, \tag{A.107}$$

then there must exist a nonsingular matrix S which connects the two sets according to

$$\gamma'_\mu = S\gamma_\mu S^{-1}. \tag{A.108}$$

The proof is given by Pauli, and can be found in [13–15]. γ'_μ need not be hermitian, and S will not be unitary in general. The matrix S which connects two sets of Dirac matrices according to Eq. (A.108) is uniquely determined by the two sets for an arbitrary numerical factor. It is customary to choose the arbitrary multiplication factor so that

$$\det S = 1. \tag{A.109}$$

Then S is unique except for a factor of ± 1, $\pm i$.

Lorentz Transformations of the Dirac Wave Equation

Now, we study the Lorentz transformation properties of the four-component Dirac wave function introduced in section A.9. A homogeneous Lorentz transformation is defined as a coordinate transformation

$$x'_\mu = a_{\mu\nu} x_\nu \tag{A.110}$$

with the condition that Eq. (A.110) leaves the length of any position vector unchanged

$$s' = (x'_\mu x'_\mu)^{1/2} = (x_\mu x_\mu)^{1/2} = s. \tag{A.111}$$

This condition requires the following relations for the transformation coefficients

$$a_{\mu\nu} a_{\mu\rho} = \delta_{\nu\rho}, \tag{A.112}$$

which are called orthogonality conditions. Using these conditions, we find from Eq. (A.110)

$$x_\nu = a_{\mu\nu} x'_\mu \tag{A.113}$$

and

$$a_{\nu\mu} a_{\rho\mu} = \delta_{\nu\rho}. \tag{A.114}$$

If A is the 4 x 4 matrix with components $a_{\mu\nu}$, then Eqs. (A.112) and (A.114) are given by

$$A^T A = 1 \tag{A.115}$$

where A^T is the transposed matrix of A. The determinant of (A.115) is

$$\det(A^T A) = (\det A^T)(\det A) = (\det A)^2 = 1$$

so that

$$\det A = \pm 1. \tag{A.116}$$

The space which we are studying now, with three real and one imaginary coordinates, Eq. (A.97), is called the Minkowski space. The transformation A should preserve the reality of the coordinates x, y, z, t. From this we must have

$$a_{ik} \text{ real}, \quad i, k = 1, 2, 3; \qquad a_{i4} \text{ and } a_{4i} \text{ imaginary}, \quad i = 1, 2, 3;$$
$$a_{44} \text{ real}. \tag{A.117}$$

The plus sign of $\det A$ will apply for the transformation continuous with the identity. This class of transformation is called the proper Lorentz transformation. If $\det A = -1$, A is continuous with the space reflection

$$x_j' = -x_j, \qquad x_4' = x_4, \tag{A.118}$$

which has $\det A = -1$ and $a_{44} = 1$, or with the time reflection

$$x_j' = x_j, \qquad x_4' = -x_4, \tag{A.119}$$

which has $\det A = -1$, and $a_{44} = -1$.

The Dirac wave equation should be covariant in the sense that, when the Lorentz transformation of Eq. (A.110) is applied, the equation

$$\left(\gamma_\mu \frac{\partial}{\partial x_\mu} + \frac{mc}{\hbar}\right)\psi(x) = 0 \tag{A.120}$$

leads to equation

$$\left(\gamma_\mu \frac{\partial}{\partial x_\mu'} + \frac{mc}{\hbar}\right)\psi'(x') = 0. \tag{A.121}$$

Here we have the same expression for γ_μ, since they are adopted simply to write down four equations for the components of ψ as a single matrix equation. For the transformation in which $a_{44} > 0$, the covariance can be obtained by the transformation

$$\psi'(x') = \Lambda\psi(x), \tag{A.122}$$

provided that the matrix Λ is appropriately chosen. It is assumed that Λ has an inverse. (This should be proved.) And it has no coordinate dependence.

From Eqs. (A.121) and (A.122), we have

$$\left(\gamma_\mu a_{\mu\nu} \frac{\partial}{\partial x_\nu} + \frac{mc}{\hbar}\right)\Lambda\psi = 0, \tag{A.123}$$

where we have used

$$\frac{\partial}{\partial x_\mu'} = \frac{\partial x_\nu}{\partial x_\mu'} \frac{\partial}{\partial x_\nu} = a_{\nu\mu}^{-1} \frac{\partial}{\partial x_\nu} = a_{\mu\nu} \frac{\partial}{\partial x_\nu}. \tag{A.124}$$

If we multiply Eq. (A.123) by Λ^{-1} on the left it becomes the same as Eq. (A.120), provided

$$\Lambda^{-1}\gamma_\mu a_{\mu\nu}\Lambda = \gamma_\nu, \tag{A.125}$$

or

$$\Lambda^{-1}\gamma_\mu\Lambda = a_{\mu\nu}\gamma_\nu. \tag{A.126}$$

The existence of Λ and Λ^{-1} is evident from the following. Let us denote

$$\gamma'_\mu = a_{\mu\nu}\gamma_\nu, \tag{A.127}$$

Then the γ'_μ make a set of Dirac matrices, since

$$\begin{aligned}\gamma'_\mu\gamma'_\nu + \gamma'_\nu\gamma'_\mu &= a_{\mu\lambda}\gamma_\lambda a_{\nu\rho}\gamma_\rho + a_{\nu\rho}\gamma_\rho a_{\mu\lambda}\gamma_\lambda \\ &= 2a_{\mu\lambda}a_{\nu\rho}\delta_{\lambda\rho} = 2\delta_{\mu\nu}.\end{aligned} \tag{A.128}$$

From Pauli's fundamental theorem, there exist Λ and Λ^{-1} such that $\Lambda^{-1}\gamma_\mu\Lambda = \gamma'_\mu$.

The transformation matrix can be explicitly obtained by considering the infinitesimal Lorentz transformation

$$x'_\mu = x_\mu + \epsilon_{\mu\nu}x_\nu, \qquad \epsilon_{\mu\nu} = -\epsilon_{\nu\mu} \ll 1. \tag{A.129}$$

The transformation matrix is given by

$$\Lambda = \exp\left(\tfrac{1}{4}\epsilon_{\mu\nu}\gamma_\mu\gamma_\nu\right). \tag{A.130}$$

This can be seen, e.g., in the book by Rose [15].

Adjoint of ψ

The hermitian conjugate of the equation

$$\left(\gamma_\mu \frac{\partial}{\partial x_\mu} + \frac{mc}{\hbar}\right)\psi = 0 \tag{A.131}$$

is

$$\left(\frac{\partial\psi^\dagger}{\partial x_k}\gamma_k - \frac{\partial\psi^\dagger}{\partial x_4}\gamma_4 + \frac{mc}{\hbar}\psi^\dagger\right) = 0, \tag{A.132}$$

where γ_μ is hermitian and $\psi^\dagger$ is

$$\psi^\dagger = (\psi^{1*}, \psi^{2*}, \psi^{3*}, \psi^{4*}). \tag{A.133}$$

A more symmetrical form is obtained by introducing the adjoint of ψ,

$$\bar{\psi} = \psi^\dagger\gamma_4. \tag{A.134}$$

Multiplying the equation of $\psi^\dagger$ by γ_4 on the right, we have

$$\left(\frac{\partial \bar{\psi}}{\partial x_\mu} \gamma_\mu - \frac{mc}{\hbar} \bar{\psi}\right) = 0. \tag{A.135}$$

This is the adjoint Dirac equation and the covariance of this equation will be shown as follows. First, we give the adjoint of the transformed field $\bar{\psi}'(x')$

$$\bar{\psi}'(x') = \psi'^\dagger(x')\gamma_4. \tag{A.136}$$

Using Eq. (A. 122), we have

$$\bar{\psi}' = (\Lambda\psi)^\dagger \gamma_4 = \psi^\dagger \Lambda^\dagger \gamma_4 = \psi^\dagger \gamma_4 \Lambda^{-1} = \bar{\psi}\Lambda^{-1}. \tag{A.137}$$

Here we have used the lemma

$$\gamma_4 \Lambda^\dagger \gamma_4 \Lambda = a_{44}/|a_{44}|. \tag{A.138}$$

The proof of this lemma is elementary, but lengthy; it can be found in the literature [15]. Substituting Eqs. (A.124), (A.126), and (A.136) into Eq. (A.135), we have

$$\frac{\partial}{\partial x'_\mu} \bar{\psi}\Lambda^{-1}\gamma_\mu \Lambda - \frac{mc}{\hbar} \bar{\psi} = 0. \tag{A.139}$$

Multiplying from the right by Λ^{-1}, we have

$$\frac{\partial}{\partial x'_\mu} \bar{\psi}\Lambda^{-1}\gamma_\mu - \frac{mc}{\hbar} \bar{\psi}\Lambda^{-1} = 0. \tag{A.140}$$

This indicates that the adjoint equation is in fact covariant, and the adjoint Dirac spinor is transformed is this way;

$$\bar{\psi}'(x') = \bar{\psi}(x)\Lambda^{-1}. \tag{A.141}$$

Bilinear Quantities in ψ and $\bar{\psi}$

There are sixteen different forms of bilinear quantities, because ψ and $\bar{\psi}$ each have four components for which the transformation properties are given in Eqs. (A.122) and (A.141). They are arranged in five groups;

$$S = \bar{\psi}\psi, \qquad V_\mu = \bar{\psi}\gamma_\mu \psi, \qquad T_{\mu\nu} = \frac{1}{2i} \bar{\psi}(\gamma_\mu \gamma_\nu - \gamma_\nu \gamma_\mu)\psi, \tag{A.142}$$

$$A_\mu = i\bar{\psi}\gamma_\mu \gamma_5 \psi, \qquad P = \bar{\psi}\gamma_5 \psi.$$

They behave as a scalar, vector, tensor, axial vector, and pseudoscalar, and their components are 1, 4, 6, 4, 1, respectively. That is, under the Lorentz transformation, the foregoing quantities are transformed as

$$S'(x') = S(x), \qquad V'_\mu(x') = a_{\mu\nu} V_\nu(x),$$
$$T'_{\mu\nu}(x') = a_{\mu\lambda} a_{\nu\rho} T_{\lambda\rho}(x), \qquad A'_\mu(x') = (\det A) a_{\mu\nu} A_\nu(x),$$
$$P'(x') = (\det A) P(x). \tag{A.143}$$

As an example, we will show the transformation in the case of $A_\mu(x)$ in detail [14, 15],

$$A'_\mu(x') = i\bar{\psi}(x)\Lambda^{-1}\gamma_\mu \Lambda\Lambda^{-1}\gamma_5 \Lambda\psi(x). \tag{A.144}$$

On the other hand,

$$\Lambda^{-1}\gamma_\mu \Lambda = a_{\mu\nu}\gamma_\nu, \tag{A.145}$$

and

$$\Lambda^{-1}\gamma_5 \Lambda = (\det A)\gamma_5. \tag{A.146}$$

The latter will be shown as follows. We introduce ϵ as

$$\epsilon_{\mu\nu\kappa\lambda} = \pm 1, \tag{A.147}$$

where +1 means that $\mu\nu\kappa\lambda$ is an even permutation of 1, 2, 3, 4, and -1 refers to an odd permutation. Otherwise, $\epsilon = 0$. Then,

$$\gamma_5 = \frac{1}{4!} \epsilon_{\mu\nu\kappa\lambda} \gamma_\mu \gamma_\nu \gamma_\kappa \gamma_\lambda, \tag{A.148}$$

and

$$\begin{aligned}
\Lambda^{-1}\gamma_5 \Lambda &= \frac{1}{4!} \epsilon_{\mu\nu\kappa\lambda} (\Lambda^{-1}\gamma_\mu \Lambda)(\Lambda^{-1}\gamma_\nu \Lambda)(\Lambda^{-1}\gamma_\kappa \Lambda)(\Lambda^{-1}\gamma_\lambda \Lambda) \\
&= \frac{1}{4!} \epsilon_{\mu\nu\kappa\lambda} a_{\mu\alpha} a_{\nu\beta} a_{\kappa\gamma} a_{\lambda\delta} \gamma_\alpha \gamma_\beta \gamma_\gamma \gamma_\delta \\
&= \frac{1}{4!} (\det A) \epsilon_{\alpha\beta\gamma\delta} \gamma_\alpha \gamma_\beta \gamma_\gamma \gamma_\delta \\
&= (\det A)\gamma_5.
\end{aligned} \tag{A.149}$$

Finally,

$$A'_\mu(x') = i a_{\mu\nu} (\det A) \bar{\psi}\gamma_\nu \gamma_5 \psi = (\det A) a_{\mu\nu} A_\nu(x). \tag{A.150}$$

The vector quantity V_μ is the four-vector current of a particle

$$j_\mu = icV_\mu \tag{A.151}$$

where $j_4 = ic\rho$ and ρ is the probability density. In fact, Eq. (A.152) is read

$$\mathbf{j} = -c\psi^\dagger \boldsymbol{\alpha}\psi, \qquad \rho = \psi^\dagger \psi. \tag{A.152}$$

We notice here that in our representation, $\boldsymbol{\alpha} \to -(\mathbf{p}/M)$ in the nonrelativistic limit. Furthermore, $\mathbf{j}$ and ρ satisfy the continuity equation

$$\frac{\partial}{\partial x_\mu} j_\mu = \operatorname{div} \mathbf{j} + \frac{\partial \rho}{\partial t} = 0. \tag{A.153}$$

This is easily derived from the Dirac equations for $\bar{\psi}$ and ψ, Eqs. (A.131) and (A.135).

REFERENCES

1 L. I. Schiff, *Quantum Mechanics*, McGraw-Hill, New York, 1949.
2 E. P. Wigner, *Gruppen Theorie*, F. Vieweg and Sohn, Braunschweig, 1931.
3 G. Racah, *Phys. Rev.* **62**, 438 (1942).
4 E. U. Condon and G. H. Shortley, *Theory of Atomic Spectra*, Cambridge Univ. Press, London, 1935.
5 M. Morita, R. Morita, T. Tsukamoto, and M. Yamada, *Progr. Theoret. Phys. Suppl.* **26**, 64 (1963).
6 C. Eckart, *Rev. Mod. Phys.* **2**, 305 (1930).
7 M. E. Rose, *Elementary Theory of Angular Momentum*, Wiley, New York, 1957.
8 T. Ishizu, H. Horie, S. Obi, M. Sato, Y. Tanabe, and S. Yanagawa, *Table of Racah Coefficients*, Pan Pacific Press, Tokyo, 1960.
9 G. Herzberg, *Atomic Spectra and Atomic Structure*, Dover, New York, 1944.
10 R. L. Mössbauer, *Z. Phys.* **151**, 124 (1958); *Z. Naturforsch.* **149**, 212 (1959).
11 M. Morita, *Phys. Rev.* **122**, 1525 (1961); M. Morita and R. Morita, *Nucl. Phys.* **51**, 433 (1964).
12 P. A. M. Dirac, *Proc. Roy. Soc.* **A117**, 610 (1928). *The Principles of Quantum Mechanics*, 2nd ed., Chap. XII, Oxford Univ. Press, London and New York, 1935.
13 W. Pauli, *Ann. Inst. Henri Poincaré*, **6**, 109 (1936).
14 R. H. Good, Jr., *Rev. Mod. Phys.* **27**, 187 (1955).
15 M. E. Rose, *Relativistic Electron Theory*, Wiley, New York, 1960.

AUTHOR INDEX

SUBJECT INDEX

MASATO MORITA
was awarded a PH.D. from the University of Tokyo in 1956 and is currently Professor of Physics as Osaka University in Toyonaka, Osaka, Japan. During 1956–1962 and 1966–1967, he was at Columbia University at a time of the finding of parity violation by Professor Lee-Yang-Wu.

Professor Morita is the author of "Theory of Nuclear Orientation" and of more than 100 papers and review articles. He is a member of the American Physical Society and Physical Society of Japan, and was awarded the Yukawa Memorial Fellowship (Japan) in 1952–1956, Nishina Memorial Fellowship (Japan) in 1956–1957, and the Ernest Kempton Adams Fellowship (Columbia University) in 1958–1959.

"**Beta Decay and Muon Capture**" consists of two parts: the study of properties of weak interaction and the study of nuclear structure from various phenomena (typically beta decay and muon capture) induced by this weak interaction. The book describes the properties of weak interaction which have become familiar to us only in the last 15 years, and introduces many exciting findings of these properties (including those for parity nonconservation for which the author made contributions), by using examples in beta decays. Recent developments in the peculiar properties of weak interaction are also introduced. Formulas for beta-ray spectra, beta-ray angular distribution, and beta-gamma angular correlations are calculated in a simple way with minimum prerequisite. Important formulas are collected for convenience to analyze experimental data.

Important features of the book:

- An elementary description of the theory of beta decay, and violation of various symmetry principles in the law of nature and easy derivation of formulas for beta decay
- Inclusion of topics in nuclear structure problems studied from beta decays
- A collection of important formulas and tables for use of analysis of experimental data
- An emphasis of a similarity of beta decay with gamma decay
- Inclusion of muon capture reactions

W. A. Benjamin, Inc.
Advanced Book Program
Reading, Massachusetts 01867